What Makes Us Social?

Jean Nicod Lectures

Francois Recanati, editor

What Makes Us Social?

Chris Frith and Uta Frith

The MIT Press
Cambridge, Massachusetts
London, England

The MIT Press would like to thank the anonymous peer reviewers who provided comments on drafts of this book. The generous work of academic experts is essential for establishing the authority and quality of our publications. We acknowledge with gratitude the contributions of these otherwise uncredited readers.

This book was set in Stone Serif and Stone Sans by Westchester Publishing Services. Printed and bound in the United States of America.

Library of Congress Cataloging-in-Publication Data

Names: Frith, Christopher D., author. | Frith, Uta, author.
Title: What makes us social? / Chris Frith and Uta Frith.
Description: Cambridge, Massachusetts : The MIT Press, [2023] | Series: Jean Nicod lectures | Includes bibliographical references and index.
Identifiers: LCCN 2022042937 (print) | LCCN 2022042938 (ebook) | ISBN 9780262546270 (paperback) | ISBN 9780262375481 (epub) | ISBN 9780262375498 (pdf)
Subjects: LCSH: Social interaction—Psychological aspects. | Social perception. | Social psychology.
Classification: LCC HM1111 .F757 2023 (print) | LCC HM1111 (ebook) | DDC 302/.12—dc23/eng/20230202
LC record available at https://lccn.loc.gov/2022042937
LC ebook record available at https://lccn.loc.gov/2022042938

10 9 8 7 6 5 4 3 2 1

Contents

Series Foreword

The Jean Nicod Lectures are delivered annually in Paris by a leading philosopher of mind or philosophically oriented cognitive scientist. The 1993 inaugural lectures marked the centenary of the birth of the French philosopher and logician Jean Nicod (1893–1931). The lectures are sponsored by the Centre National de la Recherche Scientifique (CNRS), in cooperation with the École des Hautes Etudes en Sciences Sociales (EHESS) and the École Normale Superieure (ENS). The series hosts the texts of the lectures or the monographs they inspire.

1 What Is Social Cognition?

This book is about how we interact with one another and what cognitive mechanisms underlie these interactions. These mechanisms are situated in our brains, and our brains are the product of evolution. The chapters in this book are grouped along the key concepts of cooperation, competition, computation, and culture. We use a Bayesian framework to situate unconscious and conscious thoughts and actions in a hierarchy of information-processing levels. We assume that automatic and unconscious cognitive processes are largely shared by humans with other animals, while deliberate and conscious processes may be more typically associated with human beings.

What do we mean by the term "cognition"? It enables us to talk about the workings of the brain in such a way that the description matches the workings of the mind. The mind is nothing mystical; rather, it is a complex system of interacting processes that can be understood in ordinary language, and hopefully in the future, in mathematical language. Social cognition is our term for capturing everything that can, in principle, explain how animals (including humans) learn from each other about the *world of objects* and how they cooperate and compete with each other in the *world of agents*. We also consider the *world of ideas*, and it is here that human culture is situated with pervasive effects on our social nature. As far as we know, it is only adult humans who can master this top level of the information-processing hierarchy, which bestows the ability to reflect on what we do and who we are. This top level provides an interface between the mind of the individual and the minds of other people.

* * *

The Richness of the Social World

Chances are that, as you read this book, your mind will only be half engaged. Chances are that your thoughts and feelings will be distracted by the daily roller coaster that makes up your social life. Our greatest happiness is being surrounded by loved ones; our deepest fear is being lost and alone. Who has not experienced ups and downs, highs, and lows in their social life? There are peaks of joyful togetherness: the memory of a beloved teacher, the touching trust of a child, the excitement of a reunion with old friends, or keeping a birthday present as a secret surprise. And there are errors with long-term consequences: misjudging somebody's motives, making an impulsive decision to enter or finish a relationship that we later regret, or telling lies and taking advantage of another person. Yet we move in this world with amazing dexterity, often without much thought. It is a place where we meet danger and misery, but also immense happiness.

Almost all the emotions that we experience daily are caused by our relationships with others: love, hate, forgiveness, grudge, pride, and regret. We can feel them and act upon them, and they may or may not preoccupy our conscious thoughts. Most of us feel less intense emotions when we engage with the physical world. The joy of warming oneself next to a fireplace is greatly enhanced if the experience is shared with someone else. Being able to talk to someone about the changing colors of a sunset creates a more vivid experience than watching the sunset alone. Our well-being depends, most of all, on the social world.

This book explores cognitive processes by which our brains and minds navigate the social world, guarding against its dangers and reaping its rich rewards. The social world is probably richer, more complex, and more varied than any other world we can imagine, and therefore less predictable than we might wish for maintaining comfort and security. We trust that as you continue reading, we can show you some reasons that make this roller-coaster experience easier to understand. But this is not a self-help book. Instead, we strive to show how experimental psychology and neuroscience have begun to redefine opposites of old, such as cognition and emotion, nature and nurture, brain and mind.

What Makes This Book Different?

There are dozens of books on social neuroscience, and even more on social psychology. As the field changes, more books have to be written. Here, we are providing a personal summary of what we have learned about what makes us social and how that connects us with our evolution and our culture. At heart, we are experimentalists, and experiments form the backbone of this book. Whenever there is sufficient evidence, we will report and discuss findings about how the brain supports our social abilities. Occasionally, we will step back from these details to paint a bigger picture that reveals the profoundly social nature of everything we do and think.

This book is the product of our collaboration during a late stage in our working lives. When we started long ago, we were working on very different topics and were mainly interested in the neuropsychology of clinical groups. One of us was working on autism (Uta Frith, 1989), and the other on schizophrenia (Chris Frith, 1992). We came to realize that both these conditions are associated first and foremost with troubling problems with social communication. We have attempted to explain them by exploring the cognitive mechanism that underlies mentalizing (the ability to use mental states to predict what an agent is going to do). We first wrote about this together thirty years ago (Frith and Frith, 1991). Gradually, we increased our collaboration, relishing the use of brain imaging that allowed us to probe the neurophysiological basis of this mechanism. This confirmed our belief that the most exciting topic to study is social cognition (Frith and Frith, 1999). It still excites us, and we hope it will also excite our readers.

Our original plan for this book was to stick closely to a series of lectures that we gave in 2014 in Paris when we were awarded the Jean Nicod Prize. However, once we began writing, we found that there was far, far more to talk about. Every week, there would be two or three new papers that looked as if they ought to be fitted in somewhere.

Figure 1.1 shows the number of papers on social cognition published each year between 1965 and 2020, the time span that coincides with our life in research. Nothing much appeared until about 1990, when the numbers began to increase slowly. In the twenty years since 2000, the number published per year has increased from 400 to 3,500.

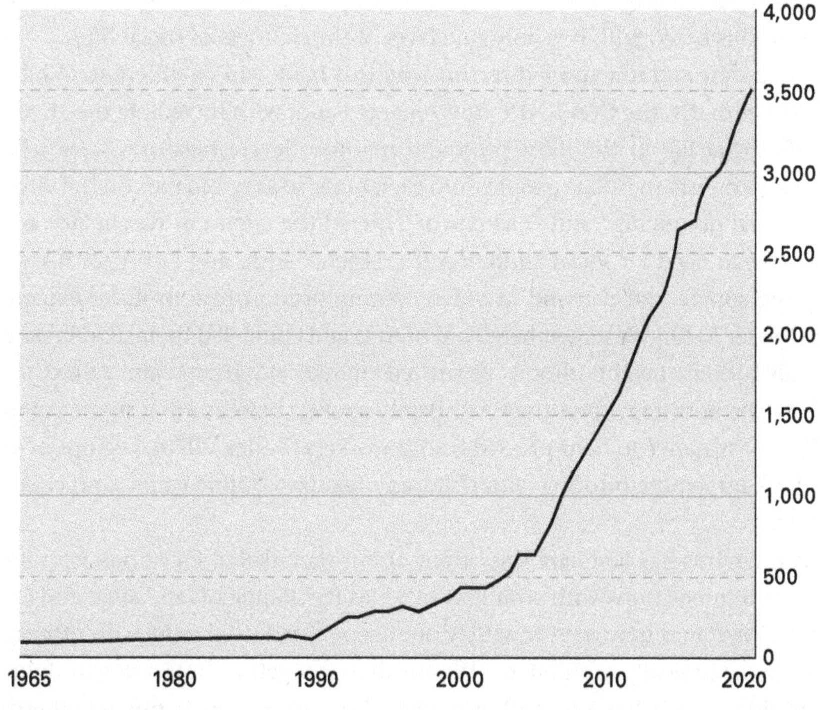

Figure 1.1
Number of papers on the topic of social cognition published per year between 1965 and 2020. Created by the authors using Web of Science.

Why was this happening? Historians will no doubt uncover a host of reasons. For us personally, it seemed to start with the possibility of connecting behavior and cognition to brain processes through theory and experiment. We were intrigued by *mentalizing*, the ability to model the minds of others to predict what they are going to do, its causes, and its consequences for our social interactions. Slowly we realized that social cognition was based on many more mechanisms, and all needed to be analyzed with an eye on evolution. This project still has a long way to go. In this book, we report the conclusions that we draw from our analysis of research done so far. Our review of the existing literature is not comprehensive, but it includes the work that we personally consider the most relevant for the future of this project.

We were surprised and flattered that Michael Gilead and Kevin Ochsner (2021) decided to celebrate the twentieth anniversary of our 1999 paper by taking stock of the rising body of knowledge since then. This resulted in an edited volume with thirty-three chapters that reflect the development of the neural basis of mentalizing. Mentalizing is now well established as a key concept in human social interaction and communication. We will have many occasions to refer to the relevant work here, especially in the later part of this book. Still, it is only one facet of the richness of social life.

There is a new and unexpected reason why this book will be different. While editing our earlier drafts, the COVID-19 virus emerged and, with incredible speed, affected everyone's social life in the most profound manner. Severe measures were taken by almost every country to isolate people from each other so as not to pass on the virus. The phrases "social distancing" and "lockdown" entered the common vocabulary and laid bare the human need for social affiliation (Dezecache, Frith, and Deroy, 2020). Almost at once, the supply and demand of video communication rose to never-experienced heights (Neate, 2020). We remembered old friends and contacted them; conference calls became ubiquitous; neighborhoods organized mutual aid groups and talked to each other from the doorstep when they had barely spoken before. An army of volunteers offered their assistance to hard-pressed health workers (Butler, 2020). Legions of grandparents took up remote tutoring, and children organized online games and chats with each other.

The coronavirus has laid bare the extent of our remarkable social predisposition to affiliate. Even interactions with strangers make us feel happier (van Lange and Columbus, 2021). We found new ways to satisfy our deep-rooted need to be with others, even when this was physically prohibited. "We are all in it together" was a slogan right from the beginning, and it has survived as a metaphor, even though this togetherness is very different from before, as we were looking at each other on screen. This misses the qualities that normally come with face-to-face interaction and are brought about by the integration of social information from different senses, including feel, touch, and smell.[1]

However, to our knowledge, the need for affiliation and the readiness to help each other during the height of the crisis have never been so evident. This crisis, always referred to as unprecedented, is deeply relevant to the substance of this book because of the issues it raises about the nature of human social interactions.

What Do We Mean by Cognition?

This book explores the cognitive processes by which our brains and minds navigate the social world, guarding against its dangers and reaping its rich rewards. For us, the term "cognition" provides a conduit through which we can connect the mind and the brain. It allows our explanations to go seamlessly from mind to brain and from brain to mind. So, whenever we say "brain," we also mean "mind."

This is not what many people mean by "cognition." They often treat cognition as the opposite of emotion and restrict it to the processes underlying deliberate, rational thinking. But we prefer a much wider definition, in which cognition is roughly equivalent to the term "information processing." And we believe that all aspects of behavior and thought, conscious and unconscious, can be understood in terms of such processing.

Our wide definition of "cognition" also provides a vocabulary that bridges the divide between human and animal abilities, similar to the definition proposed by Sara Shettleworth in her book *Cognition, Evolution, and Behavior*: "Cognition refers to the mechanisms by which animals acquire, process, store, and act on information from the environment" (2010, 5).

It is worth pointing out that cognitive processes refer to phenomena that are mostly unconscious. They are very hard to study, as they have to be inferred rather than directly observed in behavior or physiology. Just looking at the data is not enough. We need hypotheses that can be tested. The aim is to seamlessly link behavioral evidence on the one hand, and neurophysiological evidence on the other. Both sources of evidence present difficulties. First, since the same behavior can have different causes, mere observation is not sufficient to penetrate to the underlying processes. Second, physiological processes are often studied on a fine level, such as processes happening inside single neurons. The grain that we envisage is far coarser: at the level of brain hubs, systems, and connections (Barack and Krakauer, 2021). What the most appropriate units

1. As time passed, the phrase has taken on an ironic quality. The effects of the pandemic also highlighted the divide between the Haves and Have Nots, and more generally between Us and Them.

are is something that cognitive neuroscientists are attempting to solve. At present, we think that identifying mental phenomena in the brain is a bit like identifying a city as seen from a satellite in space (see figure 1.2). Wild hypotheses about what goes on there are all we have. But this is what makes the research so exciting.

What is the social content in social cognition? It is in the information! Social information originates from living biological agents rather than physical objects,[2] and all animals use social information as much as (if not even more than) physical information. But are there entirely different mechanisms to process this social information? This turns out to be a tricky question. Intriguingly, there is some evidence that the brain makes a distinction between causation in the social world and causation in the physical world (Jack et al., 2013). Causes in the physical world include forces such as gravity and magnetism. Causes in the social world include mental states, such as beliefs and desires. Based on the findings of his study, Tony Jack suggested that there is one brain network that processes physical causes and another that processes mental causes, and these networks mutually inhibit one another.

Social cognition requires us to talk about innate predispositions that make us cooperate and compete. It also requires us to talk about cultural learning and the countless decisions we make while we carry on with our lives. Whether we believe that the purpose of these decisions is for the survival of our genes or for the pursuit of happiness, we need to have some idea of what is likely to happen next and decide what to do about it—to stay alive or to become happier. We believe that these choices are largely determined by the social world that surrounds us.

Key Concepts Used to Structure This Book

The table of contents shows that the chapters in this book are arranged into four themes: Cooperation, Competition, Computation, and Culture. In parallel, we consider how they fit into the *world of objects, of agents, and of ideas*. Starting with the world of objects, we quickly proceed to the world of agents in the first half of the book, while the world of ideas occupies the second half. But the structure that runs through all the chapters is determined by an information-processing hierarchy, and we try to match this hierarchy with ideas from evolutionary biology.

At the basic level, there are processes of which we are forever unaware; they are baked into the brain by millennia of evolution. These are built into the nervous system in

2. But we can easily (but falsely) assign agency to inanimate objects and are likely to do so for increasingly sophisticated artificial agents.

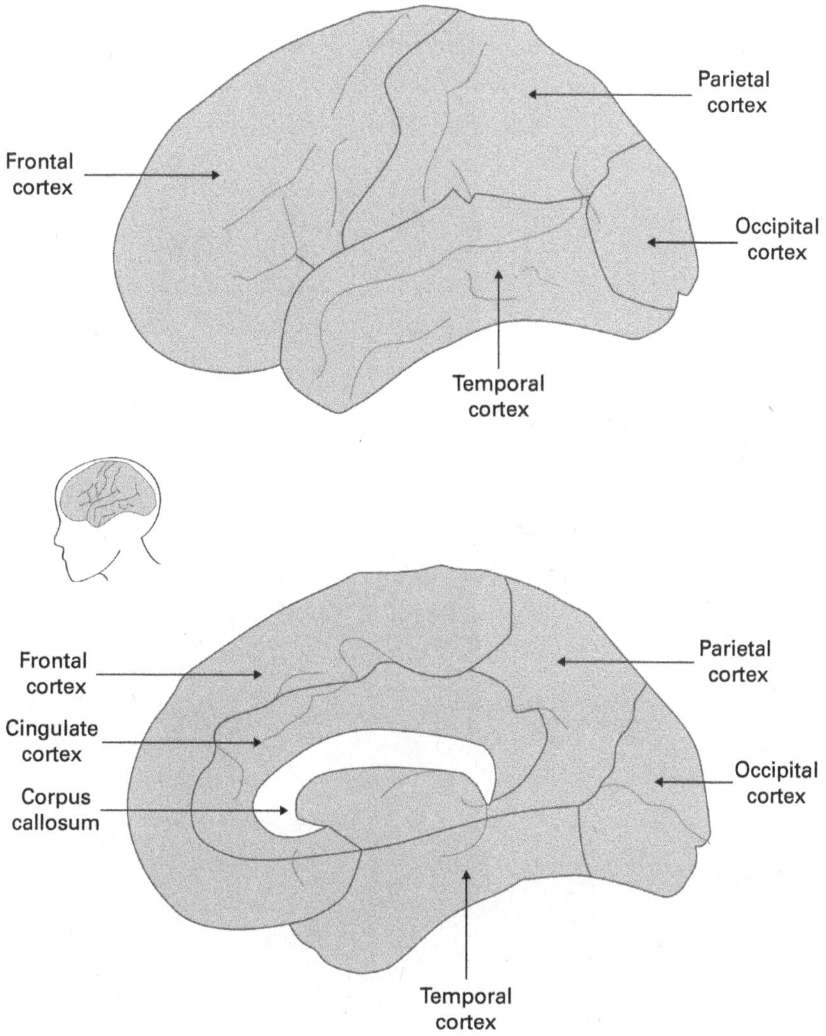

Figure 1.2a

The major regions of the human brain. The insula cortex is buried behind the temporal lobe (see figure 1.2b). The two hemispheres are joined by the corpus callosum. (Figure created by the authors.)

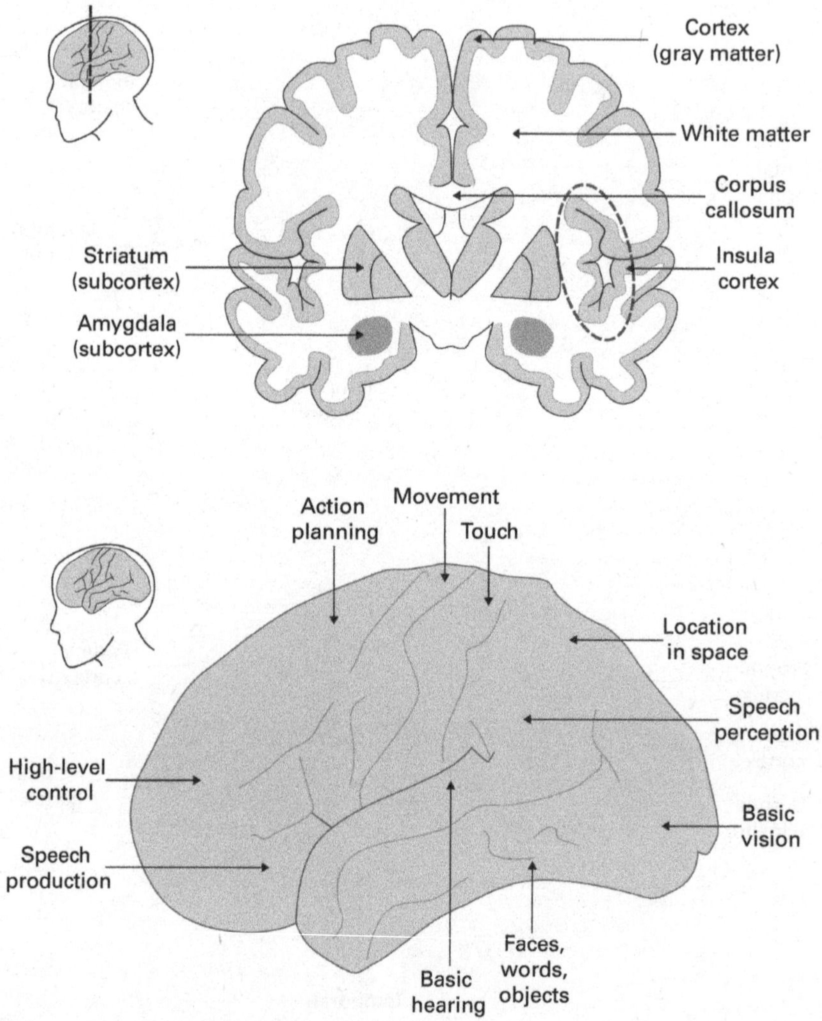

Figure 1.2b

(top) A slice through the brain showing the left and right hemispheres, insula cortex, and subcortical structures. (bottom) Left hemisphere, lateral view, showing the rough locations of some brain functions. (Figure created by the authors.)

most, if not all, living creatures to enable them to survive and even flourish in social groups. They include processes that underlie cooperation and competition, those seemingly universal features of life. When these processes come into play, we act like zombies or automata. We are unaware not only of the cognitive processes, but also of the stimuli that have affected our behavior; moreover, we are unaware that our behavior has even been affected. This is likely the case for most of our actions (Pisella et al., 2000) and emotions (Winkielman, Berridge, and Wilbarger, 2005). This bottom level has been called "type zero" (Shea and Frith, 2016).

The next two levels are roughly equivalent to Daniel Kahneman's systems 1 and 2 (2011). System 1 includes the processes by which we make rapid, intuitive, *gut* decisions. Here, we are aware of the problem and the solution, but we have no idea how the decision was reached. We just know that it is the right decision—and, sometimes, of course, it isn't! System 2, in contrast, includes processes that are conscious and deliberate. If pressed, we can give some account of how we reached the decision. This system is probably less susceptible to error than system 1, but it is often too slow to prevent us from making the error. Instead, it can help with hindsight.

We speculate that processes at the topmost level of the hierarchy are relative latecomers in evolution and are mainly observable in humans. At this level, for example, there is an interface that allows ideas to be shared among individuals via verbal instruction and formal teaching. It also opens the doors to deception and allows us to become manipulative—indeed, even Macchiavellian. Importantly, humans can create meaning and convey it to each other. This interface tends to reduce individual differences in behavior and cognition, leading to the acceptance of behavioral norms and shared beliefs. We believe that the ability to share ideas at this level fires the engine of human culture.

Cooperation

The benefit of being with others is the theme of the first six chapters. As was so plainly manifest during the coronavirus pandemic, we don't just like to be together with others; we want to move, talk, and act together. Perhaps the most important feature of our need to be with others is our willingness to sacrifice the needs of the self to the needs of the group. This is the implication of the phrase "We're all in it together." And becoming part of a group in this way has some very direct effects on our behavior.

Humans tend to do what others do and unwittingly learn from each other just by copying their actions (chapter 2). But doing as others do comes in a variety of forms and is not found only in humans (chapter 3). Many types of animals go where others go and do what others do. This kind of copying allows us to avoid the mistakes that are inevitable when we have to learn by ourselves. Copying helps us to obtain resources and stay safe. It aligns us with other members of our group at the physical and mental levels.

We copy not only the actions, but also the emotions of others (chapter 4), and this relates to one of the most valued features of our social interactions—empathy. On its most basic level, empathy is feeling the pain of another in much the same way as we feel our own pain. But there are more sophisticated forms of empathy, such as suppressing feelings of pain to appropriately help a suffering creature.

Imitation facilitates alignment, but we have an automatic drive to align, which makes it almost miraculously easy for agents to work together (see chapters 5 and 6). In these chapters, we also touch on deliberate and conscious alignment and imitation. Humans often imitate other people's actions much more precisely than is strictly necessary to meet the goal. By overimitating other people in this way, we signal a common purpose and demonstrate our attention and intention. We also become more like each other. This cements our status in the group we belong to and helps us to establish common ground, an essential prerequisite to be able to communicate with each other at sufficient depth in the world of ideas. It is not only the automatic (zombie) processes that have an important role in cooperation, but also consciously controlled (Machiavellian) processes that underlie human interactions in the world of ideas.

Competition

Alongside the basic need to be with others, and despite the many advantages that arise through learning from and working with others, there are also difficulties in being with others: there is perpetual competition. In the next five chapters of this book, we consider the many processes involved.

The most basic kind of competition, found throughout the living world, is between predator and prey. To survive this competition, creatures must be able to recognize other agents and predict what they are going to do next (chapter 7). Making such moment-by-moment predictions about future movements can be based on the physical principles associated with objects, such as the laws of motion and the constraints imposed by the body. However, for predictions over the longer term, it is useful to consider the goals of our opponents. Most animals can recognize goal-directed agents from their rational behavior (e.g., avoiding obstacles), as well as their consistent preferences. Humans, and perhaps some other species, can make even better long-term predictions about the behavior of others if they make inferences about their intentions and beliefs.

However, competition is not restricted to predator *versus* prey. There is fierce competition among groups within many species, including humans. Why do we so easily divide ourselves into groups that are hostile to each other (chapter 8)? We believe that splitting starts with predispositions with which we are endowed by evolution. For instance, we have a predisposition to affiliate, so we automatically align ourselves with others—first and foremost with our own families, but not with strangers. As a result of this affiliation, we become more similar to each other, behave in the same way, talk in the same way, and even think in the same way. At the same time, we become increasingly different from all those people who live elsewhere or lead another way of life. We create ingroups and outgroups.

In chapter 9, we ask why we put such high value on having a good reputation. We can obtain a good reputation if we can suppress our selfish urges to the benefit of the group. Our reward is a gain in status. We are now seen as valuable members of our group, and we are likely to be chosen as trustworthy partners. The benefit for the group is that it becomes more cohesive and can compete better with other groups. We discuss intuitive impressions of trustworthiness and compare them to information that we can gain through gossip. The power of gossip in building and maintaining reputations is not to be underestimated. Gossip trumps our own evaluation of another person's trustworthiness even though this evaluation comes from firsthand experience. This experience is often based on only a few encounters and can therefore be unreliable.

We situate the ability to mentalize (chapter 10) at the heart of social cognition, having been our pet subject for many years. Mentalizing facilitates the manipulation of hidden states, such as beliefs and desires. And through such inferences, it becomes possible to use deception and persuasion, which is a particularly advantageous strategy in competition. Mentalizing, according to our current view, takes two forms—one implicit and shared with other species, the other explicit and uniquely human. The explicit form has allowed people to drive up competition to never-dreamed-of heights in outthinking opponents. Knowing this, we must be constantly vigilant with each other. Unfortunately, no elegant computational theory exists to explain the mechanisms underlying mentalizing—yet. On the other hand, we have a great deal of information about the brain regions that support mentalizing.

In chapter 11, we consider some of the negative consequences of competition. Mentalizing is a distinct advantage in competitive situations, such as when we deceive others by implanting false beliefs. Vigilance is needed to combat such deception. We revisit intergroup competition and the pervasively negative evaluations of outgroups. While prosocial tendencies emerge early in development, they are not applied when interacting with outgroups. Top-down control can be used to overcome this behavior, but unfortunately, it can also be used to justify bad behavior and demonize outgroups. As for competition within groups, we often will behave selfishly if we think that we can get away with it. However, there is a spectrum of selfishness, varying from extreme altruists to extreme egoists, and it is possible that in the long run, this variation may be important for the survival of the group.

Computation

Chapter 12 takes on the task of proposing computational mechanisms for at least some of the basic cognitive processes needed to make sense of social interactions. These mechanisms give a reasonably good account of how animals, including humans, might learn about the world, predict behaviors from moment to moment, and identify causes. But these mechanisms are much less capable of predicting what individuals are going to do next in the long term. In this chapter, we strongly rely on the idea that the brain is a prediction engine. This is the basis of the hierarchical information-processing system that plays a big role in our book.

We believe that it is important to at least gesture toward the computations that underlie the cognitive processes that we are discussing. Computational descriptions will always be more exact than those that depend on words and will provide parameters that can be mapped directly onto brain activity (O'Doherty, Hampton, and Kim, 2007). In addition, results from studies of the brain provide important information as to which computational models are more plausible (Pulvermüller et al., 2021).

We find that the conscious-unconscious distinction (also referred to as "explicit-implicit") exposes a rift that goes right through all attempts to study our mental life. Our everyday folk psychology assumes that all that happens in our mind is within our conscious awareness. A century or more of psychological studies has proved this to be a grave, but understandable mistake. Under the hood, there is a mass of machinery that is processing information, but it remains so hidden that we totally ignore it. We have no idea how we predict a ball's trajectory through the air (Reed, McLeod, and Dienes, 2010). We don't even recognize that we are making a prediction. We just catch the ball.[3]

The pioneering nineteenth-century physiologist Hermann von Helmholtz was the first to recognize this. He suggested that our brains make countless inferences and, most of the time, this happens outside our consciousness. From the start, this idea has made people uncomfortable.[4] Isn't making inferences the prerogative of the rational

3. It should be clear that the unconscious processing that we are talking about has nothing to do with the Freudian unconscious, the id.

4. This includes von Helmholtz himself: "Later I avoided that term, 'unconscious inferences', in order to escape from the entirely confused and unjustified concept—at any rate so it seems to me—which Schopenhauer and his disciples designate by this name" (von Helmholtz, 1878/1971, 220).

mind? Isn't the unconscious dominated by primitive urges? But could it be a mistake to assume that our unconscious processes are basically irrational and selfish, while our conscious ones are good, rational, and altruistic? Could it be the other way round?

We subscribe to the belief that the brain is a prediction engine because it explains a whole range of behavioral phenomena and provides insights into the working of cognitive processes, as well as their neural underpinnings. This approach is often referred to as "Bayesian," in homage to the mathematician Thomas Bayes (1701–1761), whose work inspired a modern view of how the brain might work. Actually, this view dates to von Helmholtz and his insights into how perception works (von Helmholtz, 1866/1962). He recognized that the evidence from our senses is never enough to determine our perception of the world. We also need to have prior expectations about the world.

This is where Bayes's mathematical work comes in. His equation shows how best to combine our sensory evidence and our prior expectations in order to infer what is out there in the world (Bayes, 1763/1958).[5] This idea provides a basic mechanism for the idea of the brain as a prediction engine. It has proved appealing to both neuroscientists (Friston, 2008) and philosophers (Hohwy, 2013; Clark, 2015).

Figure 1.3 demonstrates the effect of a prior expectation that exerts a strong influence on how we interpret a visual image. Disconcertingly, this presents us with the realization that we cannot rely on the fact that what we see is simply "what is out there in the world" that is conveyed from the eye to the brain in a bottom-up fashion. The shape of the dots is interpreted very differently, turning from convex to concave and vice versa, according to the direction they are viewed from.

The elegance of the Bayesian formulation allows us to apply the same principles at increasingly abstract levels in a hierarchy of predictions. So, for social cognition, the same approach can be applied to the way in which we "read" intention from movement (Kilner, Friston, and Frith, 2007), as well as to the way in which our beliefs and behaviors are molded by culture (Chris Frith, 2014). In each case, it includes increasingly abstract and general levels of prediction that apply at higher levels of the system.

This correspondence is illustrated in figure 1.4. Here, we see a hierarchy with three levels with increasingly abstract content as we move up. At the bottom of the hierarchy, the system is interacting with the world of objects (i.e., keyboards and screens). At the top of the hierarchy, the system is interacting with culture, the world of ideas and norms. Prediction errors are sent up from lower to higher levels, while upper levels act on lower levels to resolve the errors. So, for example, when the wrong character appears on the

5. Why was this process of interest to Bayes, a clergyman? Possibly he wanted to argue against David Hume, who suggested that no evidence would ever be sufficient to change his prior belief that miracles cannot occur.

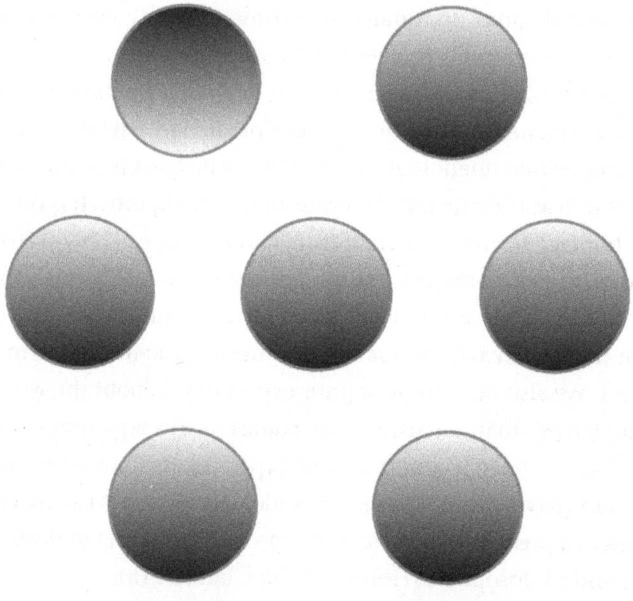

Figure 1.3
The prior expectation that light comes from above: In this image, you see six buttons and one hollow. But if you turn the page upside down, you will see six hollows and one button. This is because you expect light to come from above, so hollows will be dark at the top and buttons will be dark at the bottom. (Figure created by the authors.)

computer screen (a prediction error), signals from the upper level modify functioning at the lower level so that typing is slowed. At the top level, we can imagine an editor vetoing the title of the article "The Social Brain: Allowing Humans to Boldly Go Where No Other Species Has Been" because it flouts the cultural norm that prohibits split infinitives.[6]

In this chapter and subsequently, we argue that the top of the hierarchy is not, as one might think, the individual, but rather the culture that influences the individual in innumerable ways.

Culture

In the last part of the book, we continue to use the framework of the hierarchy of information processing to understand the world of ideas of humans. Here, we are concerned

6. There actually is a paper with this title (Frith and Frith, 2010), which managed to get past the editor.

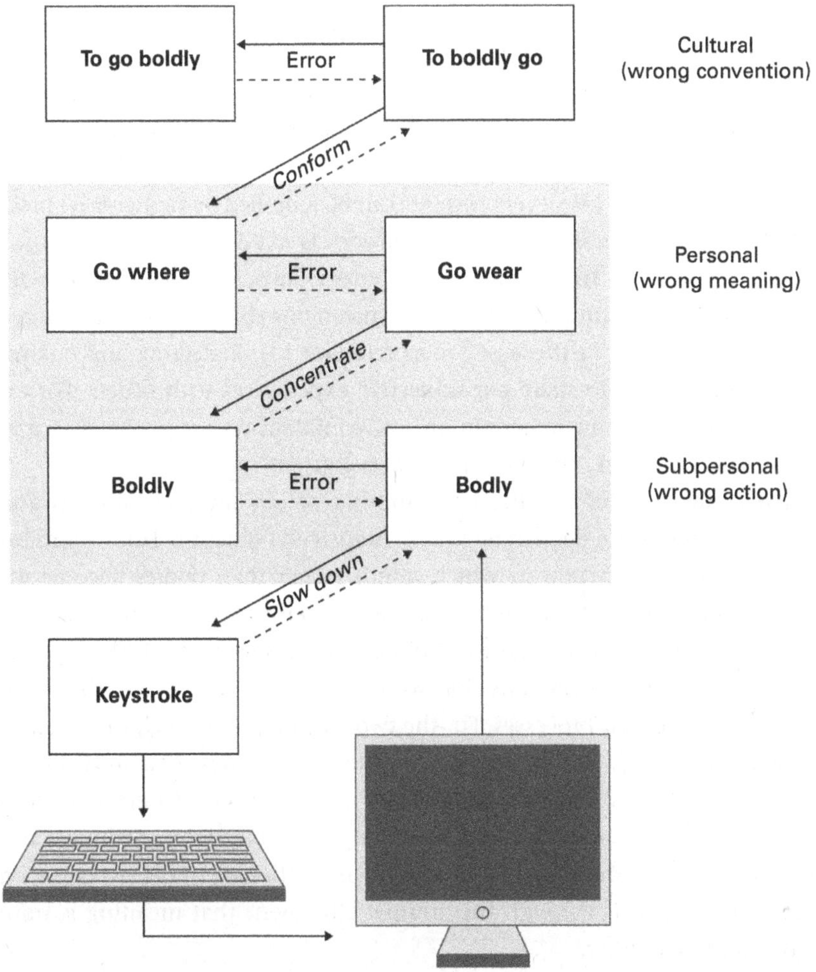

Figure 1.4
Sketch of a hierarchy of control in the mind/brain: In this hierarchy of control, the mind/brain (shaded area) interacts with the physical world of objects (keyboards, screens) at the bottom and with the mental world of culture at the top. The hierarchy has three levels. At the lowest level, there is an interaction with the world of objects via vision (letters on the screen) and action (keystrokes). At the middle level, there is an interaction between perception and prior knowledge to create more abstract entities, such as words. At the highest level, there is an interaction between the individual and culture. At each level, there is an interplay between expectations (from above) and evidence (from below). Expectations in the left column are compared with evidence in the right column. If there is a mismatch (prediction error), action is taken to reduce the error. At the bottom the error is caused by hitting the wrong keys (bodly), in the middle by choosing the wrong word (wear), and at the top by using the wrong style (to boldly go). (Figure created by the authors inspired by figure 2 in Friston [2005]).

not simply with modeling the social world, but with modeling the modelers. The levels in the hierarchy do not act independently of each other, and information can flow up and down. As we suggest in chapter 13, the low-level automatic and unconscious processes are coupled with "minders" that monitor their functions. If problems occur, the minders can send signals to higher-level systems, which then take action to resolve the problem. In this way, unconscious low-level processes can be modified by higher-level processes.

At the top of the hierarchy in our framework is explicit metacognition, a special form of social cognition and a human superpower. Conscious mentalizing is an example of explicit metacognition, and by this we mean our ability to reflect on how minds work, whether our own or those of others (chapter 14). Reflecting and talking about our thoughts enable us to share our subjective experiences with others. For example, we can tell each other how uncertain or how confident we are in the choice we have just made. This enhances our ability make decisions (chapter 15).

Explicit metacognition is critical for another of our uniquely human abilities—namely, to communicate deliberately (i.e., ostensively) (chapter 16). Communication is an example of joint action in which minds rather than bodies become mutually aligned. On the one hand, internal signals about our private cognitive processes are converted into a public form that can influence others (Sperber, 1996). On the other hand, we can convert instructions that we receive from others into a form that can change our own thought processes. For the two-way process of reciprocal communication, alignment and adaptation are critical. There is the radical claim that we cannot know what we are thinking or talking about unless we have interacted with others (Davidson, 1991). This point is made succinctly by Robert Frost: "You can never tell what you have said or done until you have seen it reflected in other people's minds" (Lathem, 1966, 71). It is through this mutual alignment that meaning is transferred from one mind to another.

Once we can convey meaning to each other, our ability to teach goes up a notch (chapter 17). Human beings can go beyond simple demonstrations and coaching; now we can teach through sharing experiences explicitly. Ideally, there is mutual adaptation between teacher and learner, just as there is for communication more generally. Thus, the teacher can correct misunderstandings, gauge the precise state of knowledge of the learner, and provide the information that the pupil is ready to receive. In everyday life as well, we use verbal instructions to enforce norms. In the physical world, we can help people avoid danger (Don't eat rhubarb leaves; only eat the stem). In the mental world, we can help people avoid confusion (Keep left).

Culture can be thought of as a body of knowledge, opinions, skills, norms, and other things that humans learn from other humans by imitation and teaching (Richerson and Boyd, 2005). Many animals, including apes and meerkats, learn specific ways of doing things through imitation and implicit teaching (Thornton and Clutton-Brock, 2011; Whiten et al., 2009). Humans, however, have a uniquely cumulative form of culture that is embodied in a mental world of explicit knowledge, opinions, and rules of behavior (chapter 18).

But how do individual mental representations of opinions, norms, and other elements of society become widely distributed in a group; and conversely, how do group representations come to affect the ideas and behaviors of the individual (Sperber, 1996)? We believe that the ability to share experiences is the foundation of cultural evolution—the oil on the wheels that propel advances of the individual and of the group. Hence, we argue that explicit metacognition, the human ability to report and discuss our subjective experiences, is the key process (Shea et al., 2014).

Through this process, it is not only our behaviors that become aligned, but also the way we think and how we experience the world. From this alignment, cultural practices and institutions emerge. Science in particular advances through our ability to share with each other our models of how the world works. These models make predictions that can be tested against reality. Those leading to more accurate predictions are chosen, so our models get better and better.

We will also discuss (chapter 19) the intimate relationship between consciousness and communication and show how explicit metacognition is critically involved in our sense of agency and our feeling of being responsible for our actions. Even young children know that when they do something wrong, claiming that "It was an accident" is a good excuse. Most people believe that we deserve punishment for bad things done deliberately, of our own free will, rather than not on purpose (Nahmias et al., 2005).

The concepts of free will and responsibility, of selfishness and altruism, of ingroups and outgroups, present tough problems. Yet they may be key to unlocking one of the trickiest questions about our ability to live together: can we become better social beings? We will argue that discussions about how the mind works can generate a cultural consensus that, in turn, modulates behavior and, of necessity, brain function. This provides a measure of hope that we might eventually get along with each other better than we do now.

Our Grand Aim

The study of human social cognition casts a strong light on the mind-brain problem. Social cognition in humans is all about people interacting with each other and with their culture. Claims that the brain is involved in such interactions imply, for some, that people are just machines and culture is irrelevant. One of our main aims in writing this book is to challenge this misunderstanding.

In this book, we explore how social cognition is enabled by the brain. We want to lay bare what is social about it. Simply being with other people is not enough, although the presence of other people affects us profoundly. Indeed, other people need not even be present. We can address them with intensely felt letters, and we can read about and imagine social interactions in stories. The coronavirus has surprised many of us about how well we can manage without physical contact.

There have been many attempts at explaining our social nature without any reference to the brain or to the countless similarities between all social animals. Here, we will try to elaborate how biology, via basic survival mechanisms and evolution, and culture, via teaching and technology, prepare us to navigate the social world. We believe that the human brain has to play a central role in our quest because the brain is the arena where biology and culture meet and enable us to build our social lives. This is why we put social cognition at the center of what it means to be human.

I Cooperation—Benefits of Being with Others

2 Learning from Others

This is the first of five chapters dealing mainly with cooperation. All animals, including humans, can learn more by copying than by direct experience. Copying allows us to benefit from the experience of others. For example, we learn where food is by following others. We learn about the value of objects by observing whether other members of our species approach or avoid them. We are automatically attracted to where they are looking. In this way, the goal that they approach can become our goal. By copying others and following their gaze, we can learn what are the important stimuli among the many that otherwise might overwhelm us. Thus, our conspecifics provide a filter for information through which we can learn about the world of objects more quickly and more accurately than we could on our own.

We also benefit from priors or predispositions that, as a result of evolution, automatically direct us to what is most important. One of these is the predisposition to do what other agents do and achieve their goals by attending to their gaze. Learning itself does not have to be learned. The mechanisms required for social learning seem to be the same as in learning more generally. However, humans do not learn indiscriminately and have developed additional kinds of social learning, such as formal teaching, which allow us to enter the world of ideas and accumulate culture.

* * *

From early in life, we need to learn about the world of objects. What is safe to eat? How do you start a fire? How can you keep dry in the rain? So why should this kind of learning be discussed in a book on social cognition? We can learn about the physical world by ourselves, but mostly we learn about this world from other people. We are not necessarily learning about social aspects of the world, but the way we learn is intensely social.

We learn from others all the time, mostly without being aware of it. In this chapter, we want to draw out one thread from the many complex processes involved—let's call it the *Zombie thread.* It refers to the unconscious processes that we find in spontaneous learning and behavior. We will later consider a very different thread, to do with conscious learning in humans. Let's call that the *Machiavelli thread.*[1] The name does

1. This is named after Niccolo Machiavelli (1449–1527), who advised the Medici prince to act deliberately and calculatedly, under the assumption that he would always meet deception.

not imply something evil; rather, it implies that it operates via calculated actions and presumes that deception is possible.

The Advantages of Learning by Observing Others

In the 1960s, the social psychologist Albert Bandura (1965) claimed that social learning is unlike other forms of learning because it does not need trial-by-trial reinforcement. Trial-by-trial learning is a basic mechanism that is still studied extensively in animal experiments. It depends on direct experience. We perform an action, and if the outcome is good, we repeat it. If the outcome is bad, we try another action.

Bandura pointed out that, when we learn from others, we can have "no-trial learning." He proposed that, rather than learning from our own direct experience of rewards and punishments, we can learn by witnessing the behavior of others. Such vicarious learning can be seen in all animals, from fruit flies to meerkats to humans.

Consider the red-footed tortoise. These are not the most social animals. They live lives of almost complete isolation, apart from the brief interactions necessary for reproduction. And yet they can learn to perform a difficult task, a detour problem (see figure 2.1), simply by observing and copying another tortoise that has already learned how to do it (Wilkinson et al., 2010). Without a model to copy, the tortoise would have to make many unsuccessful attempts and may never even be able to reach the goal at all.

The finding is remarkable in showing that social interest is not a necessary precondition for social learning. Nor is the experience of maternal care a prerequisite for becoming a social learner. Folk beliefs that children must be socialized before they can learn by observation seem rather questionable in this light.

The vivid demonstration of observational learning in a nonsocial reptile stimulated us and others to think about the basic requirements for this type of learning that does not require conscious processes. *How* do we learn from others? What are the basic tools through which we exploit the knowledge of others and need not learn afresh? Just by inspection of random cases, we can conclude that the major tool is copying. If we go where others go, we are likely to arrive in places where there is food and water, and hopefully an absence of predators. If we eat what others eat, then the result is likely to be nutritious rather than poisonous.

Learning about Nice and Nasty Objects

At its simplest, social learning, as in the case of the red footed tortoise, allows us to learn about the world of objects. If you learn about these objects from others, you can

Figure 2.1
A tortoise learning a detour problem: The tortoise can see the lettuce, but to reach it, it has to move away from it, around the barrier. A naive tortoise needs over thirty sessions of training to learn the task. But when a naive tortoise is allowed to observe a trained tortoise, it can master the task in only a few trials (Wilkinson et al., 2010).

not only save time, but also avoid pain. If you depended on your own experience, you might well not survive. Learning to avoid a poisonous snake works best if you can observe another member of your species doing the avoiding.

In a series of ingenious experiments, Susan Mineka and colleagues demonstrated that lab-raised rhesus monkeys acquired a fear of snakes very quickly by observing a video where wild-reared monkeys showed fear toward a toy snake (see, e.g., Mineka and Ohman, 2002). Interestingly, the monkeys did not learn to be afraid of a toy flower, which the crafty experimenters had inserted into a video in place of the snake. There seems to be a predisposition to learn to be afraid of nasty things like snakes. As we shall see, learning is not enough on its own. We need to have predispositions about what to learn.

Mineka's monkeys demonstrated fear conditioning through observation. This exceptionally important mechanism has also been demonstrated in humans (Olsson, Nearing, and Phelps, 2007). There are computational models that do a good job of explaining how this indirect value-based learning works, and we also can point to a useful review by Andreas Olsson, Ewelina Knapska, and Björn Lindström (2020).

The Wisdom of Sticklebacks

Just as we can learn what things are nasty by observing others, we can also learn about what things are nice and where to go to find them. We see this happening in an elegantly simple experiment with nine-spined sticklebacks, small predatory fish found in rivers and ponds. But here they are in big glass tanks in the lab. An individual stickleback was isolated and learned that food could be found on the left side of a tank, but not on the right side. After a while, the fish would always swim to the left when given the choice. After a delay of seven days, the stickleback observed other fish feeding on the right side of the tank. Promptly, it abandoned its previously learned preference and now swam to the right (van Bergen, Coolen, and Laland, 2004). Simple observation of others overturned a preference previously gained by individual learning (go to the left) in favor of the opposite preference (go to the right).[2]

Examples of social influence on foraging behavior have been observed in many other animals (Galef and Giraldeau, 2001). When looking for food, a good strategy is to go where the others go (Rieucau and Giraldeau, 2011). This phenomenon is well understood by restauranteurs, who seat early arrivals in the window to attract more customers.

It Pays to Be a Copycat

Copying pervades social learning. But beware: copying can mean many different things. It could mean that one agent is following the same goal as another, or that one agent is performing the same sequence of actions, regardless of the goal. Actions can be copied as faithfully as possible, as in mimicry. But when copying means adopting the same goal as others, it's emulation. This entails choosing different actions as a means for getting to the same goal (e.g., Charpentier, Iigaya, and Doherty, 2019).

These differences have been revealed in many studies where animals learn which actions to perform by observing others (Huber et al., 2009). An ingenious method, which works well with birds, monkeys, and humans, is the use of puzzle boxes (see figure 2.2, Thorndike, 1898). These boxes contain rewards and can be opened by a series of more or less complex actions, often with more than one solution.

Chimpanzees will imitate a demonstrated sequence of actions to gain access to food in a puzzle box (see Whiten et al., 2009). In one charming experiment, wild mongoose pups learned how to open modified Kinder eggs by observing an adult mongoose

2. We should not take from this example that habitual behavior is easy to overturn through social learning. In the fish, memory of the location may likely have faded over the seven-day interval.

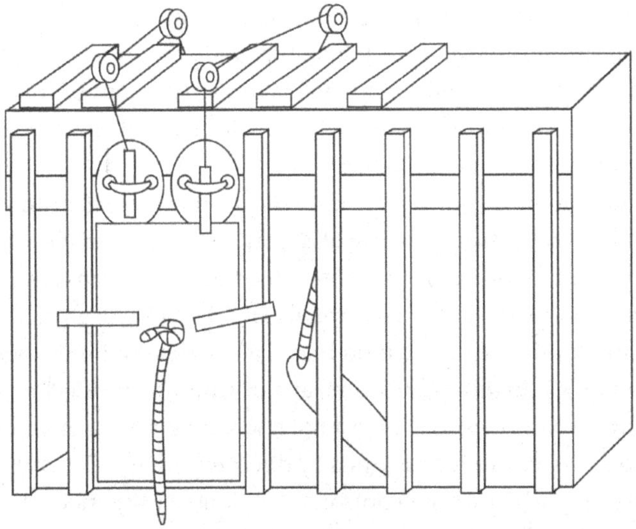

Figure 2.2
A puzzle box: The door is held in place by a weight suspended by a string. To get out of the box, it is necessary to depress a treadle, pull a string, and push a bar up or down (Thorndike, 1898).

(Müller and Cant, 2010). Some of the adults were first taught by the experimenters to open them by smashing them on the ground, while others were taught to bite them open. Both actions were equally effective at getting access to the food inside the container. However, the action chosen by the pups was determined by what they saw the adult do, demonstrating the attraction of copying.

Such imitative learning can also be seen in human infants from around the age of one year (e.g., Carpenter et al., 1998). By this age, infants will imitate useful but arbitrary actions when they see an adult interacting with a novel object. This comes as no surprise to parents, who find that their babies quickly learn that they can press buttons on remote controls and phones to access videos or images that they like.

Observing others is a very efficient way of learning about the world. But is it actually better than doing it your own way and learning by trial and error? Kevin Laland and colleagues (Rendell et al., 2010) put the question to the test. In a highly original study, they devised a competition for computers and their programmers in which the goal was for artificial agents to acquire the most adaptive behavior in a complex environment. They attracted programmers to take part in the tournament by promising a reward for the winner. The programmers had to come up with their own strategies, and it turned out that there was a clear winner.

The most successful program was one that relied almost exclusively on its agents copying other agents rather than learning by trial and error. Interestingly, the copying of other agents' strategies was indiscriminate, taking no account of their awkwardness or elegance. Copying was outcome focused and selected strategies solely on the basis that they had worked for another program.

Why was indiscriminate copying so successful in the context of this tournament? Apart from avoiding errors that are an essential part of trial-and-error learning, the copying agent took advantage of the fact that the observed agents had already selected the actions that they had found to be most beneficial. They effectively and inadvertently acted as a filter to provide the information that was most useful for the observing agent.

Clearly, copying is a highly adaptive means of gaining knowledge (Rendell et al., 2011), but it may not always work. For example, some skills, such as applying the right force to the piano keys, can only be learned by direct experience (Yuniarto, Gerson, and Seed, 2020). Likewise, when we are confronted by a novel situation, there may be no one who has the relevant knowledge, and therefore there is no one to copy. Explorers are needed who can strike out on their own, often risking great danger to themselves. So we cannot always rely on exploiting the knowledge of others. We must also be able to learn for ourselves, by trial and error.

While it is often advantageous to copy mindlessly when you can, it can sometimes lead to disaster. For example, this happens when we wrongly believe that others have more knowledge than we have. This belief can be transmitted to others through behavior, below the radar of consciousness, and then gives rise to so-called information cascades (Bikhchandani, Hirshleifer, and Welch, 1992). Each member of the group copies the others, and if asked, might justify this by the mistaken belief that these others know something that he or she doesn't know. This can lead to stock market bubbles, as well as to dangerous behavior when crossing a road, where we start to cross just because others are doing so (Faria, Krause, and Krause, 2010).

How and What Do We Learn by Observing Others?

Is a special mechanism required when we learn by observing others rather than from direct experience? Not really. When we learn about the world by direct experience, we implicitly build up associations about things. We learn that it is good to approach some things and not others. We learn about the signs that warn us of danger. And this ability to learn does not have to be learned in itself! The ability to learn is an innate predisposition and has a long history in evolution. Even the miniature nervous system of the sea slug "knows" how to associate stimuli and responses in classic association learning

(Brembs et al., 2002). The brain of a human baby "knows" how to extract patterns from speech by means of what is now called "statistical learning" (Saffran, 2003). The neural mechanisms that underlie these kinds of learning are now quite well understood. Indeed, computational scientists have achieved enough knowledge about the bare mechanics to build them into machines that manage "deep learning" (Hassabis et al., 2017). We will say more about these mechanisms in chapter 12. They are all part of the Zombie thread.

We believe that these neural mechanisms are in play whenever we learn about the world of objects by observing the behavior of others. For example, the dopamine system in the midbrain (ventral tegmentum) is activated when we receive a reward, and also when we see someone else getting a reward (in rats: Kashtelyan et al., 2014; in humans: Mobbs et al., 2009; also see figure 12.2 in chapter 12). Neural activity in the amygdala has been studied in patients undergoing neurosurgery. Neurons in this region compute expected rewards for both oneself and others (Aquino et al., 2020). One source of the power of observation is the automatic resonance that we feel with conspecifics (see chapters 3 and 4). Our empathy with the pain of others, also associated with amygdala activity, can warn us of imminent danger after observing their expressions of pain and fear (Lindström, Haaker, and Olsson, 2018). We feel their pain and associate this with the danger.

But how do we know which information we should use? There is a wealth of information available to us from the behavior of others. The cues are there for the taking. But it can be hard to know which of them are worth picking. Just as in the nonsocial world, we are always confronted with a multitude of cues, so there needs to be a mechanism to point us toward those that are important to us.

Looking Where Others Look

One mechanism that helps us to exploit the knowledge of others is to pay attention to where they are looking. It might be somewhere nice! We can identify what they are looking at by attending to the orientation of their heads and bodies, or more subtly, the direction of their eye gaze. This is an entirely automatic process that has proved its worth through millennia of evolution.[3] We can think of eye gaze as a freely available filter that screens information for us. Being alert to where others are looking can pay off handsomely.

3. Simply looking where others look is not an example of joint attention. In joint attention, mutual gaze acts as a signal indicating that we should look at something together because the object is of common interest.

Gaze following plays a major role in many aspects of social cognition (see Stephenson, Edwards, and Bayliss, 2021 for a review). It has also been demonstrated in ravens, goats, dogs, and nonhuman primates (reviewed in Zuberbühler, 2008). Infants reliably follow the gaze of another person from around one year of age (Flom and Johnson, 2011), greatly increasing their opportunities to learn about the world.

We have compelling proof that gaze following is part of the Zombie thread—namely, that it happens even when it is a disadvantage, as shown in this neat experiment. Andrew Bayliss and Steven Tipper (2006) asked people to detect a target that could be either on the left or on the right of a screen. Before each trial, a face was shown, which might be looking either to the left or to the right. People were much slower to detect the target when the person on the screen looked in the wrong direction (i.e., away from the target). Participants were unable to stop themselves from following the misleading gaze, and they were unaware that, in one condition, the gaze was consistently pointing in the wrong direction. However, intriguingly, when asked afterward, they made a socially highly relevant observation: they rated the person who looked in the wrong direction as less trustworthy.

Automatic gaze following is possible only for agents who have a distinct head with eyes—and the more visible, the better. A unique feature of the human eye is the white sclera that surrounds the iris, which makes eye gaze highly visible and allows observers to judge rather accurately where a person is looking (Kobayashi and Kohshima, 2001). In nonhuman primates, the dark coloring of the sclera makes it more difficult to infer the gaze direction. Nevertheless, nonhuman primates have been documented to follow the eye gaze of other animals quite readily. It remains to be seen to what extent the change in color of the sclera was driven by advantages of deliberate information transfer between humans, a feature of the Machiavellian thread. This thread is also seen in our use of eye gaze as a deliberate signal to alert another observer to pay attention to a target, or we deliberately avert our gaze if don't want other observers to know where we are looking.

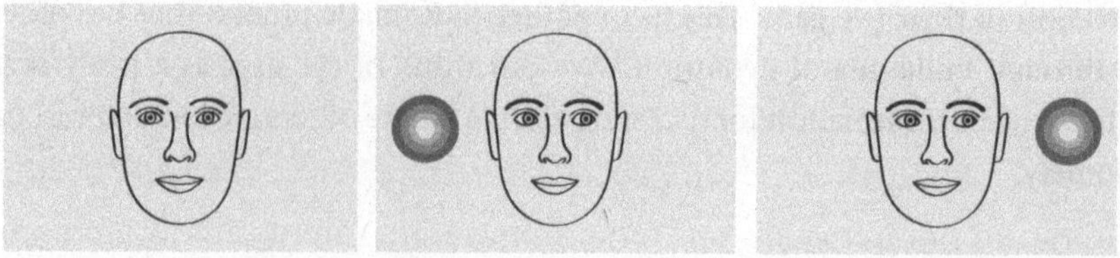

Figure 2.3
Eye gaze following: we automatically follow other people's gaze. So we can spot a target more quickly when they are looking toward it (Bayliss and Tipper, 2006).

Is there a specialized region in the brain that is responsible for responding to gaze direction? Studies of rhesus monkeys have helped to answer this question (Ramezanpour and Thier, 2020). A patch of neurons on the floor of the superior temporal sulcus (STS) is found to be uniquely sensitive to gaze direction in another monkey. Different neurons, in a nearby region, respond to the identity of the monkey.

Learning from others by observing where they look is undoubtedly a very useful trick. But how do we know to home in on eye gaze rather than on any of the other cues that abound in our environment? If we started to learn by taking in information at random, it would take far too long to avoid imminent danger. Fortunately, we don't have to start from scratch since we arrive in the world with a whole set of internal priors and biases. These set us on a fast track, so that we learn what we need to learn to thrive in our niche. An example mentioned earlier is that monkeys are predisposed to learn to fear snakes, but not flowers. We need to take a short detour here to explain why we need these biases and what they mean.

Priors, Predispositions, and Preference Settings

Nobody now thinks of the brain-mind as a blank slate. But the opposite idea—the brain as a bundle of instincts with fixed action patterns—is not right either. In the words of Kevin Mitchell (2018), the brain is not hardwired, but it is prewired.

For the brain to function as a prediction engine from the start (see chapter 12), it must be endowed by evolution with a number of priors or biases. After all, to make a prediction, you need some prior information. Giorgio Vallortigara (2021) has provided detailed evidence for innate preference settings for facelike objects in birds and mammals. Human infants too are predisposed to look at facelike objects from birth (Johnson et al., 1991). The built-in prior expectation here is that faces are a useful source of information. There are likely many other predispositions, which are necessary to get learning off the ground. We like the metaphor of starter kits for cognitive mechanisms, complete with factory settings that enable rapid learning. Just like a new washing machine, the factory settings may need to be adjusted to the demands of a given environment (e.g., the hardness of the local water). It would not be surprising if resetting is time limited by sensitive periods in development, which result in the almost indelible phonology of our mother tongue and the bane of a foreign accent.

We can now begin to answer the question of why we humans are so adept at learning from others by following their gaze. For this, several priors and mechanisms come into play: First, there is a prior that "knows" what counts as another agent; second, a prior that "knows" which agent to follow; third, a prior that "knows" to orient toward the agent's face and eyes. The orienting behavior is probably mediated by an ancient

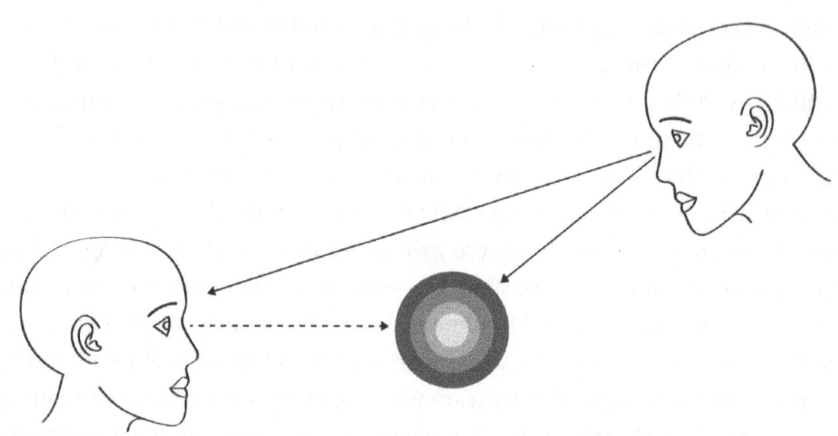

Figure 2.4
The geometric problem with gaze following: To follow the actor's gaze and look at the same target, the observer cannot simply copy the eye movements of the actor. The observer (on the right) must use the head and eye positions of the actor (on the left) to work out the target of the actor's gaze and then look at that target. (Figure created by the authors.)

subcortical route (Morton and Johnson, 1991). Orientation to the face and eyes enables gaze following, but the observer must also be able to learn that such following is likely to lead to valuable information. Here, lateral interparietal neurons seem relevant because they compute the value of social and nonsocial cues, but only when the information is relevant to decisions about gaze orienting (Klein, Deaner, and Platt, 2008).

For the observer to figure out what an agent is looking at, it is not sufficient simply to imitate the eye movement, since the agent and the observer will be in different spatial locations. The observer must take account of this difference in viewpoint (see figure 2.4). The gaze-following patch identified in the superior temporal sulcus of the monkey (Ramezanpour and Thier, 2020) plays a major role in this geometric computation.

The ability to take in viewpoints that differ from one's own can involve more than differences in spatial location. Viewpoints can differ according to knowledge, group affiliation, and culture. Taking account of such differences is a complex skill that is probably not available to all animals. We will only briefly touch on this topic here, but we will return to it in later chapters.

Who Should Be Copied?

It is not just a matter of choosing the right cues, such as faces, when we learn by observing others. We also have to choose the right people. We humans do not learn randomly from whoever is near us. Rather, we copy the people who we perceive to be knowledgeable and competent. We need to learn *about* agents in order to choose the best agent to learn *from*. We need role models. This is especially the case if, as with adolescents, experience is limited and preferences are not yet firmly established (Reiter et al., 2021). At this stage in life, we desperately need to learn the appropriate behavior and what is on trend for our group. How else can we decide what to put on our playlist or what type of clothing to wear? For the most part, this happens via Zombie learning, even though this might be vigorously disputed by those who have just begun to adopt the norms and conventions of their peer group.

We should observe and learn from people whom we judge as competent, no matter who they are. But we have a built-in predisposition to trust the members of our ingroup more. And we prefer to learn from those whom we trust. From the age of four months, children already expect imitation to be associated with affiliation (Powell and Spelke, 2018) and, by seven months, they expect members of the same social group to behave in a similar manner (Powell and Spelke, 2013).

This predisposition is not always for the good. It is likely responsible for the bias that distances ourselves from outgroups (see chapter 8). We learn from the ingroup, but we do not learn from an outgroup. This is a pity, as the outgroup might have found better solutions to a problem that we are still struggling with. And it is here that we would do well to follow the Machiavelli thread by deliberately suppressing the tendency to devalue people who are not in our group.

Is It Only the Observer Who Benefits from Transfer of Information?

Even though we can only speculate here, we are quite excited by the idea that the transmission of information among humans has evolved to be qualitatively different from transmissions in other animals. This idea is drawn from the writings of several of our favorite thinkers. To understand this idea, we need to distinguish between cues and signals. A cue is a feature of the sender, there for the taking, and of benefit to the observer, not the sender. A signal, however, is a promise, along the lines of "Hey—this is something that should interest you,"[4] of benefit to both sender and receiver.

To unpack these thoughts, we can assume that in many animals head or gaze direction is a cue. It serves information transfer and benefits the observer, not the sender. Presumably, the mechanism by which the cue is detected by the observer evolved first

4. As we shall see later, at higher levels of the information-processing hierarchy, signals can be used for deceptive purposes.

to take advantage of the information. It is likely that the corresponding mechanism by which the cue is sent (e.g., the change in the color of the sclera), evolved later to bring an advantage to the sender as well (Diggle et al., 2007; Scott-Phillips, 2008), turning the cue into a signal.

Use of signals that are of benefit to both sender and receiver can be seen in many animals. But humans take signals a step further: we exchange information deliberately and with selected individuals only. We use ostensive signals when we do so (e.g., calling someone's name, touching them, or subtly catching their eye). This is known as "ostensive communication" (see chapter 16), a huge step up from indiscriminate information transfer and part of the Machiavelli thread. As a result, we can learn from others by formal teaching (see chapter 17) in a far more targeted and effective way than by informal observation. Here, we are perhaps quite close to Bandura's one-trial form of social learning, but we see it in a very explicit form: I tell you what to do; just follow the instructions; please read the manual. Are other animals able to learn from others in this way? We rather doubt it.

What Is Special about Social Learning?

We doubt that special mechanisms are needed. All the principles of learning theory, as derived from a solid corpus of empirical studies, also apply to learning from others. However, it is difficult to shake the belief that there must be something special about social learning in humans.

For one thing, humans look upon culture as their supreme achievement, and culture would not exist if we did not learn from others. But is cultural transmission the sole reason for the overwhelming success of human culture? Take tool use, for example. While humans have a long history of inventing and perfecting tools, tool use has been documented in many animal species (Whiten, Caldwell, and Mesoudi, 2016). We have seen delightful videos of otters that crack shells by using stones as hammers. We know that groups of chimpanzees use sticks to extract food. Furthermore, their use is interestingly different in different groups, and this precisely points to social learning. Of course, affirming this ability in chimpanzees begs the question why they have not achieved a cumulative culture, as humans have.

There are several possible differences that mark social learning in humans. For example, there could be a difference in the weight that is given to social learning compared to personal experience. At least one experiment suggests this possibility. Vale et al. (2017) found that when a novel task was first learned and then demonstrated by a competent model, chimpanzees largely stuck with their own way of doing the task. In contrast, four- to five-year-old human children were far more influenced by the social demonstration of the task than by their own prior experience. The authors conclude that the children's learning was more malleable. They allowed their previously acquired behaviors to be replaced with new and often superior solutions to the task in hand. This happens when more weight is given to social learning.

Another difference is the extended period of childhood in humans, which provides masses of opportunities of learning from trusted others. Compared to humans, other animals grow up more quickly and hence have fewer chances to learn from those around them. Nevertheless, even fruit flies, that live for only about forty days, have sufficient time to learn from others. Indeed, their learning ability has been studied extensively so it can be linked to genetic factors. Remarkably, there is evidence for conformist social learning in their mating preferences, which are influenced by the prevailing behavior of other fruit flies (Danchin et al., 2018).

The evolutionary anthropologist Robert Boyd made a strong case for cultural learning as the basis of the success of human adaptation in many environments, from jungle to desert, from arctic to tropic (e.g., Boyd, Richerson, and Henrich, 2011). Cultural learning is like social learning, in that it has to be effective within an individual's lifetime. However, unlike social learning, cultural learning can evolve over generations and can be transmitted over long distances of time and space. With cultural leaning, we enter the world of ideas. It is this kind of learning that will determine the highest goals that we can pursue. These are topics we will discuss in the last part of this book.

3 Mirrors in the Brain

What is the basis of the universal predisposition for copying in the mind/brain? We assume that there are representations that connect the realms of perception and action and reach across the self and the other. Mirror neurons embody such representations. First identified in the prefrontal cortex of monkeys, they have also been identified in the human brain.

Copying is under the control of higher-level processes that determine whether it happens and when. This control, which need not be conscious or deliberate, allows animals to align with each other and be continuously ready to perform the most appropriate action in a given circumstance. Hence individuals can act as one, enhancing protection from predators. Furthermore, copying promotes affiliation with other members of the group, and ostracism leads to increased copying.

Imitation of actions is complemented by emulation of goals. Emulation is a form of social learning that allows animals to share the same goal, even if the actions they have available are different. Overimitation is a special form of social learning where even unnecessary actions are imitated. This is observed in humans and functions as a tool to forge group identity. Overimitation is situated at higher levels of the information-processing hierarchy. So are other conscious and deliberate forms of copying and emulation, as practiced, for example, in rituals. They are fundamental to the creation of human culture.

* * *

Friendly aliens visiting our planet would surely be eager to learn about our world. They would learn easily about the world of objects because the laws of physics would be familiar to them from their home planet. However, they might well be astonished at the world of agents. What they might report when they get home is that most of the animals on Earth congregate and imitate each other. They might conclude that the basic unit of life was not the individual animal, but the group.

From a distance, it might seem like a fusion of many into one, followed by splitting into different groups and sometimes clashing violently. The pervasiveness of acting as others do and following the same goals as others in the group is a startling phenomenon. It is performed Zombie fashion, and we take it for granted. As we established in chapter 2, it is the remarkable ability to copy that enables us to learn so easily from others. But how we do it?

Key Concepts: Mimicry, Emulation, Overimitation

The term "imitation" is rather vague and can be applied to very different types of behavior. Three forms of imitation—mimicry, emulation, and overimitation—have been much explored by researchers of social behavior in many different animals. There has long been debate about what the underlying mechanisms might be and how they might originate. These are popular, if as-yet-unresolved topics, which have stimulated a large number of experiments and publications. Fortunately, we can refer the reader to a review by Antonia Hamilton and colleagues (Farmer, Ciaunica, and Hamilton, 2018).

So, what are the key concepts? The term "mimicry" has been used to refer to the process by which a nice-tasting insect, for example, evolves to look like a dangerous, nasty-tasting insect, thereby deterring predators (see figure 3.1). In this chapter, however, we use the term to refer to the copying of actions rather than visual appearance. We can see mimicry of actions as a means to blend in with the environment formed by other members of our group.

This definition of mimicry in members of a group, while hinting at what might drive it, gives no clue as to the underlying mechanism. Moreover, there is something of a question mark about how exact the action copy has to be to qualify as mimicry. Faithful

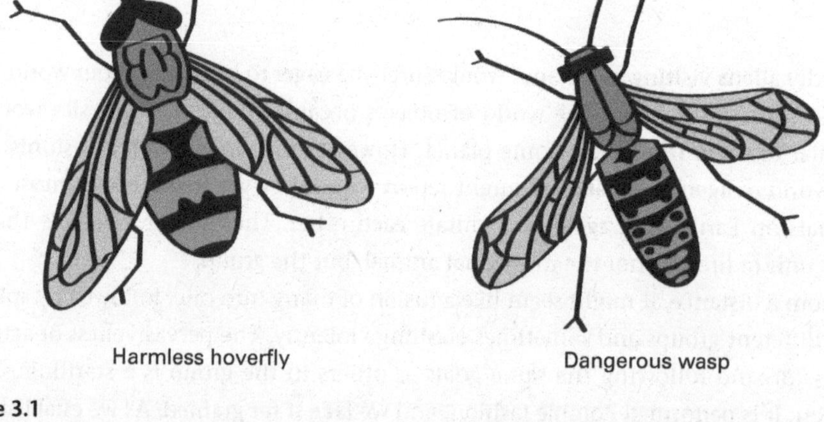

Figure 3.1
Mimicry: A harmless fly looks like a dangerous wasp. (Figure created by the authors.)

copies can apply only rarely, as there are bound to be differences in structure in the bodies of individual organisms, depending on age and nutrition, among other things. Just as even imperfect color matches with the target can protect an insect, even imprecise mimicry can provide measurable and positive advantages (Sparenberg et al., 2012).

One form of very imprecise mimicry is emulation. Here, the goal of the action is more important than the action itself. Emulating depends on identifying the goal of a conspecific and striving to get it. For example, an adult may pick up a book with one hand, but a child who would like to have this book might use both hands. The same goal is achieved, but the means to achieve it are different.

Overimitation is a form of copying too, but it is almost the opposite of emulation. Here, it is the action that is copied even in the absence of a goal. For us, overimitation is an endlessly enthralling topic. It goes far beyond the utilitarian view that copying is good because you don't make the same mistakes that others made or that you would have made when you used trial-and-error learning. Overimitation is strongly developed in humans and may even be unique to that species. Rituals are examples with undisputed cultural significance.

Ritual movements that are performed in communal ceremonies, such as sitting, standing, or kneeling in a church, are actions that have no particular goal other than to define you as a member of that community. But even in private, for example, when we cook a favorite meal, we may use utensils and perform actions that we may have observed our mother doing, even if they are irrelevant to achieving a good outcome.

The experimental study of overimitation is relatively recent and took off from seminal studies by Andrew Whiten and his colleagues at St. Andrews University. These studies used the puzzle box paradigm (see figure 2.2 in chapter 2) in an ingenious new way. Participants, whether children, adult humans, or other animals, had to learn by observation how to get a reward in the box (Whiten et al., 2009). However, the demonstrator performs some strange actions in the process of opening the box. He or she might use a wand to push out a wooden bolt, use this to tap the empty upper compartment three times and then pull out the round plug in the center of the door and use the wand to retrieve the reward. The first part of this procedure is entirely irrelevant for obtaining the reward. Moreover, it can easily be seen to be irrelevant. Interestingly, while young children happily imitated the irrelevant actions, chimpanzees did not. They obtained the reward as quickly as possible (Horner and Whiten, 2005; also see Hoehl, Makinson, and Whiten, 2019 for a review).

Experiments showed that children persisted even in situations when this was against their own interest, as when they were in a competition to get to the reward first, and even when they were warned about the lengthy procedure (Lyons et al., 2011; Lyons,

Young, and Keil, 2007). This strongly suggests that the behavior is part of the Zombie thread. But it is not just children who overimitate. The copying of the irrelevant actions actually increases with age, such that adults perform the task with even more emphasis on performing the irrelevant actions (McGuigan, Makinson, and Whiten, 2011).

What do we gain from this irrational and apparently automatic behavior? The answer will preoccupy us throughout this book. Here, we can briefly state that by copying actions that are not necessary, we are able to achieve and even emphatically demonstrate our alignment with a group. Thus, a novice is likely to perform the actions of the already initiated with particularly zealous devotion. The founder of the academic discipline of sociology, Emile Durkheim (1912), argued that the faithful copying of actions is a striking feature of human culture and the reason for the existence of rituals that enhance group cohesion.

All forms of imitation, whether mimicry, emulation, or overimitation, do not happen randomly with any model, or on any occasion. We humans, like most other animals, are quite selective about whom to imitate and when, even when we are not aware of making such a selection. But we can become aware of what we are doing and why. We may even reflect on the irrational nature of our behavior but continue with it nevertheless.

But *how* do we copy actions? We have to somehow create a bridge between what we see someone else doing and doing the same thing ourselves. How do we know what pattern of motor activation will make our action look like, or sound like, that of the person we are imitating (Brass and Heyes, 2005)? How do we link our motor system with our perceptual system?

Links between Perception and Action

The idea that actions are intrinsically linked to perception goes back, at least to the nineteenth century, when William James, in his ideomotor theory of action, claimed that "every mental representation of a movement awakens to some degree the actual movement which is its object" (James, 1890, p. 526). The implication is that observing, imagining, or in any way representing an action excites the motor programs used to execute that same action. This is the mechanism that cognitive psychologist Wolfgang Prinz has termed common coding (Prinz, 1984). A recent review (Hardwick et al., 2018) confirms that this notion has stood the test of time. Perceiving, imagining, and performing actions are represented in the brain in overlapping premotor and parietal networks.

James suggested that the links between perception and action are built by association learning. When people initiate movements, they learn what happens next. Eventually, they build an expectation about what they are going to see, hear, and feel after initiating various acts. This expectation is equivalent to the *forward model* in more recent accounts of motor control (Wolpert, Ghahramani, and Jordan, 1995). James also suggested that these associations could be used in reverse order. When we see limbs moving in a certain way, we have an expectation about the motor program that initiated these movements. This is equivalent to the inverse model: the action we need to initiate to achieve the desired outcome.

Perception-action links have frequently been studied outside a social context, and this suggests that they apply not only to social objects, but also to action maps, a topic that we will discuss in chapter 5. Action maps are hidden automatic action plans that are elicited by the mere appearance of objects, like reaching with your right hand to pick up a cup whose handle is on the right side in front of you. Seeing another person moving a cup also automatically activates motor plans.

The extraordinary fact that actions executed by the self and by others are interchangeable drew surprisingly little attention. It was assumed that the links that we acquire between perception and action apply indiscriminately, and thus enable us to copy the actions of others. However, a big change was underway in our understanding of this baffling process.

Mirror Neurons

In the early 1990s, when we were beginning to think about social cognition, a serendipitous discovery was made, which had an enormous influence on us, and indeed everyone interested in cognitive neuroscience. The discoverers were Giacomo Rizzolatti and his colleagues at Parma University (di Pellegrino et al., 1992). They were studying motor actions by recording the firing of single brain cells in monkeys. They were particularly interested in cells of the area known as F5, in the monkey premotor cortex, as they had previously found that these cells fire when the monkey performed a particular action, such as grasping a peanut (Rizzolatti et al., 1988).

One day, it so happened that a researcher in the lab picked up a peanut while watched by a monkey. To his amazement, he noticed that the recording apparatus attached to the monkey's brain started whirring because of a sudden burst of activity in exactly the same neuron that had been active when the monkey himself was grasping a peanut. Following up this observation, the team was able to identify other neurons in the premotor cortex that fired when a monkey performed a particular movement, and also when the monkey saw someone else making the same movement (see figure 3.2). These became known as "mirror neurons" (Rizzolatti et al., 1996; Ramsey, Kaplan, and Cross, 2021).

Figure 3.2
Mirror neurons. The same nerve cell fires when the monkey picks up a peanut (left) and when the monkey sees the experimenter pick up a peanut (right). Redrawn from figure 1 in Rizzolatti and Fabbri-Destro (2008). Copyright 2008, with permission from Elsevier.

The concept of mirror neurons had a dramatic impact on neuroscience and beyond. It confirmed the idea that there are *representations* of actions in the brain that combine the sensory and the motor components of action. Moreover, these representations are also activated when the actions are performed by others. Thus, mirror neurons embodied a direct link between the self and the other and provided a new way of thinking about learning through observation.

The 1990s were also the time when brain imaging in humans started to take off, with early claims emerging about a network of brain regions underlying mentalizing. Social neuroscience was no longer a pipe dream (see e.g., Frith and Frith, 1999). Generalizing from monkey brain to human brain, researchers started to speculate that, whether we are aware of it or not, our mirror neurons will be activated when we observe other people, and this will prime us to do what they do and sometimes actually make us perform the action (Fadiga et al., 1995)

Our friendly aliens who were baffled by the ubiquitous tendency of animals to copy each other, might contrive transplants of such cells to become a bit like us. But could they then simply copy our actions without any history of learning and experience? According to Celia Heyes (2011), association learning is essential to build up mirror neurons in the first place. As James had suggested, certain observations and actions are repeatedly encountered and become glued together. The aliens would be well advised to first invest in a mechanism that allows them to learn by association.

There is indeed empirical evidence that association learning can lead to the creation of mirror neurons. An example is a study in which monkeys were trained to use reverse pliers. This is a strange tool that goes against the grain because it is opened by closing the hand and closed by opening the hand. No wonder the monkeys needed many trials to learn the correct movements. After the monkeys had eventually learned to use the pliers to pick up food, neurons were found in the premotor cortex that fired when the monkey used the reverse pliers and when the monkey observed them being used by someone else (Umiltà et al., 2008).

This evidence certainly suggests that mirror neurons can be created from scratch when needed. But do they have to be? Not necessarily. Mostly, animals do not have the luxury of being trained in the safety of labs. Mostly, they have to learn very fast to survive. We have made the point already that there are predispositions and priors prewired in the brain as a result of evolution. At the very least, priors are needed to direct attention to relevant stimuli. For instance, Marco Del Giudice, Valeria Manera, and Christian Keysers (2009) pointed out that human infants preferentially attend to their hands and showed that their perceptual-motor system is optimized to provide the right kind of input for setting up perception-action representations of manual actions, such as grasping (see also Werchan and Amso, 2021).

Inevitably, the early excitement about mirror neurons led to controversy. While some researchers claimed that mirror neurons could explain almost all aspects of human social cognition (e.g., Oberman and Ramachandran, 2008), others considered them to have a very minimal role (e.g., Hickok, 2009). However, to explain how we align with others in our group and how we learn from them, the concept of mirror neurons occupies a unique place. The urge to copy the action of another individual plays a key role in the world of agents.

What Processes in the Brain Are Involved in Imitation?

In humans, it is only possible to study the properties of single neurons by invasive techniques, but noninvasive imaging studies suggest that there are several mirror systems in the human brain, which are likely to involve thousands of neurons. Brain regions, corresponding to areas where mirror neurons are found in the monkey brain, inferior frontal cortex and parietal cortex, are activated when people perform an action and when they see the same action being performed by someone else (Rizzolatti and Craighero, 2004). However, because the activity of so many neurons is lumped together in brain-imaging data, how can we be sure that the same neurons are involved in the perception of the action and in its performance?

Fortunately, a technique, known as "repetition suppression," can help. Repetition suppression exploits the following fact: If a group of neurons in the brain has sprung into action in response to a particular stimulus, then the next time the stimulus appears, the neurons will respond less. They are suppressed. Think, for example of an occasion where you listened to the same message so many times that you stopped hearing it. Repetition suppression is not only observed when a person performs the same action twice, but also when action observation follows action performance (or vice versa—see Dinstein et al., 2007 and Kilner et al., 2009). These results suggest that the same neurons are being activated by both perception and action.

It is always important to remember that localized brain regions, such as those containing mirror neurons, never work in isolation. They are part of a much bigger system. Mirror neurons are constantly being activated, but we don't copy each other all the time, and we need to explain this. Here, we need to return to our hierarchical processing system. How is it embodied in the human brain? We know there are massive top-down connections originating in frontal and parietal regions, which meet bottom-up connections originating in sensory parts of the cortex. These two major information-processing streams in the brain must communicate with each other to maintain a fine balance from moment to moment. How well does the incoming stimulus fit the prior expectation? How important is this for survival? How opportune is an instant response? Should mirror neurons, as soon as they are primed, pass the command to other neurons to initiate and execute the action? Should the action be inhibited instead?

The top-down information-processing streams of the brain are constantly busy, but they do not always have the upper hand compared to bottom-up processes. Control cannot always be exerted, simply because there are capacity restrictions. If the executive parts of the brain are busy supervising other tasks, such as maintaining data in working memory, automatic imitation may be triggered for no other reason than that it was not inhibited (van Leeuwen et al., 2009). Patients with frontal brain lesions have problems inhibiting automatic imitative responses and may not even be able to do so when explicitly instructed (Lhermitte, Pillon, and Serdaru, 1986). This phenomenon appears to be entirely independent of more general problems with inhibiting overlearned responses (Brass et al., 2003). Imitation must be looked at in its context, which is typically social (Farmer et al., 2018).

Antonia Hamilton has proposed an elegant model, STORM (which stands for "Social Top-down Response Modulation"), in which imitation is integrated into a top-down control network (see figure 3.3). This control guarantees that not everyone and not everything is imitated (Wang and Hamilton, 2012; Hamilton, 2015). So our friendly aliens, so interested in transplants, would have to embed mirror neurons into this complex control network. But once this was accomplished, it would become clear that this is not sufficient, as everything is connected with some other regions of the brain, not just the cortex. Nothing less than a total brain transplant would do.

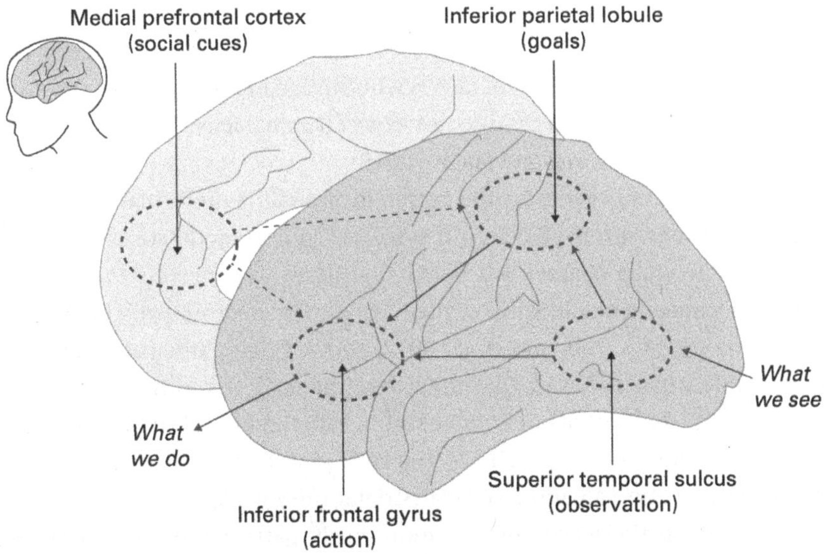

Figure 3.3
A brain system for imitation: The medial surface of the right and lateral surface of the left hemisphere. The movements we see are analyzed in the superior temporal gyrus. Goals are represented in the inferior parietal lobe. Actions are planned in the inferior frontal gurus. To emulate someone's behavior, the route goes from observation to goal to action. To mimic someone, we go directly from observation to action. The medial prefrontal cortex, among other regions, modulates the function of this system: inhibiting inappropriate behavior and directing attention to the appropriate social cues (Hamilton, 2008).

When and When Not to Imitate

So, when should we imitate? Even at the Zombie level, the brain has to weigh the costs and benefits when deciding whether it is desirable to imitate the behavior of others or to ignore it. For example, a stickleback doesn't bother to follow other sticklebacks when it knows for certain where the food is (van Bergen, Coolen, and Laland, 2004).

The system must also make decisions about *who* and *what* to copy. This has been considered in some detail in the behavior of nonhuman animals when they are learning from others (Laland, 2004; Hoppitt and Laland, 2013). Human decisions will be affected by preexisting unconscious biases regarding which particular person should be imitated. A person of our own gender and age? Of the same status? (Kendal et al., 2015; Wood, Kendal, and Flynn, 2012)? Imitating someone of higher status might be the result of calculated decisions via the Machiavelli thread, so as to affiliate with people who might be advantageous to us later. However, we should bear in mind that those we imitate might not reciprocate by imitating us. This asymmetry is observable by the age of five months. Infants expect that characters who engage in imitation will approach the people whom they imitated. But they do not expect the characters who were imitated to approach their imitators (Powell and Spelke, 2018).

These various effects depend on the hierarchical brain system that modulates the activity of the mirror neurons. But mirror systems can also modulate activity in other systems. In the previous chapter, we noted that items can increase in value simply because we see someone else looking at them. Cognitive neuroscientist Mathias Pessiglione and his colleagues (Lebreton et al., 2012) showed that participants rated objects as more desirable when they saw that they were the goal of another's action. This is known as *goal contagion*, part of our Zombie realm. When they looked at what happened in the brain, they found that activity in the mirror system (parietal-frontal regions) triggered activity in the valuation system (striatal-frontal regions). The stronger the connectivity between these systems, the more likely participants were to show goal contagion. This, then, is how we can acquire the values of others via mirroring without any need for explicit verbal communication.

Sometimes, at least in humans, our desire to imitate others is strong enough to overcome aversion and is pursued even at high costs. This is particularly true for the complex rituals that are often associated with religious ceremonies and military displays. These can involve lengthy and strenuous singing, praying, and marching. Extreme rituals may even involve self-inflicted pain, such as caused by body piercing. But they also promote cooperation among the participants and empathy and admiration from the audience for those who undergo such grueling rites (Xygalatas et al., 2013).

The Power of Imitation

We hope that we have convinced you that imitation takes many forms and depends upon a complex system involving many brain areas, following only the Zombie thread. For many, if not most creatures, there seems to be a pervasive disposition toward automatic mimicry, a continuous readiness to do what others do when in their company. All the members of a herd, swarm, shoal, flock, group, tribe, or festival audience are ready to act as one. It must be worth a lot to be always primed to carry out the actions of others or chase the same goal as others!

In humans, automatic mimicry shows its Zombie credentials by being hard to suppress, even if it produces a competitive disadvantage. A compelling example of this was shown in a study by Celia Heyes and colleagues using the rock, paper, scissors game.

Here, the tendency to make the same hand movement as your partner often happens, even though this is not a winning strategy (Cook et al., 2012). A similar result was obtained for a two-player arcade game, where the faster players could not stop themselves from imitating slower opponents (Naber, Pashkam, and Nakayama, 2013). And, as we have seen, competitive disadvantages are discounted even in the case of effortful overimitation.

Why is imitation so important to animals? It forges individuals into groups, and being part of a group conveys many advantages. For example, a shoal of fish can protect its members better from predators, and indeed some predators use the strategy of dispersing the shoal so that they can pick off individuals. Seen in an evolutionary context, it makes sense that an individual animal does what others do, and it makes sense that they have an enduring mechanism for doing so.

For humans, there are additional advantages. For example, as Farmer et al. (2018) point out in their comprehensive review, unconscious mimicry and emulation enhance our understanding of others. Plenty of examples can be found in the domain of speech and emotions, where better comprehension is achieved by partners who are automatically aligning with each other in their choice of vocabulary, syntax, and prosody (Pickering and Garrod, 2007). Even imitation of a speaker's accent can improve the comprehension of the listener (Adank, Hagoort, and Bekkering, 2010). This is an example of imitation at a much more abstract level than simple movements, and it is still a Zombie thread.

A Tool for Affiliation

We imitate because we need to affiliate, but perhaps we affiliate because we imitate. Perhaps the most important role for imitation in humans is in the creation of an emotional attachment between the members of a group. Mimicry increases liking and prosocial behavior in people who are being imitated, with the proviso that they don't know that they are being imitated (Kulesza, Dolinski, and Wicher, 2016). If they do notice, the Machiavelli thread comes into the fore, which changes everything. Suspicion lurks, and more often than not, imitation is perceived as mockery. In contrast, people who have been imitated without being aware of it, feel more rapport with their partner (Lakin and Chartrand, 2003). They are also more generous and helpful to others immediately afterward (van Baaren et al., 2004; but see Hale and Hamilton, 2016). This effect is already observed in eighteen-month-old infants (Carpenter et al., 2013).

Automatic copying and its positive effects on affiliation have been studied extensively in young children (Over and Carpenter, 2013). For instance, children tend to put more trust in those who have copied them previously. Five-year olds can infer affiliative relations from observing imitative interactions in videos. They naturally assume that individuals imitate only those who are nice, and they apparently attribute prosocial motives to characters who imitate. This holds even for much younger children (Over et al., 2013; Nielsen and Blank, 2011).

If young infants are already expert at attaching social significance to imitation, we can be sure that not a lot of practice is needed to link affiliation with imitation. Lindsey Powell and Elizabeth Spelke (2018) demonstrated this with the aid of animated characters who were interacting with each other and sometimes imitated each other. Just like older children, the infants expected that characters who engaged in imitation will approach and affiliate with characters whom they imitated. In another study, infants aged six months reacted with smiling and approach behaviors after they experienced being imitated. Conversely, if adults suppressed their emotional mimicry, the infants' social responsiveness decreased (Sauciuc et al., 2020).

Clearly, imitation even at a very early age is loaded with social meaning, and for this ability to get off the ground, not much practice is needed. Young babies not only understand mimicry, but they also display it in their facial movements. This behavior critically depends on eye contact, confirming that the social context is paramount in eliciting copying behavior (de Klerk, Hamilton, and Southgate, 2018).

Belonging to Your Group—the Project of a Lifetime

When you are imitated by your conversation partner, Zombie-style, you may feel that you rather like this nice, friendly person who is listening to you and clearly understands you. When you depart, you may feel just a little happier, and it is not surprising that if you are then asked to make a small donation to a charity, you will be a little more generous (van Baaren et al., 2009). But it is the converse that clinches the connection between mimicry and attachment and liking: we show less mimicry toward people whom we dislike or people who belong to an outgroup (Stel and Harinck, 2010).

A prominent theory proposes that this kind of mimicry serves as a social glue (Chartrand and Bargh, 1999; van Baaren et al., 2009), which could be a mechanism for bonding with other members of our group, liking them better, and trusting them more. This would provide a secure basis for cooperation. Imitation is a signal that I am to be trusted.

Imitation does not just make us more generous and helpful. It is also a sign of group membership. By imitating, we indicate our group identity. We see this, for example, in the rapid propagation of fashion and taste among like-minded people. But there are far more subtle examples that pervade our cultural life and create group identities. As the British psychologist William McDougall (1926, p. 352) wrote: "Most Englishmen would scorn to kiss and embrace one another or to gesticulate freely, if only because Frenchmen do these things."

Group membership is not guaranteed. Individuals who have been shunned (ostracized) by their group will increase the extent to which they imitate members of their own group, but not members of other groups (Lakin, Chartrand, and Arkin, 2008). The experience of ostracism also increases the fidelity of imitation (Watson-Jones, Whitehouse, and Legare, 2016). These effects occur even when it is not us ourselves, but another person, whom we have observed being excluded. Five-year-old children (Over and Carpenter, 2009), as well as toddlers aged thirty months, were found to mimic facial expressions more strongly after they had observed third-party ostracism (de Klerk et al., 2020).

We want to be accepted as members of our group, and the way to foster and maintain this acceptance is to imitate the behavior of the others. But we still can't tell which came first, mimicry or the need for affiliation. Studies to date suggest an intimate, even circular, relationship between mimicry and affiliation, such that each can act as cause or effect. Mimicry causes affiliation, and wishing for affiliation causes mimicry (Lakin and Chartrand, 2003; Stel et al., 2010)

Alignment with our group is important for personal development since we have a deep-seated need to belong. However, at different ages and stages in life, we will probably move to different groups and may belong to more than one group. Somebody will insist on customs and conventions about what we must do to fit into a group, such as prescribing a dress code. Such conventions not only create group identities, but they also take away the pain of choice and the uncertainties created by meeting strangers. On the one hand, they lessen the fear of sticking out from the crowd, and on the other, they provide a cover for cheats who are not actually group members.[1] It is not surprising that conventions can become very elaborate and secretive.

Deciding whom, when, and how to imitate creates uncertainty, and this uncertainty probably reaches a peak during adolescence, when self-identity is being reassessed. Parents are often horrified by their children's choices of what to wear, what music to listen to, or what hairstyles to adopt, and feel powerless to influence these choices. The overwhelming importance of fitting in with your peer group during adolescence has been highlighted by Sarah-Jayne Blakemore in her book *Inventing Ourselves* (Blakemore, 2018). However, the powerful social values of imitation and affiliation, as well as their dreaded shadows of mockery and exclusion, seem to regulate our social lives throughout the whole of our life.

1. One example occurs in the classic G. K. Chesterton Father Brown story "The Queer Feet."

Automatic and Deliberate Imitation

The contrast of the Zombie and Machiavelli threads is vividly illustrated by the distinction between automatic and deliberate imitation. We usually trust Zombie mimicry, but anything to do with the Machiavelli thread occurs under the shadow of suspicion and needs vigilance. Only the most naive person will think that others can *always* be trusted. Mimicry can imply "I like you," or "I am like you." It can also mean, "You are ridiculous—just listen to how silly you sound." While these two threads have strikingly different effects, there is also a gray border in between. This happens when imitation is used as a communicative signal. Signaling may or may not be consciously initiated. It may be simply triggered by the presence of an observer.

The Canadian psychologist Janet Bevin Bavelas and colleagues (1986) conducted an ingenious and revealing experiment to investigate just this problem. They created a scenario where participants (observers) were sitting in a waiting room, supposedly waiting to be called into a testing room. However, the real experiment happened while they were waiting: An actor passed through with a heavy weight, which he pretended to drop on his foot, with corresponding signs of pain, manifest in pronounced wincing. The actor was instructed in advance whether to make eye contact with the observers or not. The critical question was whether the observers would wince too, as a result of automatic mimicry. The scenes were recorded on video and later painstakingly analyzed by students who were unaware of the real purpose of the experiment.

The results showed that the presence of wincing mimicry in the observer depended critically on the presence of eye contact between the actor and observer. Observers winced more when the actor was looking at them. The result fits with the finding that young infants showed facial mimicry only when eye contact occurred (de Klerk et al., 2018). The decoders of each video also provided subjective interpretations to the extent that they thought the observer showed caring. This was also significantly related to the amount of eye contact with the victim. To generalize freely: If you want to be seen as a caring person, then when a victim of an accident is looking you in the eye, rev up your facial response. But if they don't, and if nobody sees you, why bother?

In this chapter, we have emphasized the similarity of almost all social animals in the way that they copy each other's actions and goals. This reveals the urge of animals to align with their conspecifics. Motor mimicry and goal emulation are behavior patterns that exist throughout many if not all species. The benefits are so massive that even the top-down control systems in the human brain do not always have a chance to inhibit the urge to copy. In humans, the presence of other people and the wish to be like them affect each and every one of us at different stages of our lives and at different levels of conscious awareness.

4 Sharing Emotions

It is not only actions that can spread from one agent to another through unconscious mimicry, but also sensations and emotions. The transmission of negative feelings, such as fear and disgust, is vital for avoiding danger in the world of objects and agents. However, all emotions, negative or positive, are designed for sharing and act as signposts for negotiating the social world. Social laughter is a perfect example. Laughter is highly contagious and provides an automatic means for bonding with members of one's group. The way that people laugh together indicates how connected they are. The conscious awareness of emotions in humans is remarkably dissociated from their unconscious physiological effects. Thus, individuals differ in the way that they interpret their inner feeling states. With the use of top-down control, humans are expert at feigning emotions and using them to manipulate others.

Empathy is a highly complex and sometimes misunderstood concept. At the more basic levels of the information-processing hierarchy, sharing in feelings such as misery and pain is present not only in humans, but also in other animal species. At a higher level, humans can monitor and control these feelings. Consciously controlled empathy can foster reflective compassion. However, it can also lead to morally dubious outcomes, such as aggression toward individuals who are unjustly held responsible for inflicting harm. Nevertheless, a certain level of control and inhibition of basic levels of empathy is essential if we are to help others in distress rather than merely experience their pain.

* * *

Emotions in the Brain

We are continuously immersed in waves of emotions, and we confess that on the whole, we find this trying, even exasperating. We don't just experience our own emotions—we catch them from others around us. Emotions can flood us in torrents or they can be so faint that we can barely catch them. Emotions can cruelly reveal our ugly side, as in hate, jealousy, and contempt. But we admit that they also have a positive side. There are love, forgiveness, and empathy, and without these feelings, life would be dull and gray. We can learn to regulate emotions so as not to be overwhelmed by them, although it may take a lifetime to do so.[1]

All emotions, whether positive or negative, convey useful information, and this information often comes from other people. Body language and facial expressions guide us to learn about the value of objects. Emotions are both private and public. We both mirror and respond to the emotions of other people as they respond to and mirror our own, and we do this even before we become aware of the emotion we are experiencing. Unconscious mimicry for emotions occurs in much the same way as it does for motor actions.

Are social emotions special? There is little reason to think so. The evolutionary biologist Randolph Nesse (2004) proposed a phylogeny of emotions and illustrated it by a tree (see figure 4.1). We adopted this image to show that even the most sophisticated social emotions, as listed in the tips of the branches, can be understood as arising from two basic roots: opportunity/desire and threat/fear.

As indicated in the diagram, desire gives rise to approach behavior, while fear gives rise to avoidance behavior. This marks the big divide between positive and negative emotions. As we discussed in chapter 3, in the world of objects, approach and avoidance depend on the value of the objects that elicit each behavior. If the value is high (nice), we approach. If the value is low (nasty), we avoid. There is probably also some reverse influence, such that extremely positive values elicit desire and extremely negative values elicit fear.

The pleasure of food, sex, addictive drugs, and sustained states of happiness can produce strikingly similar patterns of brain activity. They involve many brain regions (the orbitofrontal cortex, the insula cortex, and subcortical structures, including the nucleus accumbens and the amygdala). The same regions also mediate aversive emotions, such as fear, pain, and disgust (Berridge and Kringelbach, 2015). However, positive and negative emotions diverge sharply in the actions they trigger. The circuits involved must be foolproof to work in Zombie mode, since confusion or delay could be fatal.

Nevertheless, the brain structures involved in processing nice and nasty stimuli are subject to regulation from the top of the hierarchy of the brain's information-processing system. These include the prefrontal cortex, and particularly the orbitofrontal cortex (Roelofs et al., 2009; Ochsner, Silvers, and Buhle, 2012). Thus, fear can be suppressed and actions can be modified and inhibited depending on context.

Emotions have physiological effects, which can readily be studied in animals, but in humans, there is an added dimension. We can reflect on our emotions and then

1. We still suffer when we read the comments of "Reviewer 2" about our latest paper.

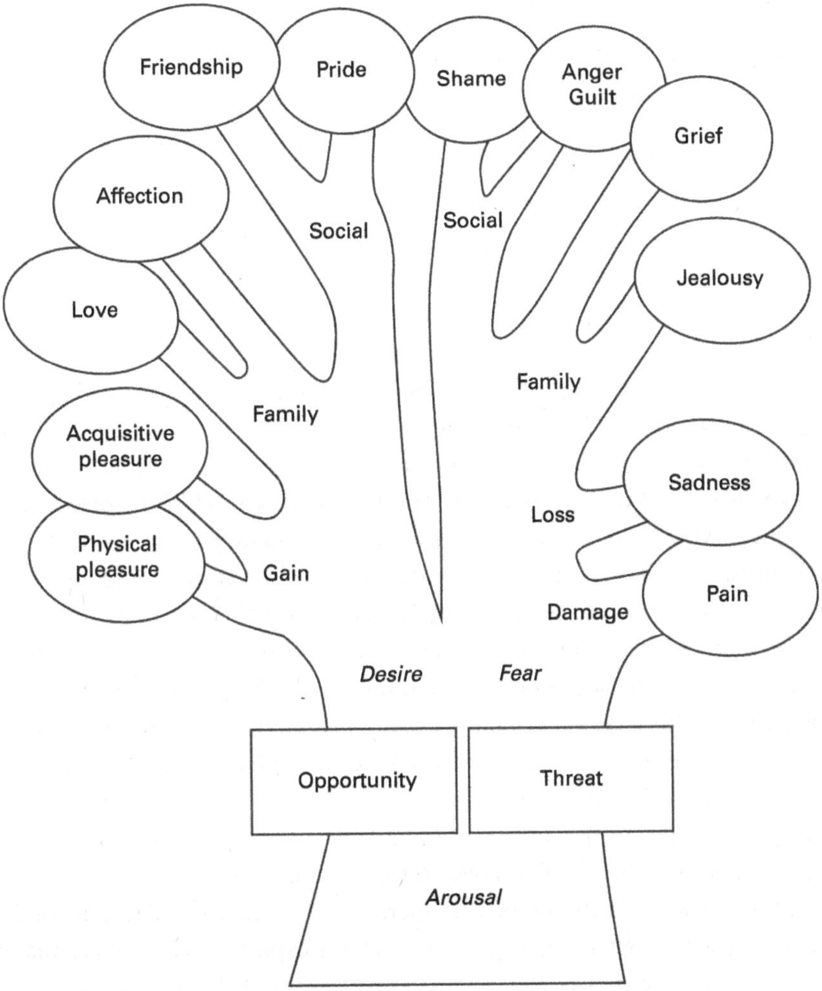

Figure 4.1
The phylogeny of emotions. Redrawn from figure 2 in Nesse (2004). Copyright 2004, with permission from the Royal Society.

report our feelings to others. John Lambie and Tony Marcel (2002) pointed out that the physiological arousal associated with emotions is one thing and the awareness of this arousal another. This distinction seems relevant to understanding the condition known as "alexithymia" (literally "no words for emotions"). Alexithymic individuals, by their own account, are unable to read their own emotions and frequently misidentify them. Does their brain not recognize emotions at a basic level, or is their problem situated at a higher level?

To learn more about the neural basis of responses to emotionally arousing stimuli, we used the hierarchical information-processing framework in a brain-imaging study that we carried out together with Giorgia Silani, Geoff Bird, and Tania Singer (Silani et al., 2008). Our participants were autistic adults and a neurotypical control group. All completed an alexithymia questionnaire.

We showed participants pictures with pleasant, unpleasant, or neutral content while they were in the brain scanner. We asked them to report (using a slider on an analog scale) the emotional effect that the picture had on them. This required conscious access to the emotion aroused. Brain activity during this task was contrasted with a task where we showed the same pictures but asked simply how colorful they were. Here, the emotionally arousing effect of the image would be incidental. In fact, we found that the amygdala was activated by unpleasant pictures in both tasks, confirming that this region responds automatically to emotional stimuli. This was the same for all participants, regardless of their alexithymia score. At the lowest level of the hierarchy, individuals with alexithymia processed their emotions adequately.

At the intermediate level, activity in the anterior insula cortex reflected the degree of unpleasantness felt by the participants. The anterior insula cortex has a special role in the representation of feelings since it receives interoceptive signals—that is, signals about the physical state of the body, including touch, smell, and pain (Craig, 2002). This suggests that this region plays a role in bringing bodily states into awareness. Individuals with high alexithymia scores showed weaker activation at this level. This fits with the idea that indeed, their awareness of emotions is rather feeble. There is increasing evidence that alexithymia is a problem in interpreting physiological information from the body, such as heart rate (Brewer, Cook, and Bird, 2016).

At the highest level in the processing hierarchy, the medial prefrontal cortex (see figure 4.2) was active when participants had to introspect on their emotional experience (compared to colorfulness). This is consistent with a role for this region in self-reflection (Ochsner et al., 2004). In other words, this region tapped a level of processing, where the self is monitoring its own inner states. The autistic participants in this study showed weaker activations in this region. This fits with the idea that they have difficulty in mentalizing (see chapter 10).

Shared Sensations and Shared Bodies

The way that we experience our own emotion is important, but what about the emotions of others? In the last chapter, we described mirror neurons, part of a neural system that provides a direct link between actions of the self and of others. A whole host of

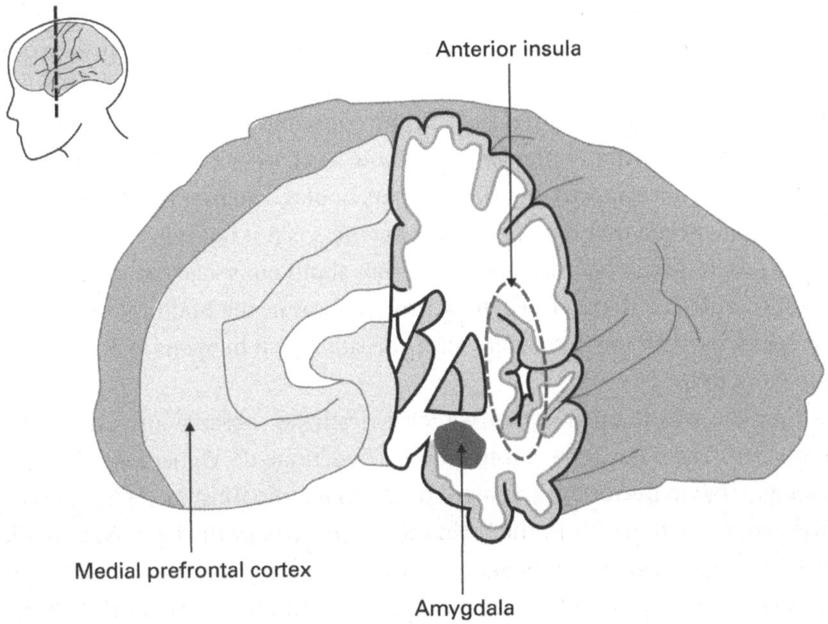

Figure 4.2
Brain regions associated with three levels of emotion. Here, we see the medial surface of the right hemisphere and a slice through the left hemisphere. The amygdala responds automatically to unpleasant stimuli. The anterior insula cortex is associated with awareness of emotions. The medial prefrontal cortex is concerned with self-reflection.

studies now exist indicating that emotions of the self and others are also directly linked via mirror systems (see, for example Keysers and Gazzola, 2009).

To get a more concrete idea of what this means, we can look at subtle touch sensations, which can also be shared between people. A small number of individuals have been found who say that they feel touch on their own body in exactly the part of the body where they see another person being touched. This is a form of synesthesia that involves the experience of touch rather than the more common experiences of color when seeing or hearing words (Ward, 2013).

Sarah-Jayne Blakemore and Jamie Ward, in collaboration with Chris Frith, studied one of these people, C., a woman who had been unaware that the vast majority of the population do not experience similar shared sensations (Blakemore, Ward, and Frith, 2005). They found that the brain region (in the somatosensory cortex) that became active when she was touched on a particular spot (e.g., her left cheek) also became active when she saw someone else being touched on the left cheek. Remarkably, however, she and other similar synesthetes were not quite as exceptional as it might appear. Ordinary people also showed activation in the somatosensory cortex, when they observed another person being touched. The difference was that this activation did not trigger activity in those brain regions that support the conscious awareness of the touch.

This finding demonstrates that, at a very basic level, we are all affected by the sensation of touch when it happens on other people's bodies, whether we like it or not. Most of us may feel relieved that our conscious awareness is not bothered by this. However, we should take note of what this example reveals about our social nature: we are deeply and bodily connected to other people. Mirror systems in the brain make sure that our body prepares itself for the immediate future because what happens to others may soon be happening to us.

Nothing like this happens when we, or synesthetes, observe an inanimate object being touched. There has to be another body. Frédérique de Vignemont (2014) argued convincingly that to mirror the actions of others, our brain must factor in the body that is performing the actions. Thus, the brain's representation of the body is central for the whole range of mirroring phenomena. This representation appears to have a default setting for sharing bodies with similar agents.[2] In line with the idea that the body acts as a crucial point of reference, emotions are often located in particular body parts. We use many expressions referring to the heart. But we also know what it means to feel weak in the knees, and when our stomach turns. An encounter can be hair-raising, spine-tingling, breathtaking, and so on.

The representation of our own body is something we take for granted, and it is anchored not only in our actions, but also in our perception of touch, and in our experience of emotions. Moreover, this representation is constantly modified by observing other people. The boundary between ourselves and others at this deep level is more blurred than we might think, and it makes sense of some odd but everyday mirroring phenomena.

Contagion and Resonance

Robert Provine (1989) found that roughly 50 percent of adults will yawn during or shortly after watching a video showing people repeatedly yawning. Likewise, we tend to involuntarily salivate if we see somebody biting into a lemon. Florence Hagenmüller showed people a video of a person cutting and eating a lemon and measured the

2. This poses an interesting problem for our future life with robot assistants. It remains to be seen whether we would mirror their actions or their expressly designed emotions.

amount of salivation that the observers produced by means of a cotton pad in the mouth (Hagenmuller et al., 2014). The study also found that the amount of saliva produced correlated positively with self-reported empathy. However, it has been argued that yawning and salivating are not straightforward examples of emotional mirroring (e.g., Schürmann et al., 2005), but should be considered as automatically released physiological responses through contagion.

Contagion is a metaphor derived from disease processes involving a replicating pathogen, such as a virus. The metaphor emphasizes the notion that our brain, like our body, can unwittingly catch something from the behavior of another person. Yawning, salivating, and itching fall into this category. Perhaps the trigger stimulus can serve a useful purpose in helping us to react adaptively to an environmental stimulus, be it lack of oxygen (yawn) or the presence of irritants (itch).

Resonance is a closely related concept, but here, the metaphor is from music rather than disease. Many instruments, such as the sitar and viola d'amore, have sympathetic strings, which will automatically resonate in response to the sounds being emitted by the playing strings. Resonance too is an involuntary response, but it is a response that we do not usually want to shake off. And it is everywhere. Emotional resonance seems to happen when our control system is relaxed, such as when we are immersed in reading a story, following an opera, or watching a film. Many a tear has been shed by viewers when Bambi lost his mother. We are outraged when a protagonist is treated unjustly, embarrassed when they commit a faux-pas, proud when a hero succeeds against formidable odds, and elated when the baddies get their comeuppance.

The observation that we can be moved to helpless tears or laughter when watching a stylized animation of fictional events or reading a novel suggests that we are responding to quite abstract qualities. Thus, to explain resonance, we should not confine our search to subtle facial or bodily cues, nor to pheromones that waft between bodies. There are also cultural factors that power up the sharing of emotions.

The death of Diana, Princess of Wales, in 1997, resulted in a huge wave of grief and mourning spreading just like a physical wave through the population in the United Kingdom and beyond, as she had been a universally beloved figure. The intensity and duration of this wave took many by surprise and remains a phenomenon that can be best explained through contagion. Interestingly, hardly anyone expressed doubt over whether the shared emotion was authentic, precisely because it was both spontaneous and intense. We confess, however, that we ourselves are rather sad that "keeping a stiff upper lip," a cultural norm from a previous age, is no longer considered appropriate.

Advertisers have capitalized on the idea that the spreading of emotions at the unconscious level is more effective than at the conscious level. Instead of using testimonials, they use images of people smiling, wearing glamorous clothes, eating exquisite food, and sipping exotic cocktails in order to prime you to covet and buy their products. As far as we know, this seems to work.

Emotional contagion and resonance vividly imply an uncontrollable and invisible spreading of information from one individual to another. However, as theoretical concepts, they have yet to prove their worth (Dezecache, Jacob, and Grèzes, 2015). It makes perfect sense to assume that it is the mirror systems that do the heavy lifting in all situations when agents perform the same actions, pursue the same goals, and display the same emotions.

How Do We Mirror Facial Expressions?

How is it even possible to mirror what happens on another person's face when we don't have a mirror to check out how accurate we are? Such mirroring may not be observable with the naked eye, but it can be captured using electromyographic recording. Dimberg, Thunberg, and Elmehed (2000) measured the electrical activity in two major facial muscles that are known to be involved in producing happy and angry faces (the zygomatic major muscle and the corrugator supercilii muscle, respectively). Almost uncannily, the same muscles are activated in the face of the observer as in the face of the observed. And this happens even when the observer is not aware of the emotional expression. Using the same methods, the same result was found in infants aged only eleven months (Addabbo et al., 2020). Like adults, the babies responded by mirroring the activity of a face presented in a video, activating the same muscles as those used by the model.

Carina de Klerk, Antonia Hamilton, and Victoria Southgate (2018) investigated facial mimicry in babies just four months old. They showed them videos of actors who either opened their mouths or raised their eyebrows. The babies mirrored these expressions, and this was verified by electromyographic recording of activity in the corresponding face muscles. But the most exciting finding was that the babies only showed evidence of mimicry when the actors made eye contact. This flags up a mechanism beyond association learning and beyond mirror neurons. As we shall see in chapter 16, this mechanism may explain how perception-action representations across self and other come to serve ostensive communication.

What does the automatic sharing of facial expressions mean? Are we merely reproducing bodily expressions of emotion? Are we also experiencing the emotion? We could ask people. But if the effect is occurring Zombie fashion at the unconscious level, they would not be able to tell us. However, thanks to neuroimaging, we can look at the corresponding activity in the brain. The question, analogous to the original mirror neuron studies of actions, is straightforward: Do the same brain areas become active when we experience an emotion as when we see someone else experiencing that emotion? This question resulted in a whole series of brain-imaging studies (for a review, see Bastiaansen, Thioux, and Keysers, 2009), and fear proved the most tractable emotion.

The feeling of fear is known to be elicited by the expectation of a painful shock, and this is associated with activity in the amygdala. Neuroimaging studies have shown this and have also shown that the same activity is elicited by the presentation of a fearful face (Morris et al., 1996). Remarkably, this was true even when the participants were prevented from becoming consciously aware of the fearful face (see figure 4.3). This was done by presenting a neutral face that immediately followed the fearful face, a procedure that results in masking the first face (Whalen et al., 1998).

The same approach was applied to the feeling of disgust. Again, the same brain region, here the anterior insula cortex, was activated by the feeling of disgust, induced

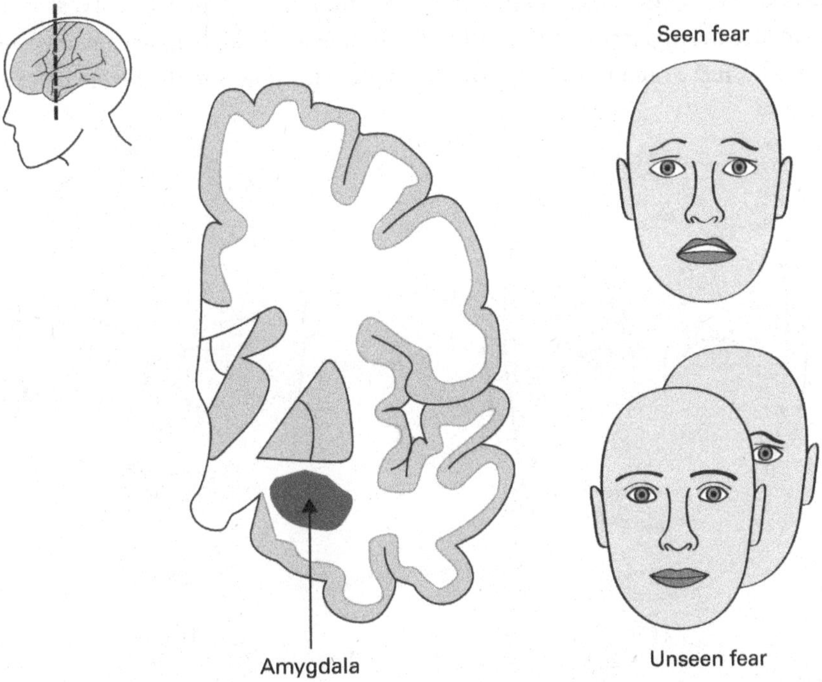

Figure 4.3
Unconscious response to a fearful face: a slice through the right brain hemisphere. A fearful face elicits activity in the amygdala, even if you're not aware of seeing it (Whalen et al., 1998).

by a noxious odor, as by observing the facial expression of disgust (Wicker et al., 2003). There is no need to invoke the Machiavelli thread to explain these results.

Now, we take a step back to take a broader view. What advantages does any kind of mirroring confer? At the risk of overstating the case, we suggest that it gives animals the ability to transport themselves a split second into the future. If so, then, simply by automatically monitoring other agents, they gain an edge in being ready to fight, flee, or approach. Even tiny trips into the future can help any organism to survive better.

To apply this specifically to humans, we can return to the example of the face. The sight of a fearful face is a clear signal of danger in the environment that calls for preparation. Likewise, a face expressing disgust indicates the presence of a noxious substance that should be avoided. The expression of fear is characterized by wide open eyes and flared nostrils (see figure 4.4). This means that the visual field becomes larger and nasal volume increases, which improves the likelihood of detecting visual and olfactory cues about the source of danger (Susskind et al., 2008).

Nearly opposite facial movements characterize disgust. Here, the facial expression involves closing up the nose and mouth. Thus, imitating an expression of disgust is helping to reduce the impact of noxious stimuli. These actions can help us only if we respond very rapidly and without thinking. This is no place for slow and deliberate reflection.

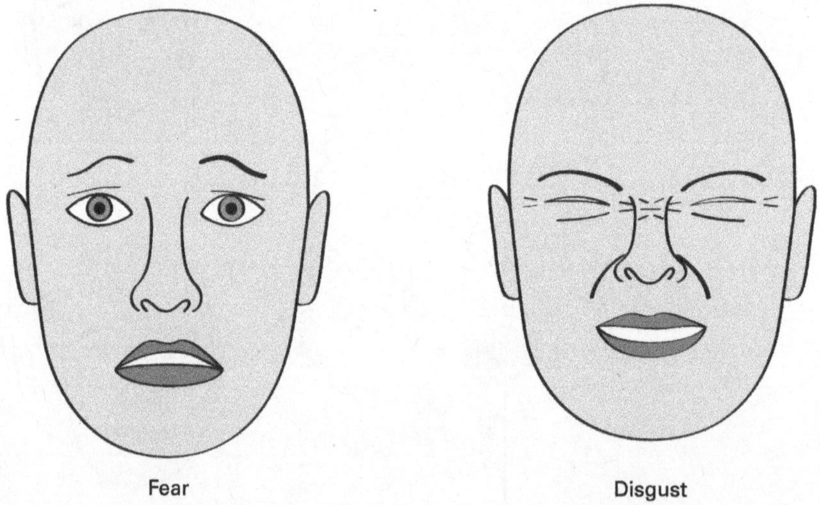

Figure 4.4

Larger visual field	Smaller visual field
Increased nasal volume	Decreased nasal volume
Increased vigilance	*Reduced impact of noxious stimuli*

Redrawn with permission from figure 2 in Susskind et al. (2008). Copyright 2008.

Emotions in the Unconscious and Conscious Realms of the Mind

We humans reflect on our emotions, and we love to talk about and experience them—even our negative emotions. A striking example is our love of ghost stories and horror movies (speaking for ourselves), and the love of extremely violent video games (speaking for others). But we also love to absorb romantic stories, hilarious stories, and very tragic stories. What is going on here? Are we giving our emotions a good workout from time to time? Is it that the unconscious and conscious levels of the brain's information processing use emotions to indulge in crosstalk?

Joseph Ledoux and Richard Brown (2017) point out that we should not take it for granted that conscious and unconscious fear responses map tightly onto each other. Indeed, events in the subcortical fear network and the conscious experience of fear do not show a one-to-one match. The amygdala network controls defensive responses and performs this job satisfactorily throughout many animal species. But the subjective experience elicited by fearful stimuli is a separate matter. This suggests that Zombie and Machiavelli threads do not necessarily entertain a lot of crosstalk.

Ledoux and Brown (2017) provide evidence from neuropsychological patients who have suffered damage to the amygdala, the core of the fear network. These patients should no longer feel fear. However, this is not the case. Even if the amygdala is damaged bilaterally, the subjective emotion of fear is not abolished (Feinstein et al., 2013). Thus, patients with bilateral amygdala damage no longer showed the normally expected neurophysiological response to the inhalation of carbon dioxide, a life-threatening procedure. But astonishingly, at the conscious level, the inhalation of carbon dioxide did evoke fear and panic attacks in these patients.

A dissociation of bodily expressions and subjectively experienced feelings is also obvious when we watch actors portraying great emotion. We may accept the signals in their faces and body movements to temporarily convince ourselves that they truly feel the emotions they portray, and we even resonate with their displays. However, the Machiavelli thread tells us that facial expressions can be deceptive and people can express emotions that they do not actually feel. Conversely, we can use a poker face to suppress the expression of truly felt emotions.

Laughter—the Social Glue That Binds

Experimenters have studied negative emotions, like fear and disgust, for some time, but a more recent chapter was opened up by the study of laughter. Laughter is one of the brain's innate response patterns, and it has been studied in rats (Panksepp, 2007) and in apes (Davila Ross, Owren, and Zimmermann, 2009) as well as humans. Laughter is easily transmitted between people, so it is often labeled infectious.

Sophie Scott, doyenne of the neuroscience of laughter, concluded from her studies that laughter should not merely be seen as a bodily expression of happiness, but above all as a social phenomenon (e.g., Scott et al., 2014). More precisely, laughter happens much more often in the presence of other people than when we are alone. In a group, if one person laughs, then the other members are very likely to join in. What is the point of social laughter? Sophie Scott is clear about the answer: laughter provides a powerful glue that binds members of a group together.

Laughter is a means to preempt and deescalate all those negative emotional experiences, which are bound to arise in difficult social interactions. Thus, laughter is particularly frequent during conversations, estimated at the rate of five laughs per ten minutes (Vettin and Todt, 2004). This is far higher than most people realize. Laughter in conversations typically occurs after banal statements rather than jokes, and it comes from the speaker rather than the listener. Laughter after one's own utterance might be used unconsciously to avoid misunderstandings, especially when the conversational partner does not react to the utterance in a way that one expects.

The brain systems underlying spontaneous laughter have been revealed through studies of patients with epilepsy and stroke, where involuntary laughter can be a consequence of pathology (Wild et al., 2003). This work has identified an involuntary system including the amygdala and parts of the brainstem, and a second, voluntary system originating in medial prefrontal regions. The overlap of these two systems with those identified for expressions of negative emotions is striking. Are they the engines for binding us together, willy-nilly?

It turns out that listeners readily distinguish genuine spontaneous laughs from posed laughs. People can detect emotional authenticity in laughter, and there are individual differences in this ability (Neves et al., 2018). People, who rate themselves as having a higher degree of empathy are better at detecting authentic emotion. These possibly supersocial people also report greater emotional contagion, and they are more likely to join in laughter.

Listening to other people laughing also reveals that we can interpret the quality of the laughter as a signal between conversation partners that tells us something about their relationship, and this signal seems to be the same across many languages and cultures (Bryant et al., 2016). Laughter between friends is more spontaneous than between strangers. Indeed, if you want to judge the extent of liking between partners, listening to their laughter gives you more clues than listening to their speech (Bryant, Wang, and Fusaroli, 2020).

The Bewildering Concept of Empathy

When we laugh with others, is this an example of empathy? Actually, we seem to reserve the term "empathy" for negative emotions. Empathy with another person who is suffering pain or sadness has become a favorite topic in social cognitive neuroscience (Zaki and Ochsner, 2012). Nowadays, we believe that many of our problems could be solved if only there were more empathy in the world.

We all think that we know what we mean by "empathy," but the word has many different interpretations (Zahavi, 2014). We might think that "empathy" means the same as "sympathy," but with more feeling. But looking at definitions is only touching the surface of the meaning. As Zaki and Ochsner (2012) and other authors have pointed out, empathy is a multidimensional concept. It implies orienting toward others' needs and experiencing personal distress when observing others in pain. It sometimes also implies perspective-taking, such as appreciating that different people (and animals too) might feel pain differently and might need help in different ways.

A basic form of empathy can be thought of as unconscious resonance or mimicry of the emotions of others (de Vignemont and Singer, 2006). It has at least two distinct functions. One is to prime us to deal more successfully with future events by predicting feeling states in others (Bernhardt and Singer, 2012). The other is to facilitate prosocial action (Decety et al., 2016).

When people report high levels of empathic concern, they are more likely to help someone. And such predictions can also be made on the basis of brain activity in the anterior insula cortex in response to seeing a loved one suffer (Hein et al., 2010). Manipulations that enhance empathic reactions to the distress of others can also increase cooperation in social interactions. For example, after such manipulations, people behave less selfishly in games like the prisoner's dilemma (Rumble, van Lange, and Parks, 2010). Empathic concern for others, as shown in signs of distress at hearing a baby crying, is documented from the age of three months (Davidov et al., 2020), with considerable variability in individuals, predictive of later prosocial behavior.

Frederique de Vignemont and Pierre Jacob (2012, 2016) have provided an in-depth analysis of the various meanings of empathy, from contagion to complementary response, such as bursting into tears at the sight of somebody crying as opposed to helpfully offering some tissues. Clearly, an optimal form of empathy is more than mimicry. This is an opportunity for the Machiavelli thread to be on the side of the angels by calculating which actions are most appropriate to lessen the hurt of another person.

Empathy for Pain in the Brain

According to this analysis, empathy at a very basic level means contagion by another person's distress. This Zombie thread produces numerous physiological signs, such as tears, facial expressions, and body posture. How can this phenomenon be studied experimentally? Obviously, it matters who the other person is. Since imitation is not indiscriminate, it would seem a good idea to choose as participants couples who are close to each other. Empathy for pain is an ideal target, as the brain's pain matrix has already been explored, so the neural responses for pain in oneself can be compared with those responses that occur when responding to pain in a loved one.

Tania Singer with colleagues, including Chris Frith, used a daring paradigm involving couples, where both partners received transient but painful electric shocks—obviously with their consent (Singer et al., 2004). The shocks were administered while one partner was lying in the scanner and the other was sitting beside it. Brain responses to a participant's own pain could be measured and compared to what happened when the person knew that the friend was experiencing pain.

The results of this comparison showed that brain regions that receive signals from the skin concerning touch and heat, as well as pain, only responded to the body's own pain (see figure 4.5). In contrast, the anterior insula and anterior cingulate cortices responded both to the body's own physical pain and to a signal that a painful stimulus was being applied to the partner. Furthermore, the intensity of the response in these exact regions correlated with self-reported empathic concern. But why does this happen in these particular brain regions?

The results suggest that the neural basis of empathy depends on brain regions where pain is represented being decoupled from sensations that are caused by damage to the physical body. Such decoupling is also a feature of social pain. Here also we have an experience of pain in the absence of sensations relating to physical damage. The social pain associated with the distress of being rejected by others also activates the anterior insula and anterior cingulate cortices (Eisenberger and Lieberman, 2004).

As we saw earlier in this chapter (see figure 4.2), the anterior insula is known to receive signals about the physical state of the body (interoception), including pain, and is probably required to bring these signals into consciousness (Craig, 2002). It is this conscious experience of emotion that is under voluntary control (de Vignemont and Singer, 2006). At this level, we can not only regulate our own emotions, but also vicariously experience the emotions of others.

The precise role of the anterior cingulate cortex is yet to be fully understood. This region is immediately adjacent to brain regions concerned with the planning and execution of actions and has a role in guiding future action selection (Walton and Mars, 2007).

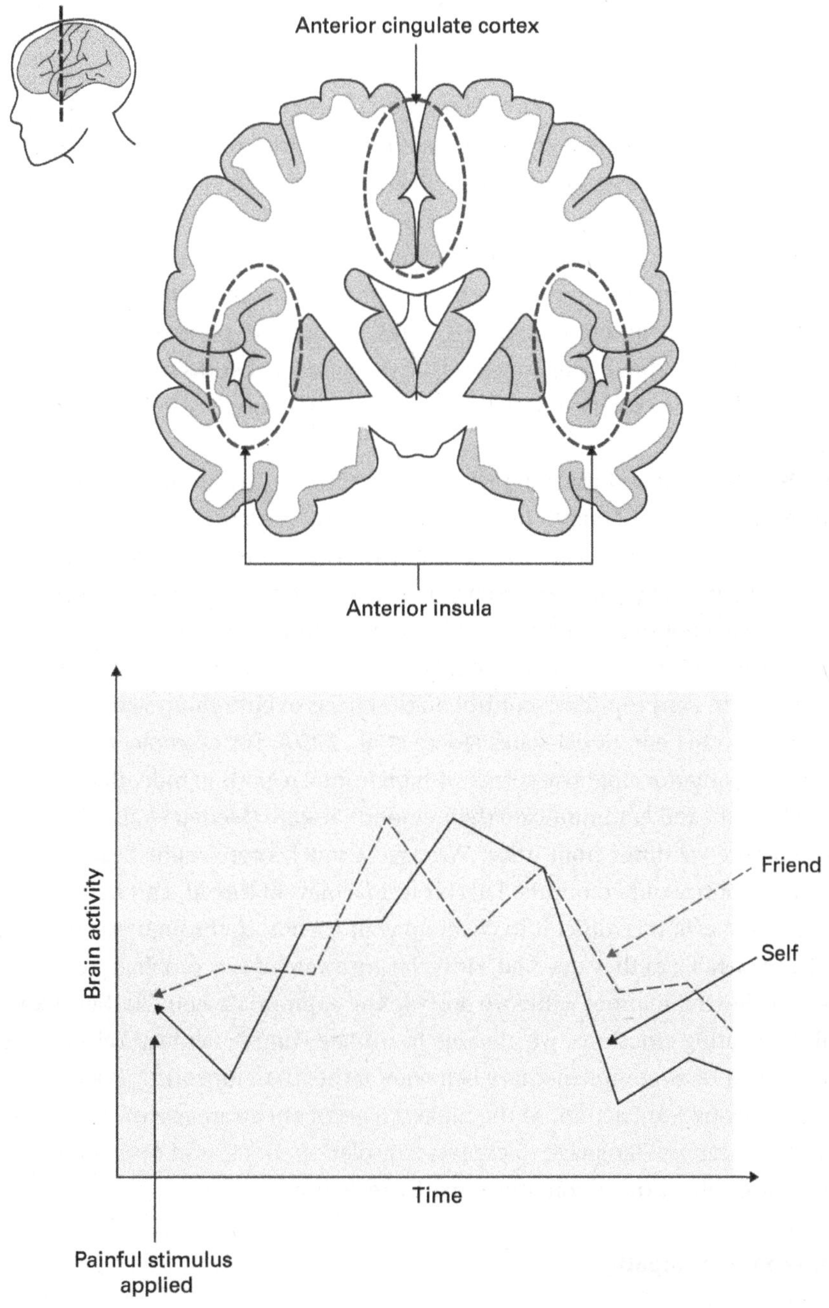

Figure 4.5
Empathy for pain in the brain: The anterior insula and anterior cingulate cortices are activated by pain. The same regions are activated when we know that a friend is in pain. Redrawn with permission from figure 2b and 2c in Singer et al. (2004). Copyright 2004, AAAS.

Pain, whether it occurs in the self or the other, is a signal that action is required. We suspect that the anterior cingulate cortex plays a role in prioritizing the most appropriate actions (see Ullsperger, Volz, and von Cramon, 2004). We will talk about action priority maps in chapter 5.

We tend to think of empathy as a rather special human experience that is very likely modulated by our cultural environment. But responses to the pain of others and the related activity in the anterior cingulate cortex can also be observed in mice (Smith, Asada, and Malenka, 2021). This suggests that the neural networks supporting this basic form of empathy are of such high value to social animals that they have been conserved in evolution.

Both the prefrontal and the anterior cingulate cortices play an important role in the monitoring and control of emotions. Top-down control is exerted by these regions upon structures where emotions are generated, such as insula cortex and amygdala (Ochsner et al., 2012). Prefrontal regions are needed to control the balance between emotion and thought, so that emotional responses can be damped down or enhanced when appropriate. Within this system, the anterior cingulate cortex acts as a critical hub, enabling prefrontal systems to exert top-down control on the effects of bottom-up signals concerning pain, fear, and other emotional states (Joyce et al., 2020). For example, there is greater coupling of the anterior cingulate cortex with prefrontal regions in individuals with good ability to identify and communicate their emotional state (Meriau et al., 2006).

This is where we differ from mice. We have a much more sophisticated ability to monitor and control our emotions. This is our Machiavelli thread, and here it sheds all connotations of evil. It is difficult to cheer up your friends, if, through excess empathy, you feel as miserable as they do. And, since we are aware of this problem, we can try to behave in a cheerful manner while we provide the appropriate help. Rather than automatically mirroring emotions, we are able to initiate complementary behavior. As we shall see in chapter 6, complementary behavior, rather than mirroring, is a key requirement for successful joint action. At the highest level of our awareness of emotion in self and others, we can use language to express consolation, hope, and optimism. This is a significant upgrade of the automatic empathy response.

The Other Side of Empathy

Paul Bloom (2017) aroused a passionate debate by pointing out that empathy is not always what it is made out to be. It is not always a force for good.

There is no guarantee that empathy, at a basic or at a higher level, will always lead to positive outcomes. Showing empathy to a victim does not motivate universal kindness, nor does it inhibit aggression. As we shall see in chapter 8, automatic empathic resonance does not extend to outgroup members. Thus, empathy for members of our ingroup (e.g., our fellow citizens who might become unemployed) can unleash violent and cruel actions against members of an outgroup (e.g., the foreign migrants who supposedly are taking away their jobs). Empathy can be used to motivate a blame game, seeding resentment and arousing aggression toward those seen as offenders (Decety, Echols, and Correll, 2009).

Once we can monitor and control emotions, they can be manipulated to achieve selfish ends in devious ways. For example, imposters persuade their victims to give them money by appealing to them on the grounds that they need emergency medical treatment. Another problem with empathy is that it can lead to burnout. Empathizing with the suffering of another person can lead to strong feelings of distress and other aversive emotions in the observer. Rather than helping the person in distress, such emotions can motivate us to avoid such situations. The alternative is to encourage compassion rather than empathy (Singer and Klimecki, 2014). Compassion promotes positive feelings concerned with valuing other people and caring for them. It is possible to recognize their distress by acting with kindness without having to share their pain.

But what would it be like to live without empathic resonance? Psychopathy is generally believed to be a condition that provides a prime example of this. Psychopaths are characterized by callous, unemotional traits and are generally thought to have markedly reduced emotional resonance with other people's suffering (Marsh et al., 2013; Viding and McCrory, 2019). One strong indication is that they show reduced physiological responses to images of people in distress (Blair et al., 1997).

Lack of resonance also extends to positive emotions. Essi Viding, with Eamon McCrory and colleagues, used Sophie Scott's paradigm to elicit spontaneous laughter. They found that boys at risk for psychopathy did not fall under the spell of contagious laughter (O'Nions et al., 2017). However, even if people with such traits lack automatic emotional resonance, they might be able to use higher-level systems to apply deliberate and helpful forms of empathy.

If psychopaths are the prototype of individuals who are unable to spontaneously share emotions, then there is an even more extreme version—namely, robots. Robots in the popular imagination have no emotions. We feel thoroughly entitled not to empathize with them and do not feel guilty in exploiting artificial intelligence (AI) agents (Karpus et al., 2021). Emily Cross et al. (2016) have reviewed results from brain-imaging studies in which participants observed human actions or robot actions. The action observation network, which roughly corresponds to the mirror neuron network, was activated when observing both human and robot actions. However, the robots were not treated as agents with feelings. The mentalizing network (see chapter 10) was activated only when participants believed that the agent was human. So far as emotions are concerned, robots are the ultimate outgroup.

5 The We-Mode

Most animals, including humans, have a basic need to affiliate with others. We like to be together, and there are simple and automatic rules of alignment that enable us to behave as a collective (whether called a group, a flock, a swarm, a shoal, or a herd) rather than as individuals. This need for affiliation underpins the We-mode, where the world is represented from a collective point of view, in which the group is more important than the individual. When in this mode, individuals automatically take account of what others can see and do. We-representations can be understood as maps of what we, as a group, can currently see and do. These maps are very malleable since they will be constantly changing depending on where we are and who we are with.

In human societies, frames of reference, which are independent of individual viewpoints, can be effectively used as We-representations. This happens automatically in the world of agents. In the world of ideas, however, such representations often depend on group consensus (e.g., where north is) and provide a starting point for the cultural development of shared meaning.

* * *

Greta Garbo famously said, "I want to be alone," but no one really believed her. Mostly, we hate to be on our own. Like all social creatures, from the moment of birth we spend nearly all our time with other humans. Even when we are not with other people in person, we may be talking to them in our heads, or nowadays on our mobile phones. We not only like being with others, but we need others for comfort, for safety in numbers, for collaboration and competition, and for access to their knowledge and beliefs.

Like all social animals, we have an unconscious drive to affiliate (i.e., to be attached to or associated with others). In chapter 4, we discussed how affiliation is intimately connected with our automatic tendency to mirror the actions, sensations, and emotions of others in our group. In this chapter, we discuss other kinds of alignment that allow individuals to join with others to form well-connected groups that act as one entity. It is inevitable that when agents become part of such a group, they lose some of their individuality. So what are they gaining?

Alignment

The groups that we will consider here consist of agents who are doing things together, living together, acting together, or simply moving together. But even when simply moving together, how do we avoid bumping into each other? We need to align our movements. And it is not just our movements that we need to align. We need to align our goals, and even our thoughts, and through aligning our representations of the world, we can achieve much more than we can on our own. Alignment is one of the keys to group behavior. Only then is coordinated behavior possible. It is no exaggeration to say that an organism that doesn't have the ability to align with others of its kind is doomed. It will not be able to procreate, and it will be easy prey.

Alignment is directly observable in behavior, but how is it regulated? It is not obvious whether alignment stems directly from falling in line with others, bottom-up fashion, or whether it stems from following a leader. Bottom-up alignment is seen in its most striking form in flocks of birds and shoals of fish. In these conglomerations, large numbers of individuals move as one. At first sight, such magnificent mass coordination would seem to require top-down control from a choreographer. Amazingly, the synchrony can be achieved by everyone obeying a few simple rules, which can be followed automatically. In the case of a shoal of fish, for example, just two rules suffice (Ballerini et al., 2008): (1) keep close to your nearest six neighbors and (2) avoid collisions with your neighbors (see figure 5.1).

Computer simulations using such rules can generate group behavior that appears remarkably realistic (Reynolds, 1987). Synchrony results, and to achieve it, there is no need for an external clock or a leader, nor does any agent have to be aware of these rules.

Simple rules of physical alignment can create the appearance of carefully planned and strategic behavior. An example is the hunting behavior of a pack of wolves: tracking the prey, carrying out the pursuit, and encircling the prey until it stops moving. It turns out that this behavior also can be generated by two simple rules: (1) move toward the prey until a minimum safe distance is reached and (2), move away from the other wolves. All that each wolf needs to do is track the position of the prey and the other wolves (Muro et al., 2011). Their hunting behavior does not need hierarchies of leaders and followers, nor any special means of communication between each other.

For physical alignment in a freely moving group of animals, the individual agents seem to obey tacit rules about their position and movement in space. But such rules of alignment need not be restricted to moving in space. As we shall see in chapter 6, mental alignment seems the rule for people to perform a task together successfully.

Figure 5.1
Formation of a shoal of fish: a simple rule is sufficient: each fish stays close to the six nearest fish (Ballerini et al., 2008).

When freely working together, they need to have the same understanding of the nature and goal of the task. As with spatial alignment, mental alignment seems to happen automatically, Zombie style, and without deliberate effort.

Affiliation

Alignment is an essential part of affiliation. Once we are inside the safety of a group, continuous alignment increases our affiliation. If so, there is a virtuous circle that makes groups stronger and stickier and better able to reap the benefits of group protection. Through affiliation, we not only gain immediate protection, but we can also count on cooperation with other group members. We become more like each other when, as individuals, we come together in a group. We also feel loyalty toward the group and hate traitors. In this chapter, we will focus on how individual minds can merge into one unit, even if only fleetingly.

Of course, there are individual differences in groupiness. Not everyone shows the same engagement and enthusiasm for cooperating with group members. While high levels of affiliation facilitate prosocial actions, there is also a converse. One alarming observation is that young children who show less need for affiliation are at high risk for showing antisocial behavior and psychopathic traits (Viding and McCrory, 2019; Perlstein et al., 2021).

But is affiliation and merging together always best for individuals? As much as we benefit from being part of a group, we also have to pay a cost. Armies work on this principle and plan on the basis that individual soldiers will die. In everyday life too, we can experience the cost of alignment. When crossing a road, pedestrians are about twice as likely to start crossing if their neighbor has started to cross. This effect occurs even though this impulsive crossing comes at the cost of increased risk of injury (Faria, Krause, and Krause, 2010). Our drive to affiliate is so strong that the benefits must outweigh the costs.

What Do We Mean by We-Mode?

As described by Mattia Gallotti and Chris Frith (2013; also see Tuomela, 2006), We-mode is an irreducibly collective mode. It captures the idea that our view of the world changes when we are "in it together." This group viewpoint ensures that gains for the group are more important than gains for the individual (Tomasello and Carpenter, 2007). In this mode, we have much less desire to make gains for ourselves at the expense of the group (see chapter 9 for a discussion of free riding). This form of alignment, through which we share our intentions rather than just synchronizing our behavior, seems to be uniquely human (Tomasello, 2010).

We-mode is more than the phenomenal experience of each of the individuals concerned. We use the term to indicate that it allows a type of joint representation, a We-representation, that is well below phenomenal awareness. It is important for the success of joint action, as formulated and studied in the lab of Natalie Sebanz and Günther Knoblich (e.g., Sebanz, Bekkering, and Knoblich, 2006). Through We-representations, group behavior is not simply the sum of individual behaviors. For example, Kourtis et al. (2019) present evidence to suggest that actors represent the combined consequences of their joint actions. This makes it easier for the actors to understand the task requirements and to initiate their actions in a speedy way. We will say more about this lab's groundbreaking work on joint actions in chapter 6.

We-mode throws light on a rather amazing fact: people represent the common space and the objects in the space, overriding their own individual point of view. This comes in handy for groups that are equally affected by their environment. As James (1904) pointed out long ago, if one person blows out a candle, then the room is dark for everyone (565).

Merging Different Viewpoints

Because they have different viewpoints, people can often be aware of different things. For example, I can see things that you can't see because they are behind you. Merging different viewpoints seems like an advantageous thing to do for two or more people. We would all have more information. But what does this putting together entail, and how is it represented in an individual brain? Is it part of the Zombie thread? Do we have to put conscious effort into these joint representations?

Imagine that you are taking part in an experiment originally conducted by Dana Samson and her colleagues (Samson et al., 2010; also see Surtees, Apperly, and Samson, 2016). She shows you a schematic picture of a room (see figure 5.2). On the walls are one or more dots, and you are asked to report how many dots you can see. Easy, right? Now imagine that another person, shown as a schematic figure, is looking at the same room but can see only one wall. This person sometimes sees a different number of dots

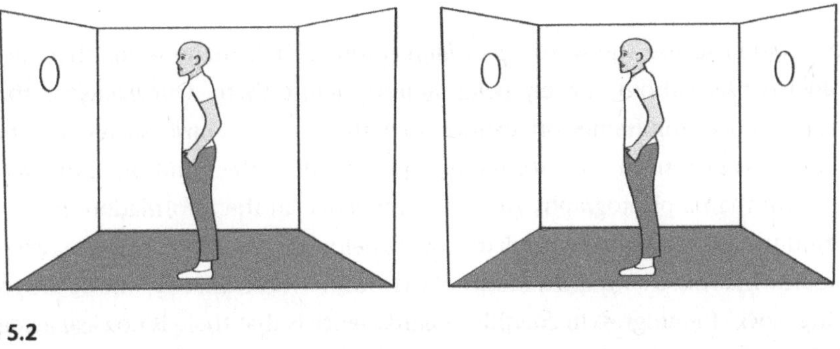

Figure 5.2
Taking account of what others see. In the left panel, both you and the other person can see one dot. In the right panel, you can see two dots, but the other person can see only one dot. This discrepancy slows you down. Redrawn from figure 2 in Samson et al. (2010). Copyright 2010, with permission, American Psychological Association.

from the ones you can see. Remarkably, when this occurs, you are slowed down in saying how many dots *you* can see.

In this task, it should not matter to you what the other figure can see. So why do you care? As the experiment shows, you cannot help it. The spontaneous adoption of another person's perspective has also been found with real people sitting at right angles to each other (Freundlieb, Kovács, and Sebanz, 2016).

Outside the lab, when two people are performing an action together, an advantage occurs from taking into account each other's viewpoint. But it seems that we do this even if we are not working together. We do not ignore what another person sees, even when this is disadvantageous—slowing us down, an indication of a truly automatic process, part of the Zombie thread. Still, does the slowing indicate that we are spending increased mental effort?

This was explored in an experiment where participants had to do a complex tapping task at the same time as the perspective-taking task. This extra task indeed took up more conscious mental effort, and it did result in a general slowing of response times. However, the participants continued to pay unnecessary attention to the figure's perspective, confirming the automaticity of this process (Qureshi, Apperly, and Samson, 2010). We suggest that this experiment is an example of the We-mode in action. The viewpoint of another actor is mingled with our own viewpoint as we form a We-representation. The mingling makes the dots that only we can see a little less prominent, and the dots that they can see too a little more prominent.

Elusive Representations and Why We Need Them

We have started to use the word *representation*, one of the many terms that cognitive psychologists use without entirely being able to define them. Our excuse is that our information-processing framework cannot do without them. Our basic tenet is that the world outside is present in some form (i.e., represented) in the mind/brain in a far more abstract form than a photograph. We can't experience all the information in the world that surrounds us, as it would completely overwhelm our capacity. Therefore, we need a sparse representation, a simplified model of the world. Unfortunately, one of the greatest stumbling blocks for progress in cognitive neuroscience is that there is no clear agreement about what representations are, or even if they are needed at all (but see Shea, 2018).

We must emphasize that representations need not be consciously held, even if we can consciously talk about them. It suits our still-primitive theories about how our mind works, to use various metaphors, models, images, tokens, or symbols. Their use is sometimes interchangeable. Often when we talk of representations, we think of objects in space and where they are in relation to us and to others. But this is not enough. If we are going to do things together, we need to represent potential actions. Actions are what all living creatures need to do, or else they might never eat or, even worse, they might be eaten themselves. And when they are with others, they need to be able to act as a group. This means that we need action to be at the center of We-representations.

Representations for Action

Let's first consider individual actions. When we see an object, such as a cup sitting on a table, we don't just acquire information about its size and shape. We also acquire information about what we can do with it. Simply seeing the cup has the effect of automatically eliciting a plan for the appropriate action. Is the cup within my reach? How can I drink from it? Is the handle on the left or the right side? In other words, the mere sight of the cup elicits a potential plan for action. This is what we mean by representations for action.

The idea that when we perceive an object, we also perceive the possibilities for action on that object, was introduced by James J. Gibson (1979). He called this the *affordance* of the object. He believed that all the necessary information was in the object and there was no need for representations in the mind/brain. We diverge from Gibson, as we believe that representations are needed to understand how we put perception into action. We assume that representations of objects and actions relating to them are computed effortlessly, without the need for conscious thought. Thus, they can activate the motor system of the brain merely at the sight of objects and without any movement occurring at the time.

Fortunately, it is now possible to detect changes in activity in motor cortex using harmless magnetic pulses (transcranial magnetic stimulation, TMS). When the motor cortex is activated, it is easier to excite, and a smaller magnetic pulse will elicit a movement. Using this technique, Cardellicchio, Sinigaglia, and Costantini (2011) showed that when we see a graspable object that is within our reach, our motor cortex is excited, not only our visual cortex (see figure 5.3). This is in line with the idea of an *action priority map*, indicating the location of all the graspable objects, a representation of the immediate environment in a form relevant for action (Klink, Jentgens, and Lorteije, 2014).

The Monkey and the Rake

We can use TMS in humans to demonstrate that the motor cortex is excited by the appearance of a graspable object. But we can get a much more detailed understanding

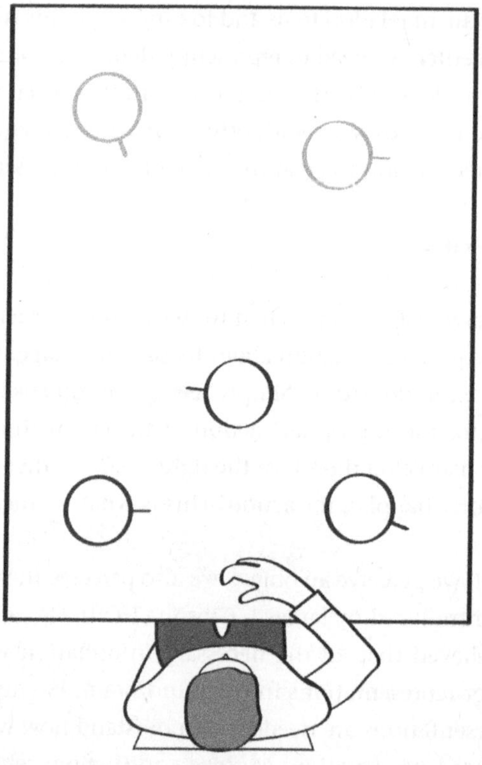

Figure 5.3
An individual action priority map: Three of the five mugs are within reach of the individual. Their presence activates the motor cortex, and they become part of the individual's own action priority map. Reprinted from figure 2 in Gallotti and Frith (2013). Copyright 2013, with permission from Elsevier.

of neural representations for action through single cell recordings in the monkey brain. In chapter 3, we highlighted the discovery by Giacomo Rizzolatti and colleagues of mirror neurons in the premotor cortex whose activity relates to the type of action seen or performed. This work also made clear that neurons are concerned with precise actions, like gripping an object between finger and thumb, while others are concerned with power grips where the object is grasped with the whole hand. While these neurons represent information about how to grasp objects, other neurons, in the parietal cortex, provide information about the location of the objects and whether they are within reach. These neurons are activated when an object is close to the monkey's hand (Rizzolatti et al., 1988).

The activity in these various neurons provides the basis for a map of the objects available for us to interact with (an action priority map). But such maps need to be very malleable, as the environment in which we act, whether physical or social, is continually changing. We can get some insight into this malleability from an inspired experiment conducted by Atsushi Iriki at the world-famous Riken Centre for Brain Science in Tokyo. He taught monkeys to use a rake to reach objects that were too far away to grasp (Iriki, Tanaka, and Iwamura, 1996).

As mentioned, there are neurons in the parietal cortex of the monkey brain, which provide maps of objects near the hand. If an object appears near the monkey's hand, then these cells start to fire. This firing activity relates to objects that can easily be reached. But what happened to the receptive field of the nerve cell after training? Exactly what Iriki predicted. As soon as the monkey learned to use a rake, the receptive field of the neurons expanded enormously, but only when the monkey was holding the rake (see figure 5.4). The action map now includes objects that are way beyond the hand, but near the top of the rake. When holding a rake, more objects become within reach, and this is reflected in the activity of the neurons.

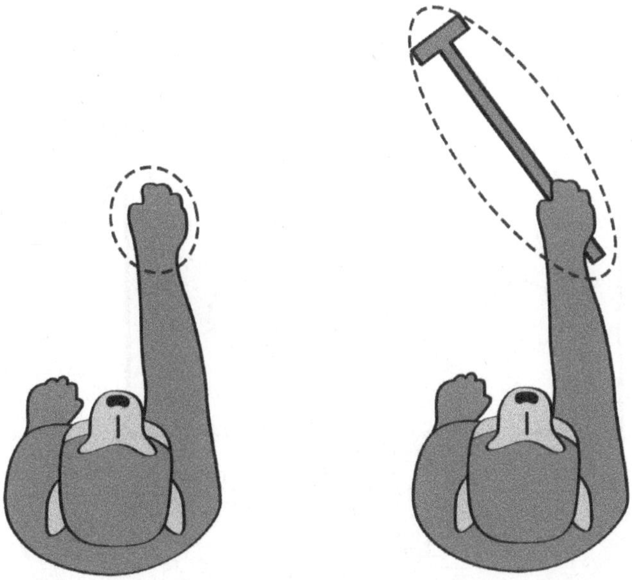

Figure 5.4
The monkey and the rake: neurons in the parietal cortex fire when an object is close to the monkey's hand. But when a trained monkey is holding a rake, they also fire when objects are close to the end of the rake (Iriki et al., 1996).

An Action Map for Groups

Imagine yourself sitting at a table with a glass and a jug of water quite far away. It is too far away for you to reach. It will not be part of your action plan. Now imagine another person joining you, as in figure 5.5. If they can reach the jug, then they could pass it to you. They put you in a position to get things. So does the jug now become part of your action plan?

Here, we return to the work of Cardelicchio and colleagues (Cardellicchio, Sinigaglia, and Costantini, 2013). They used their TMS technique to demonstrate that action maps of people indeed expand when they are in the presence of others. When you are

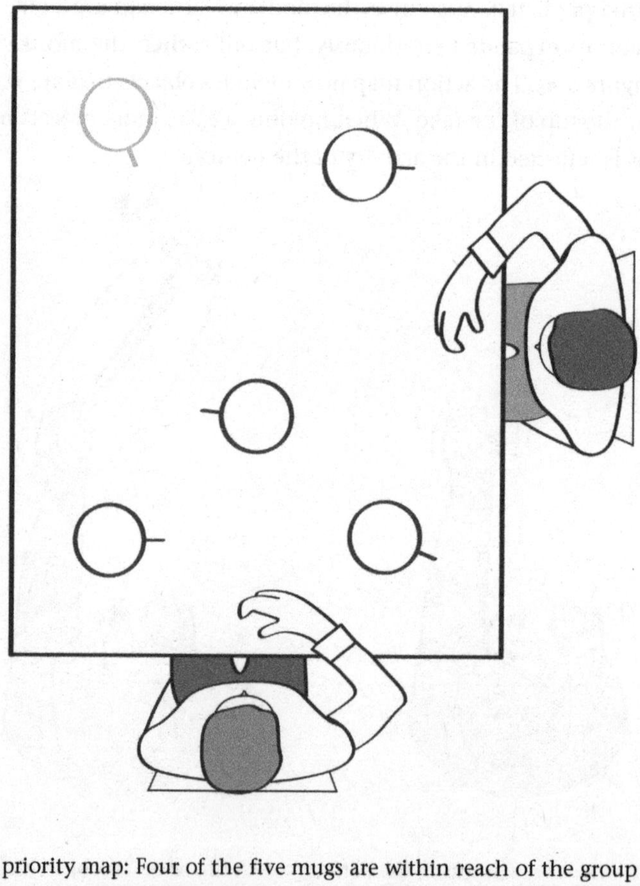

Figure 5.5
A group action priority map: Four of the five mugs are within reach of the group (i.e., both individuals). Their presence activates the motor cortex of both individuals, even though some of the mugs are out of reach. *I* can reach three mugs, but *we* can reach four mugs. Reprinted from figure 2 in Gallotti and Frith (2013). Copyright 2013, with permission from Elsevier.

alone, only objects within your reach elicit increased excitability in your motor cortex. But when another person is present, objects within his or her reach also elicit increased excitability in your motor cortex. Your action priority map is now a We-representation, including you and the companion as one, combining self-related and other-related action maps. In this sense, each person takes on the role of the monkey's rake for the other.

Just as in the example with the monkey's rake, the action space has proved to be malleable. It can be represented either for an individual or a group. The map (or rather, its representation) highlights all the objects within *our* reach. So long we are both in the We-mode, our action maps will be aligned. Each of our brains is using a representation that includes both our reaching possibilities.

We suggest that people seem to adopt these We-representations automatically when someone else is present. But does it matter who this someone else is? Do we adopt a We-representation if we are competing, or if the person is not from our ingroup? Some work suggests that the members of the group and their relationships do matter. One study showed that We-representations were inhibited when a person was excluded from a virtual ball game (Ambrosini et al., 2014). People who had been excluded did not adopt the We-mode. Their action priority map for grasping a mug did not change when another person was present.

It seems that action maps are as malleable in the social sphere as they are in the spatial sphere. It makes sense that, after having been excluded from a group, we can no longer expect help from its members, so we do not adopt the We-mode. But is this a deliberate decision or one of those purely automatic actions that our brain does behind our back? We suggest that both are possible. Adopting a We-representation, like other social processes, is subject to high-level, conscious control. In later chapters, we will add complexity, revealing that such high-level control can also be inhibited. This can happen involuntarily, usually when attention is engaged by some other urgent problem. The reason is that we don't have enough capacity to consciously control more than one action at a time. When we have enough control, we can suppress our tendency to adopt the We-mode, which would result in selfish action. Conversely, we can keep the We-mode, which would result in a dilution of selfishness and lead to altruistic action.

Mapping the Space That We Share

We have highlighted the role of We-representations as action priority maps. However, it seems very likely that collective representations of our environment are created during many situations. We-representations of value will be created during shared attention (e.g., Shteynberg, 2018), joint memory (e.g., Elekes et al., 2016) and joint-decision-making (e.g., Klink et al., 2014). For example, a We-representation of payoff values can resolve problems in games like the prisoner's dilemma (Luce and Raiffa, 1989). In this game, if players take account only of individual payoffs, then they will compete, and both will suffer. But if they both take account of the joint payoff, adopting a We-representation, then they will cooperate, and suffering can be reduced or even eliminated (Colman and Gold, 2018).

We also believe that there is an interesting family resemblance between, on the one hand, individual representations and We-representations, and, on the other hand, egocentric and allocentric spatial frames of reference (Burgess, 2006). The individual representation is equivalent to an egocentric frame of reference in the way that it links the individual to objects in his or her environment: "This is where the object is in relation to me." This representation guides you when your brain decides whether to reach for an object with your left or your right hand. You, at your conscious level, will often be entirely unaware of which choice your brain has made at the unconscious level.

What about the allocentric frame of reference? In this frame of reference, objects are related to one another rather than to the self. For example, I might represent the mug as being near the edge of the table rather than near my left hand. I might also represent the object as being to the north rather than to the left of me. At first sight, there might seem little similarity to a We-representation. There is nothing specifically social about such a representation, but the key feature of the allocentric frame of reference is that it is independent of any particular viewpoint. This means that the representation is the same for all the people present. It is effectively a We-representation, ideally suited for group endeavors. One prediction to be explored is that people will put more weight on allocentric representations when they are in the presence of others.

Representations Revisited

We hope that the concept of representations has now become a little less elusive and that we have convinced you of its usefulness. Here, we give it another try. Just as there are two ways of representing the location of an object, there are two ways of representing the shape of an object. One of these relates to action, while the other relates to recognition. Both are maps or representations that we normally expect to help us navigate the world.

A famous case study allows us a glimpse of the separate nature and location of these representations in the brain. This is the case of DF, a patient with severe visual agnosia caused by damage to her inferior temporal lobe. Her fascinating story has been told by Mel Goodale, David Milner, and others (Goodale et al., 1991). For us, the key observation is that DF could accurately grasp an object (e.g., a stick or a piece of card) and post it into an aperture by holding it in the right orientation. She performed the right action, and yet, consciously, she was unable to recognize the shape of the object that she grasped. DF was able to use the shape of an object to prepare her hand for grasping, and at the same time, she was unable to perceive or name the shape of the object, so she didn't know what it was. One representation was accessible to her, but because of her brain damage, she had lost access to the other representation.

Using shape for grasping requires an egocentric frame of reference: the nature of the grasp depends on the relation between the object and the grasper. This motor plan, however, is deeply unconscious and does not communicate with the representation of the identity of the object, which is normally accessible to consciousness and can be reported to others. Object recognition is enhanced by an allocentric frame of reference, so the object can be identified from whatever direction it is seen. DF had a specific impairment regarding the allocentric frame of reference (Schenk, 2006).

The allocentric frame of reference associated with object recognition provides a basis for a shared understanding of what objects are. Human groups take this a step further and imbue objects with meaning as well as identity. It is advantageous to the group if everyone can agree on the identity and meaning of objects such as predators or prey. However, for this advantage to be achieved, we must be able to communicate our understanding to others, topics that we will discuss in later chapters. If we can extrapolate from the case of DF, we can speculate that identifying objects is associated with consciousness. As we shall see, consciousness is critical for sharing experiences. If so, meaning can be seen as a We-representation created by culture. We will return to this possibility in chapter 16.

This point reminds us of the mapping and merging of spatial viewpoints. For instance, in an allocentric viewpoint, we can represent objects as being toward the north. But for this to work, there needs to be a permanent knowledge of what "north" means. Furthermore, there needs to be a group consensus as to where north is, something that deliberate teaching can bring about.

These considerations bring us close to our main message, which we hope will become clearer in the final chapters of our book. But first, we need to lay down some basic stepping stones—fundamental mechanisms that make us social and that we are likely to share with many other animals. This chapter has hopefully provided an inkling of how representations change when we are in the presence of others. The next chapter will concern some basic mechanisms that enable us to actively cooperate with each other.

6 Joint Action

Some tasks are too big for one individual to perform and require us to work together. Such joint actions can be coordinated at different levels of the processing hierarchy, from aligning of movements to aligning of goals. Some surprising effects have been observed when two people are doing a task together. Typically, the task becomes *our* task, with tight coordination between partners, sometimes at the cost of becoming more difficult to perform. There is evidence of not just physical alignment in motor tasks, but mental alignment of different points of view in perceptual tasks. This alignment is hard to resist even when it slows individual performance.

When people are working together, complementary actions rather than identical actions are often needed. This does not produce a conflict with a predisposition to mimic when the partners align at the level of goals, as opposed to the level of motor action. Partners often need to adapt to each other, allowing for different abilities or physical circumstances. There are consistent strategies for mutual adaptations at an unconscious level of processing that make working together a fluent and efficient process. Nevertheless, explicit rules are needed to solve more complex coordination problems. These rules govern normative behavior to mitigate friction and facilitate alignment.

* * *

In the early 2000s, we visited Wolfgang Prinz, the director of the Max Planck Institute for Psychological Research in Munich. We stayed for a month or so; not only did we explore the cultural treasures of the town, but we also met Natalie Sebanz, who was Prinz's PhD student at the time. She told us about a trailblazing experiment (Sebanz, Knoblich, and Prinz, 2003). She was using a standard reaction time task, traditionally carried out by one person. The instruction was, "Press the left button if you see a white ring on the finger, and press the right button if you see a black ring." There is a little bit more to the task because the finger can point either to the left or the right (see figure 6.1). The pointing is irrelevant, but distracting nevertheless. When the finger points in the opposite direction of the button to be pushed, the reaction times are slower. The task becomes much simpler if you only have to do half of it. In this case, the instruction

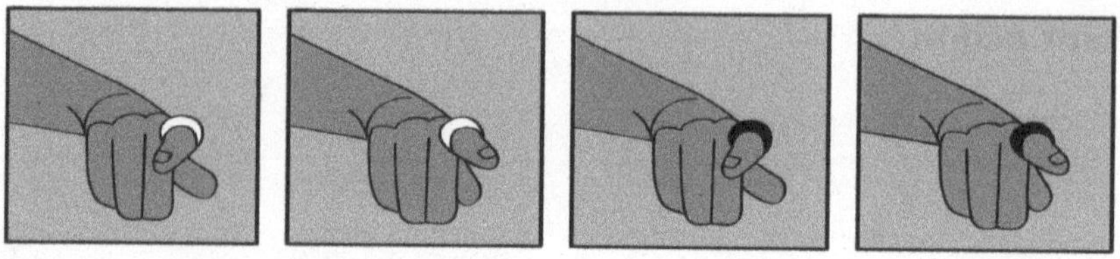

Figure 6.1
Natalie Sebanz's pointing task: Press the left button when you see the white ring. Press the right button when you see the black ring. Ignore which way the finger is pointing (Sebanz et al., 2003).

was, "Press the left button if you see a white ring." Now you do not slow down when the finger points in the wrong direction. There is only one aspect of the task that you have to worry about.

Natalie's brain wave was to have two people do this simple task together, where each one has only one aspect to attend to. One person is pressing the left button when they see the white ring. The other is pressing the right button when they see the black ring. Amazingly, here the interference comes back. The conclusion is that joint tasks are not two separate tasks done at the same time. As soon as another person is involved, we cannot help but take account of what they are doing. We automatically adopt a joint representation, a We-representation, of the task that we are doing together. In other words, our performance seems to be governed by a rule, such as, "Your task is my task—it's our task," and this rule is applied quite unconsciously—and, in this case, to our detriment. We have turned a simple task into a more complex one!

Impelled by this exciting result, Natalie Sebanz and Günther Knoblich launched a major research program to investigate the capacity to create joint representations automatically when people are acting together with others. In 2011, they set up a lab at the Central European University, then located in Budapest, which we were lucky enough to visit. There is now a mass of evidence for joint representations, particularly for fine-grained temporal coordination among partners. Joint action is supported in many tasks, from haptic coupling and shared goal representations to reading mutual expressions of emotions to tacitly obeying cultural norms and conventions (see Vesper et al., 2017; Sebanz and Knoblich, 2021, for reviews).

Physical Alignment and Mental Alignment

Physical alignment is essential for acting together, but mental alignment has equal if not greater importance. When we align physically, as in marching or rowing, this is often organized by cues from outside. The rhythm of a beat can govern our coordination. But we don't depend on such external cues. Instead, we can plan and direct coordination from the inside. We can seamlessly switch from individual work to joint work in complex sequences of joint action. And we are able to do this through automatic mental alignment.

We often need to work not just with familiar others, but also with strangers whom we will never meet again. This might vary from the simple act of coordination that is required when walking along a crowded street, to the more complex negotiations needed when we help a frazzled mother carry a baby carriage down the stairs at a subway station. The reason that these actions can be remarkably successful is that we have mechanisms that create spontaneous alignment and coordination in the short term[1].

It is worth pointing out a subtle difference in these examples. When we pass someone walking along the street, we may both have the *same* goal of not bumping into others, but it is not a *shared* goal since avoidance would occur even if the goal were held by only one of the two protagonists. In contrast, when we help someone struggling with a baby carriage, we have a shared goal since the goal will not be achieved unless we both play our part. The latter situation exemplifies joint action that involves an interaction, for which mutual adaptation and coordination is necessary (Gallotti, Fairhurst, and Frith, 2017). Of course, there are plenty of more complex examples, such as musicians playing together in an orchestra.

Prosocial Actions through Shared Goals

Successful joint actions that benefit all parties require that the people involved have joint goals. They are committed to achieving the same goal (Michael, Sebanz, and Knoblich, 2015). Such commitment is commonly brought about by prior instruction. For example, removal men are instructed to put a piano in the living room, and they are under contract to do this as safely, carefully, and efficiently as possible. This is also typically what happens in laboratory studies of joint action. Here, pairs of volunteers are given tasks to perform, with a shared goal that is clearly indicated by the experimenter (e.g., Sebanz, Bekkering, and Knoblich, 2006).

Coordinated action happens not only with instructions or carefully orchestrated experiments. Outside the lab, joint actions often occur spontaneously, especially when we are helping each other. And we may do this without giving any thought to the time and effort this involves. Such spontaneous action belongs to the Zombie thread, but

1. This has implications for working jointly with robots. The driverless cars that we have been promised are not taking any account of what we are seeing or planning. Perhaps we could accept them more easily if we could rely on some degree of mutual alignment.

in some cases, we can get a strong hint of the Machiavelli thread. If there are onlookers about, more people might be inclined to help someone with a baby carriage facing steep stairs. They might see an opportunity to increase their reputation, since helpful individuals tend to earn admiration and gain status.

Helping others can already be observed in eighteen-month-old children. In a charming experiment, available to watch on the web, Felix Warneken and Michael Tomasello (2006) showed that these infants will spontaneously open the door for a stranger whose hands might be occupied carrying books. Using ingenious paradigms, these researchers and colleagues have also shown that spontaneous helping can be observed in chimpanzees when a conspecific is trying to obtain food (Warneken et al., 2007). For such behavior to occur, the helper must not only have the motivation to help but must also recognize the goal of the other. Human infants aged six months, when observing an adult feeding herself with a spoon, anticipate with their gaze that food will go into the mouth (Kochukhova and Gredebäck, 2010). Goal recognition is tied to agents rather than to objects. One-year-olds anticipate where a toy will be placed if it is moved by a human hand, but not when the toy is moving of its own accord (Falck-Ytter, Gredebäck, and von Hofsten, 2006).

The spontaneous presence of helping behavior in other animals (see McAuliffe and Thornton, 2015) bolsters the conclusion that altruistic action could be one of the social brain's preference settings, implanted through evolution. If so, altruism may not be purely a product of cultural learning. Of course, in older children and adults, the motivation to help may well depend, in part, on deliberate thought processes, reasoning that if we help others, we can gain friends as well as a good reputation. In chapter 11, we will revisit this intriguing question. As a foretaste, we present two tantalizing attempts at an explanation.

Prosocial actions, such as helping and cooperation, are facilitated by norms that endorse them and at the same time punish antisocial behavior. The mechanism underlying these processes may involve inhibitory control of more selfish low-level processes. This control is thought to emanate from the dorsolateral prefrontal cortex (DLPFC). The alternative account starts with the idea that the preference setting in the brain is for altruism. This fits with many observations of spontaneous helping and proposes that prosocial behavior does not require effortful, top-down control of self-interest.

There are a number of brain-imaging studies that suggest that prosocial behavior can emerge from processes that are intuitive and automatic (Zaki and Mitchell, 2013). A top-down account has been proposed (Buckholtz, 2015), in which DLPFC, rather than inhibiting selfish urges, generates representations of the model-based values of rules, such as antiselfish social norms. Top-down control also modulates the effects of the lower-level (model-free) value systems instantiated in the striatum. This is of particular importance for norm-guided decision-making. Both accounts fit with the general hierarchical Bayesian approach outlined in chapter 12.

What Makes for Successful Joint Actions?

Alignment at the different levels of information processing provides a framework for the successful conduct of joint actions. This principle applies to any kind of joint action, including the simplified motor tasks typically studied in the laboratory. At the higher levels of control, there is a shared goal and a common representation of the task. At the lower levels, there is temporal and spatial coordination of movement.

For an example at a relatively low level in the hierarchy, we can consider the situation in which you and a partner are sitting next to each other and move your hands back and forth between two targets. But the task is not equal for the two of you. Your partner must move over an obstacle before hitting the target. Curiously, in this situation, you will find that you move your hands in a higher arc than is strictly necessary (van der Wel and Fu, 2015). This is somewhat costly for you, as it means some unnecessary effort, but it is easier for you to align with your partner's timing.

Alignment is advantageous for any joint action, where movements must be synchronized in time. As we have seen in other examples, an individual's actions can be slowed precisely because we cannot help but take account of what our companion is seeing and doing (e.g., Samson et al., 2010; Sebanz et al., 2003; Surtees, Apperly, and Samson, 2016). However, just as in the case of copying actions and emotions, humans are picky. We don't align with just anyone. It matters who the partner is. At the most basic level, the partner must be perceived as a biological agent. Alignment does not occur when a wooden hand rather than a human hand is seen performing the partner's task (Tsai and Brass, 2007). The partner also must be seen to be the direct causal source of the action, rather than merely intending that the action should occur through some indirect means (Stenzel et al., 2014). For the automatic alignment of movements and timing to occur, we need to believe that we have not only a live partner, but also, the right kind of partner.

Coordinating Complementary Actions

The mirroring phenomenon discussed in earlier chapters tells us that when we are watching someone else performing an action, we are primed to copy that action. This priming, which is manifest in spontaneous mimicry, is likely one of the social brain's preference settings. However, in the case of joint actions, the actors often need to make complementary rather than identical actions. If furniture movers carry a heavy piece up a flight of stairs, one may be walking forward while the other is walking backward. To maintain complementarity within joint actions, the partners need to coordinate their actions while, at the same time, they need to distinguish between their own individual roles and each other's role in the joint action (Knoblich and Sebanz, 2008).

When acting on our own, we typically adopt an egocentric frame of reference with movement trajectories computed relative to our own body. During a joint action, when the partners attend to objects together from opposite perspectives, people tend to adopt an allocentric frame of reference rather than the default, egocentric frame (Böckler, Knoblich, and Sebanz, 2011). As we argued in chapter 5, the allocentric reference frame has the advantage of being independent of any particular viewpoint, allowing easier comparison of the movement trajectories of the two partners.

Because of our preference setting to copy another agent's action, it is typically more difficult to perform an action that is different from the one we are observing (e.g., Cook et al., 2012). But this effect is eliminated when we perform a joint task with someone else, in which complementary actions are required. For example, if two people are lifting a cylinder together (see figure 6.2), they can make the same (i.e., compatible) grasping action. If, however, they are lifting a bottle, then one would be grasping the thin neck end (i.e., precision grip), while the other would be grasping the fat end (i.e., power grip). In this case, the actions would be different (i.e., incompatible).

In this joint action task, where the partners have to perform different actions, a reverse compatibility effect is seen (van Schie, van Waterschoot, and Bekkering, 2008). Now interference occurs when the partner is seen making the same action. Alignment is occurring at the level of task requirements rather than at the level of simple actions. The We-representation that we automatically adopt for joint actions depends on the nature of the task.

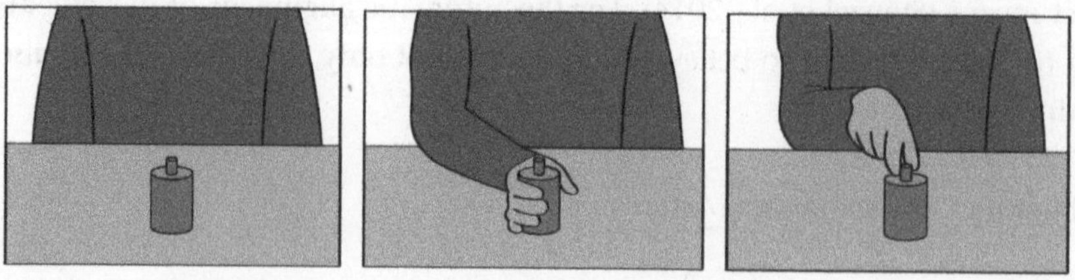

Figure 6.2
When complementary actions become compatible: to lift the bottle together, we need to make complementary actions. Redrawn with permission from figure 1 in van Schie et al. (2008). Copyright 2008, American Psychological Association.

Leaders and Followers

Whenever two people attempt to work together, they need to adapt to each other to allow for differences between them. How does one do this? One possibility is mutual adaptation. This depends on knowing something about the movements of the other, often on the basis of what has just happened. For example, in the task where two people have to tap in synch, the partner who was slower on the last trial speeds up on the next trial, while the other slows down (Konvalinka et al., 2010). The partner who adapts more can be defined as a follower, while the one who adapts less is the leader.

Leader-follower distinctions spontaneously emerge in tasks that benefit from a division of labor. Even the simple tapping task that we just mentioned has two components: first, to maintain the tempo indicated by the experimenter, and second, to keep in synch. In most cases, one member of the pair concentrates on keeping the tempo while the other devotes more effort to maintaining synchrony, thereby becoming the adapting follower (e.g., Konvalinka et al., 2014). This happens automatically, without the need for any discussion. The same effect can be observed in the coordination of musicians. In a typical string quartet, the first violin is the leader, in the sense that the other players adapt more to the leader than the leader does to them. This can be demonstrated empirically by measuring the internote intervals generated by the players (Wing et al., 2014).

Division of labor along leader and follower lines also emerges if one participant knows more about the task than the other. The more knowledgeable one becomes the leader. Leaders can automatically help followers to align their actions. They must somehow indicate to followers what the task requirements are. Remarkably, experiments have shown that they can do this spontaneously, without being aware of it, at the Zombie level. Leaders make alignment easier by reducing the variability in the timing of their movements (Vesper et al., 2011). Leaders also tend to exaggerate their movements (see figure 7.7 in chapter 7; and Vesper and Richardson, 2014). This is a signal that provides the partner with implicit cues regarding the action to be jointly performed (Sacheli et al., 2013). As the interaction develops, leaders will dynamically adjust their strategy. Fewer cues will be needed as the follower learns what is going on.[2]

Talking to Each Other as a Form of Joint Action

Successful joint actions, whatever their goals, typically require coordination at many levels of control. This is nicely illustrated in the case of a conversation between two

2. Of course, there can also be deliberate designation of leaders and followers and explicit communication between them. This adaptation belongs to the Machiavelli thread.

people (Garrod and Pickering, 2009). At the highest level, the speakers must have the common goal of understanding each other or else their conversation is not worth having. Consequently, they need to align their mental representations (Friston and Frith, 2015). Because the various levels of information processing are interdependent, there also needs to be alignment at lower levels. For example, nonlinguistic elements such as speech rate and posture will become mutually entrained, and there also will be mutual alignment of linguistic elements such as word choice and syntax (Pickering and Garrod, 2004). Evidence for this is that both speakers tend to use the same words and similar sentences, and increasingly so during their conversation. To some extent, this alignment may be governed by mechanisms of imitation. However, at intermediate levels of the discourse, the behavior of the speakers must be complementary rather than imitative (Wohltjen and Wheatley, 2021).

Alignment in verbal communication also occurs with quite subtle hidden states. When we hear a word that we cannot integrate into the preceding sentence, this unexpected word elicits activity in an electroencephalogram (the N400 component), which is a characteristic electrophysiological marker of semantic difficulty (see Kutas and Federmeier, 2011). This effect was shown to be present already in infants aged fourteen months (Forgács et al., 2018). Their brain response indicated that they noticed a semantic incongruity in the words that two speakers used in a conversation.

A beguiling experiment went one step further. Here, participants had to read text that sometimes included implausible sentences ("The girl had a little beak"). As in previous experiments, the N400 marker tracked the semantic oddity. However, the marker disappeared when the sentence was read in a context that made it plausible ("The girl was dressed up as a canary for Halloween"). Astonishingly, the marker reappeared when the participant knew this context, but a person sitting next to them did not (Jouravlev et al., 2018). This finding adds weight to the assumption that we continuously engage in modeling the minds of others and are sensitive to differences between what they know and what we know, and that this happens automatically.

Playing Music Together as a Form of Joint Action

A study conducted at the Institut de recherche et coordination acoustique/musique (IRCAM), the Advanced Music Institute in Paris (Aucouturier and Canonne, 2017), investigated the interactions of pairs of free-jazz improvisers. One member of each pair was assigned to play in a manner that conveyed one of a set of five interpersonal attitudes: caring, insolent, disdainful, domineering, or conciliatory. The other partners in these pairs performed well above chance in their ability to recognize which attitude was

Figure 6.3
Listeners with different knowledge: only one listener knows that the girl "was dressed up as a canary for Halloween" (Jouravlev et al., 2018).

being expressed (M = 64 percent, where chance is 20 percent). Musicians who listened to the recordings were also well above chance on recognizing the attitudes, but this ability dramatically increased when listening to both players (36 percent when hearing only one player, as opposed to 54 percent when hearing both players).

When hearing both players, features of the interaction emerge that are not available when hearing just one player. For example, the caring condition was characterized by mirroring and complementary play. In contrast, in the insolent and disdainful conditions, there was much less simultaneous playing. The condition being conveyed also affected leader-follower relationships (measured with Granger causality).[3] When players conveyed dominance, they became leaders of the interaction, while when they

3. Here, Granger causality (Bressler and Seth, 2011) is used to measure the degree to which the behavior of one player can be predicted from the behavior of the other player.

conveyed caring, they became followers. When disdain was being conveyed, there was no causality at all: they ignored each other.

In this study, we see how nonverbal cues do not just modulate interactions in a joint action task but can also convey subtle information to outside observers. The theatre gives vivid illustrations of our appreciation of different levels and types of alignment. For instance, audiences find dance more aesthetically pleasing when there is greater coordination between the dancers, as indexed by Granger causality (Vicary et al., 2017).

Conflicts and How to Avoid Them

The work reviewed in this chapter demonstrates that there is more information in a joint action than is conveyed by the actions of each individual. What we find particularly interesting, however, is the realization that observers need not be aware of the nature of the information, and yet it can affect them deeply. We have focused on experiments on joint action in humans because they target putative cognitive processes that underlie the complexities of joint action. We still don't know how the trade-off between the costs of modifying actions and the benefits of interaction success has been solved (Pezzulo, Donnarumma, and Dindo, 2013). It can be viewed as joint action optimization, where the coactors automatically optimize the success of the interaction and their joint goals rather than only their part of the joint action. It is remarkable that it works. However, there are clearly advantages of explicit coordination, which we will briefly consider here.

Two people are passing each other in a narrow passage. How can they avoid bumping into each other? Simple as the problem seems, it poses a puzzle. Each person has two options: dodge left or dodge right. Looked at from a bird's-eye view, they should dodge in opposite directions, but how can this be achieved without them reading each other's minds to discover which way they are intending to dodge?

How do groups, herds, or hunting packs coordinate their actions so smoothly and effortlessly? One way to achieve a good outcome is the spontaneous adoption of the leader and follower roles. If one agent makes his move sufficiently earlier than the other, then a collision is likely to be avoided. The first to move takes the role of leader, and the other person now has sufficient information to choose his or her move appropriately (Braun, Ortega, and Wolpert, 2011). But sometimes each might wait for the other to go first; and other times, both try to make the first move. These are not efficient strategies for smooth passing.

Another way, for humans at least, is to follow prior rules or conventions. A simple convention for solving the passing problem is "Keep left." Arbitrary rules that have been agreed upon in advance and have become part of convention for a society are powerful forces for coordinating activities of all kinds. Through conventions, people can coordinate their activity without the need for constant negotiation and explicit agreement. This is because their expectations are aligned so each individual believes that everyone will follow the rules (Gintis, 2010). Conventions can be very arbitrary and local. For example, when riding on an escalator, people in Tokyo stand on the left (passing on the right), while people in Osaka stand on the right. Some time ago, people in Copenhagen entered buses in the front, in Aarhus at the back. Such conventions are themselves examples of the need of people who live together to align with each other. At first, these are conscious and explicit strategies that have to be communicated and then followed. But eventually, they become automatic habits.

The Benefits of Conventions

There are more complex problems than having to pass each other in a narrow passage, on an escalator or at a bus entrance. Even moral dilemmas can sometimes be solved by rules and conventions. To study this process, coordination games have been used. One such game is called, tongue in cheek, the "battle of the sexes." A husband and wife are the two players whose aim it is to spend time together, but they have different preferences. One prefers football, the other opera.[4] They can be together if they decide to both watch football or both go to the opera. But one of them will be less happy.

A tragic outcome of this problem is told in a famous short story by O. Henry "The Gift of the Magi" (1905). In this tale, a poor young couple decide to surprise each other with a wonderful Christmas gift. This is what they do: The husband sells his watch to buy a set of combs for his wife's beautiful long hair. Meanwhile, the wife sells her hair to buy a watch chain for her husband. This was the worst possible outcome as far as the game is concerned. However, there is a beauty in their choices, in that they revealed their deep, mutual love.

Going back to the less tragic but incompatible preferences for football and opera, there is a way to achieve coordination so that, in the long term, clashes can be avoided. The best outcome for both partners can be obtained when they both follow a convention that each player knows that the other will follow. For novel problems, the best way to achieve this is through an external advisor who tells the players what to do. External randomization is another solution, known as Aumann's correlated equilibrium (Aumann, 1987). Our advice would be that they should take turns: go to the football one week and go to the opera the next.

4. Given that the game dates from the 1970s, you can guess who prefers what.

Social conventions act as priors that are instilled in us just by being exposed to our culture. They act as preference settings, and individuals acquire the habit of following them (Gintis, 2010). The normative preferences tell you how you should behave, even if they are sometimes trumped by selfish preferences. But it helps because flouting social norms has a bad effect on your reputation (see chapter 9). Explicit rules and sanctions are necessary when the cooperation is costly and when selfish people are likely to take advantage of conventional practices. They may argue that they "were only following the rules" or "weren't doing anything illegal" when the rules are still new and not yet watertight.

How do we start to follow newly set up conventions? The case of the COVID-19 pandemic provided the opportunity to answer this question. It proved possible to predict how well social distancing rules were followed by individuals. Using data from 115 countries, it appears that people distanced most when they thought their close social circle was doing so as well (Tunçgenç et al., 2021).

Conventions need not be explicit and external signals facilitating collaboration need not be explicit either. Nudges may indirectly remind people of conventions that exist for the common good, but are often disregarded out of laziness (Thaler and Sunstein, 2008). For example, trash cans are placed prominently in picnic spots, and people are encouraged to throw their garbage into the trash can rather than leaving it where they were eating. Even though nudges don't always work, it seems a good idea to make the better choice as obvious as possible by using prompts.

We need to coordinate even before we collaborate, and rules and conventions are excellent means of doing so. They avoid anarchy and ensure that the common interests of those working together remain a priority. Unfortunately, as we all know, not everyone in our group will follow the norms, and this is something that must be dealt with. For instance, we expect group members to monitor each other's behavior.

If norms for good behavior are followed by group members, major rewards can be reaped. We can see this in the history of guilds and corporations. In the more successful of these, members are monitored to see how strictly they obey the rules. This makes learning within their bounds very effective, such as by apprenticeships and certificates, and hence provides excellent conditions for training and innovation (Brahm and Poblete, 2021). If the institution can prove that it upholds the rules, it can act as a guarantor for individuals to behave accordingly. We will return to the benefit of conventions in chapter 19.

II Competition—Difficulties of Being with Others

7 Predicting Behavior

This is the first of five chapters that consider cognitive processes that are relevant to competition. The most basic competition faced by all animals is between predators and prey. We all need to recognize, very quickly, whether we are confronted by friend or foe. From birth, most animals rapidly recognize other agents from their movements. Agents are not like physical objects, such as pebbles. They are self-propelled, although their movements are constrained by their physical structure. In the short term, these constraints allow their movements to be predicted. In the longer term, movements can be predicted based on their goals. Goal-directed agents behave rationally, in that they avoid obstacles and reach their goals by the most direct path available.

Human infants in their first year of life expect self-propelled agents to have goals and behave rationally to reach them. Infants also recognize that preferences determine goals and other people can have different preferences. But there is a higher level still where we can recognize agents as behaving according to their inner intentions. It is difficult to detect intentions from movements alone. Critically, we will interpret the same movements differently if we believe that we are dealing with an intentional agent. Without this assumption, intentional behavior seems irrational. This happens, for example, when people exaggerate their actions to signal their intentions. This departure from least-effort, rational, goal-directed behavior is a sure sign of that we are dealing with an intentional agent.

* * *

So far, we have talked about people and other animals being, moving, and working together. We have emphasized how cooperative we can all be. However, wherever there is cooperation, there is also competition. Between different species of animal, there is competition between predators and prey. Within species, different groups will be competing for resources. Even within groups, there will be competition among individuals, and it is crucial for us to detect and outmaneuver these competitors. So how do we detect them in the first place? [1]

1. This chapter has benefited from a prior collaboration with Johannes Schultz (Schultz and Frith, 2022).

Is It a Pebble? Is It a Bird? Is It a Man?

In the world of objects, agents stand out. They move and do things of their own accord. How do animals learn about these peculiar objects?

Do they even have to learn? Few scenes are more charming than that of newly hatched, chirping chicks in a traditional farmyard: the little balls of fluff follow their mother as soon as they open their eyes. However, it is not a miracle that they can do this. Their ability to recognize and follow a particular agent is due to a mechanism that is prewired in their brain (Vallortigara, Regolin, and Marconato, 2005). Keeping close to their mother is important for newborn birds and mammals, but the need to detect other agents remains vital during their entire lives. Movement turns out to be critical for detecting agents, and any movement in the outside world attracts our attention and is instantly classified. In the human brain, a circumscribed region in the posterior superior temporal sulcus (pSTS) (Puce and Perrett, 2003) is exquisitely tuned to detect moving agents.

Here are some types of movements that we can readily distinguish: a pebble carried by a wave to the beach—an example of movement without agency; a housefly hitting a windowpane—an example of a biological agent moving aimlessly; a dog retrieving a stick—an example of an agent showing goal-directed motion; a detective surreptitiously following a suspect—an example of an agent whose movement is guided by the underlying intention to catch the suspect unawares.

To explain the ease with which we distinguish between these kinds of observations, Alan Leslie (1994) proposed a three-level hierarchy, with self-propelled movement at the lowest level, goal-directed action at the next level, and intentional action at the top. To understand and predict the movements we see, we have to adopt the stance appropriate to the kind of agent we are dealing with (Dennett, 1971).

The movements of inanimate objects, like the pebbles on the beach, are entirely determined by the laws of physics. So we adopt the *physical stance*, which is appropriate for the world of objects. The movements of all animate agents, from flies to humans, also obey the laws of physics, but not completely. These agents have an internal source of energy, so they do not show conservation of momentum. As a result, their movement can be determined by internal states such as goals. They can avoid obstacles and move toward food. To understand this goal-directed behavior, we adopt a *teleological stance*, which is appropriate for the world of agents. At the highest level, agents, such as humans, can move on the basis of intentions caused by hidden mental states such as beliefs, which need not correspond to the actual state of the environment. Here, we adopt the *intentional stance*, which is appropriate for the world of ideas.

Who Am I? Recognizing Actions as One's Own

Creatures that were able to move emerged some six million years ago. This was associated with the development of a nervous system, and not long afterward with the appearance of what is known as "efference copy" (or corollary discharge; Crapse and Sommer, 2008) in the part of the nervous system that is concerned with motor control.

What is efference copy? We can compare it to the copy that goes into the Sent folder every time we send an email message. The efference copy is proof that a particular movement that you see was caused by you. In biology, copying is a ubiquitous (and likely low-cost) process. We have dwelled at length on how we learn from others by copying their actions. Still, isn't making a copy of every action we perform somewhat over the top? Surely, it should be obvious that any movement made by an individual has indeed been made by that creature. Except that it is far from obvious. And this is because *everything moves*.

Surprised? Just think of the illusion of self-movement created when we are sitting in a stationary train when another train is passing on the next track. It is not at all easy to decide which train is moving. This ambiguity arises because we are not active; we are sitting passively. If we are actively moving, then efference copy guarantees that we can distinguish our own movements from movements that we experience passively. If there was no efference copy, all the movements we saw would be ambiguous.

For example, if we are suddenly pushed to the left, we should compensate by moving right. But this compensation should not be applied when we suddenly move to the left deliberately. We have to know whether we caused the movement in order to take the right action (see figure 7.1). Even simpler organisms such as fruit flies use efference copy to suppress the perception of visual motion during self-propelled turns (Kim, Fitzgerald, and Maimon, 2015) and thus maintain their intended movement direction.

Just imagine for a moment what it would be like if you could *not* be sure whether an action was performed by you or another agent. This is a strange idea if you believe that an agent must have a *sense of self*. But how do you get a sense of self in the first place? We can speculate that an efference copy mechanism is one root for the sense of self. This first sense of self is only the beginning, possibly weak as well as sketchy, and perhaps only minimally conscious. It would do no more than separate what is happening inside and outside a body.

But that is not all that efference copies can accomplish. They facilitate learning! They can do this because animals learn about their environment by acting on it. For learning, it is important to know whether our actions are followed by the outcomes we were aiming for. Very often, there is a tiny difference and occasionally a very large one,

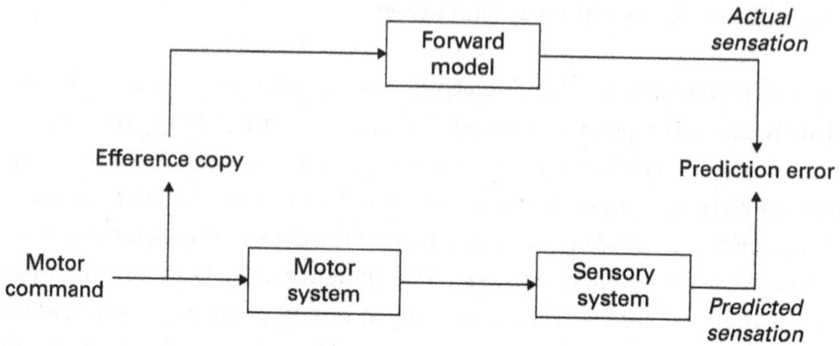

Figure 7.1
Efference copy and prediction error: A motor command causes us to move, which generates sensations of movement. At the same time, a copy of the movement command (efference copy) is used to predict (forward model) the sensation to be expected. If there is a difference between the expected and the *actual sensation*, there is a prediction error, indicating that we did not fully cause the sensation.

a big surprise: this is not the outcome that I was expecting These differences are known as *prediction errors*. They keep you on your toes. They guide learning.

As shown in figure 7.1, a prediction error can occur for two reasons. The first is that the expectation is in error and should be updated; the other is that the movement missed its goal and you need to correct it. The two errors guide your next learning step. Here is a concrete example:

1. The agent opened the blue door expecting a reward, but the reward was behind the red door. In this case, the agent should change its expectation of the reward being behind the blue door.
2. The agent intended to open the red door, but its paw/hand slipped and opened the blue door by mistake. In this case the agent should not change the expectation that the reward will be behind the red door (Parvin et al., 2018). Instead, the agent needs to refine the movement needed for opening the door.

The critical difference between the prediction errors in these two cases is the source of the evidence. In the first case, it comes from the outside world. In the second case, it comes from the inside. The efference copy mechanism helps to make this distinction.

What Is This? Recognizing Other Agents from Their Movements

Perhaps the most striking difference between living and nonliving things is that living things are self-propelled (Tremoulet and Feldman, 2000). When an object is self-propelled, its motion appears not to conserve energy. This is a violation of a law of physics called the law of energy conservation. In humans, at least, there is evidence of a heuristic used for predicting the behavior of physical objects on the basis of the conservation of energy (Sanborn, Mansinghka, and Griffiths, 2013).

Self-propelled agents deviate from these predictions, giving rise to the perception that they have an internal source of energy. They change direction and speed in the absence of any external force and before they hit a barrier. Observers perceive such striking behavior as a sign of animacy (using visual cues: Scholl and Tremoulet, 2000; using auditory cues: Nielsen, Vuust, and Wallentin, 2015).

A classic demonstration of the perception of agency comes from an experiment by Albert Michotte. The top row of figure 7.2 represents an example of a launch event (Michotte, 1963). The gray square A moves toward the black square B. When contact is made, A stops and B moves. This launch event is easily seen as part of the physical world. A and B might be billiard balls, for instance.

In the second row, a different event occurs, which is a reaction event (Kanizsa and Vicario, 1968). Instead of square B moving on contact, it begins to move just before square A reaches it, with both shapes moving simultaneously at a distance for a brief period before A stops. Observers typically report that B is reacting to A's motion—that B is escaping from A.

Michotte's experiment illustrates the three levels of Leslie's hierarchy of agency. Observers can distinguish between agents and nonagents. Over and above this, they distinguish agents showing self-propelled movement and agents showing goal-directed action. There is also a hint of intentional action when observers say that B *wants* to get away from A.

In both the launch and the reaction events, B's motion is seen as caused by A. However, in the reaction event, the interaction is interpreted as social, not physical in nature. The geometric shapes do not look in the least like animate creatures, but for the reaction event, they are perceived as animate.

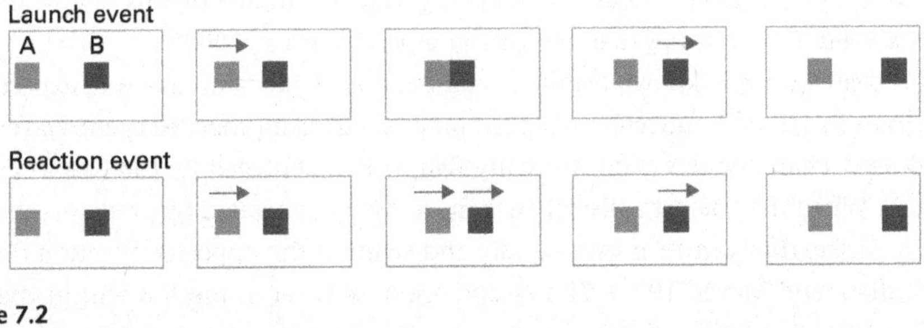

Figure 7.2
Launch events and reaction events.

These differences between inanimate objects and agents are so basic that they are likely to be built into the nervous system as factory settings, which we can also refer to as priors. Priors guide learning just as strongly as prediction errors, as they are there even before any experience. They make us pay attention to the important information in the first place. It is highly plausible that there are priors that predispose us to attend to animate agents and so ensure that social learning gets off to a good start. Given these priors, it is no surprise that infants, tested shortly after birth, prefer to look at movements that show signs of animacy (Di Giorgio et al., 2021). Nor is it surprising that nine-month-old infants readily distinguish launch events from reactive events and expect that agents will behave in a goal-directed manner (Schlottmann and Surian, 1999).

Such an innate predisposition explains why cars and trains, as well as mechanical toys with hidden batteries, greatly fascinate young children. Being self-propelled, they seem to be alive, and being goal directed, they seem to be their own agents. Children have to learn that these hybrids are self-propelled and goal directed only up to a point. Their animacy is an illusion achieved by a clever engineer.

Where Is the Agent Going Next?

At the most basic level, the future position of a moving object can be predicted from the trajectory so far. There are physical constraints on the behavior of self-propelled agents that determine how fast they can move and how sharply they can turn. These constraints create kinematic regularities that can be used to make predictions from movement trajectories. Predatory dragonflies anticipate that movement trajectories will be smooth, and focus their attention on a location just in front of a moving target (Wiederman et al., 2017). Bodily limitations, such as limb geometry, create more subtle kinematic regularities. In the case of primates, for example, there is the two-thirds power law, which relates speed to the curvature of limb movement trajectories (Viviani and Flash, 1995). Human observers implicitly take advantage of this law to predict hand movement trajectories (Kandel, Orliaguet, and Viviani, 2000).

Arms races are a well-known feature of competition. A predator may well expect prey to continue in a smooth movement, but the prey, being a self-propelled agent, can escape by suddenly changing direction. In many fish, this is achieved by the C-start escape response. When they detect a disturbance in the water generated by predators that are about to strike, they can turn very rapidly and swim in the opposite direction (Eaton, Bombardieri, and Meyer, 1977). The escape response is set in motion within the first millisecond or two for small fish, and no amount of subsequent stimulation from the other side of the fish can counter the turn. This is an example of a fixed action pattern

Figure 7.3
Taking advantage of escape responses.

(Moltz, 1962). These are stereotyped patterns of behavior that are elicited by particular stimuli and, once started, they will continue to completion, come what may.

The problem for the fish is that this escape pattern is highly predictable. The tentacle snake, *Erpeton tentaculatus*, takes advantage of this predictability. As shown in figure 7.3, the snake curves itself round the fish so that movement of its body on the left side of the fish initiates an escape response to the right (left panel). As a result, the fish sometimes swims directly into the snake's mouth (right panel) (Catania, 2010).[2]

There is another way to succeed in escaping predators. When they detect a possible predator, animals such as the cockroach (Domenici et al., 2008) and praying mantis (Yager, May, and Fenton, 1990) make escape responses in directions that are essentially random. The adaptive value of such behavior consists of being unpredictable to predators (Brembs, 2011). But we are in an arms race here, and the predator can up the game by minimizing signals that elicit escape responses in the prey.

Lions are masters of the art of avoiding sudden movements while stalking (Schaller, 2009). Predators can also disguise their movements through the choice of the approach trajectory. Dragonflies, for example, can make themselves appear stationary during aerial pursuit (Mizutani, Chahl, and Srinivasan, 2003). The attacker chooses its flight path so as to remain on the line between the moving target and some landmark point. The target, therefore, does not see the attacker move from the landmark point. The only visible evidence that the attacker is moving is its looming; a steady increase in size as the attacker gets nearer.

Humans too can see similar effects of an arms race in sport competitions. The defender can take advantage of the attacker's expectations when taking into account

2. So long as such predators are sufficiently rare, these kinds of escape response remain adaptive.

Figure 7.4
The sidestep in rugby: Using deceptive movements of his legs and shoulders, the man makes his opponent believe that he is going to change direction, but then doesn't (Brault et al., 2012)

bodily limitations. An example of this is the sidestep in rugby. The man with the ball (the prey) avoids a tackle by giving the impression that he is going to dodge to his left when he is actually going to dodge to his right (see figure 7.4).

Bodily constraints prevent a rapid change of direction. This is indicated by the trajectory of the center of mass (the waist, in this example), which provides an honest signal, indicating that the person will move to the right. However, dishonest signals are provided by the positions of the head, shoulders, and left foot. These all suggest that the movement will be to the left. Experts are better than novices at detecting these deceptive movements. Novices attend more to a source of dishonest signals (e.g., the shoulders), while experts attend most to the source of honest signals (the waist). In addition, experts wait a little longer before making their move (Brault et al., 2012).

How to Recognize Agents with Goals

To successfully predict on a moment-to-moment basis where an agent will go next, there is no need to take account of hidden states such as goals and intentions. In the short term, the biological constraints on the possible movements determine where the agent is likely to move next. But in the longer term, taking account of goals and intentions will give an advantage.

One striking difference in the behavior of goal-directed agents is that it is not restricted to fixed actions patterns, such as the C-start escape response in fish. Goal-directed behavior is more flexible. The same location can be reached by different routes, and choice behavior can change as a result of experience. Infants will even treat inanimate objects, such as wooden boxes, as goal-directed so long as they show such behavior, e.g., if the box moves to the same location by different routes (Csibra, 2008).

Infants also expect objects to be goal-directed if their behavior is contingent. If two objects take it in turns to perform actions or make noises, then we perceive them as goal-directed. Think of R2-D2 interacting with C-3PO in *Star Wars*. The little robot uses beeps and whistles instead of speech, but we know that it is communicating even if we don't know what it is saying. Human infants and adults alike will treat a shapeless object like a floor mop as a goal-directed creature if the mop moves or makes noises, apparently in response to the actions of another (Johnson, 2003). They will readily look and point to a part of the mop as its "head," as derived from its presumed orientation.

But the most important marker of goal-directed behavior is that it should be rational with regard to action. That is, the agent should take the most efficient route to the goal, always minimizing the cost of its actions. Indeed, 6-month-old infants already expect goal-directed agents to behave in this way (Liu and Spelke, 2017), revealing a sophisticated prior. This prior was first demonstrated in a groundbreaking experiment by Gergely Csibra and Gyorgy Gergely, as illustrated in figure 7.5 (Csibra et al., 1999). Infants aged 6 months expected a self-propelled ball to take a new shorter route when a barrier was removed. They were surprised if the ball continued to jump over the now-nonexistent barrier.

This, as well as a host of other studies (e.g., Falck-Ytter, Gredebäck, and von Hofsten, 2006), suggest that, for infants, movements that are goal-directed are those in which

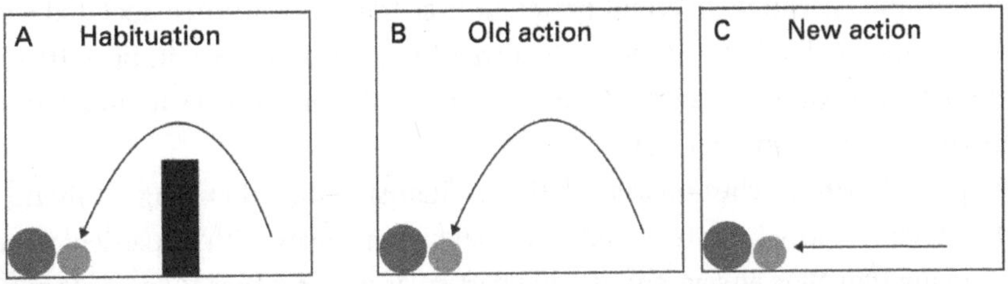

Figure 7.5
We expect agents to behave rationally: The infant repeatedly sees the small ball jumping over the barrier to reach the large ball (A). Then the barrier is removed and the small ball continues to jump (B) or moves directly to the large ball (C). The infant is surprised if the small ball continues to jump (Csibra et al., 1999).

the end point of the movement is more important than the form of the movement. For instance, six-month-old infants attend to the target of an action (which object is grasped) rather than to the form of the action (the way that it is grasped) (Woodward, 1998).

Studies exploring brain responses during the observation of simple, goal-directed actions reliably identify an action observation network (Caspers et al., 2010). This includes the inferior parietal lobule (IPL), the inferior frontal gyrus (IFG) and a swath of visual cortex including the superior temporal gyrus (STG). Interestingly, the IFG and IPL are core components of the human mirror neuron system for action (Rizzolatti and Craighero, 2004; also see chapter 3).

Activity in the pSTS is reliably elicited by biological motion (Grossman et al., 2000) and especially robustly activated by movements indicating the presence of a goal-directed agent. This is the case, for example, with a hunting scenario where people watching moving geometric shapes can detect a "wolf" that is chasing a "sheep" (Gao, Scholl, and McCarthy, 2012).

Why This Goal Rather Than Another? Learning about Preferences

So far, we have talked about how easy it is to detect goal-directed behavior and discovered some remarkable predispositions in the brain that guide how we learn about agents. But why does an agent have this goal rather than another one? We can learn an agent's preferences from tracking their choices. It is easy to learn that different agents have different preferences: Mice like cheese, bears like honey, and toddlers like going on the swing in the playground.

Preferences determine goals. But this does not mean that we should think of them as hidden mental states, like desires. Fleeting desires can only be inferred, but preferences can reliably be derived from observed behavior. They are attributes, like height or hair color, and we can learn about preferences as traits or features associated with an individual. Repeatedly seeing Princess Leia wearing a white dress will be sufficient to associate her with white garments. Thus, we can ascribe preferences to an agent without implying a hidden mental state.

Association learning relies on very basic mechanisms shared among all living creatures. By three months infants already associate preferences with a particular person after observing that they always choose one particular object rather than another (Choi, Mou, and Luo, 2018). A formal mathematical analysis (Robalino and Robson, 2012) has shown that there are considerable advantages for agents to learn about the preferences of others. When we interact with someone, we learn the rules of the interaction, but also the preferences of our partner. For example, when we trade with someone, it is to our advantage to know what sorts of things this trader values and would buy at any price. So long as these preferences are stable, we can carry this knowledge over to future interactions with that person. We don't need to relearn their preferences.

Of course, behavior is not always clear cut or reliable, but it can still be used to discover preferences. Even preschoolers can use statistical information to infer preferences from behavior. Moreover, they are able to infer that less consistently chosen objects are less preferred than those that are chosen consistently (Hu et al., 2015).

How to Recognize Agents with Intentions

What if you could learn the reasons for an agent's preferences? This would be extremely useful, since preferences can change and may not always be what they seem. Perhaps you don't really like going to art exhibitions. You pretend to like them so that you can be with your friend. Behind your apparent preference lie hidden intentions.

But how do we know when we are dealing with agents who have intentions? Are there any characteristics of movements that make agents appear intentional rather than simply goal directed? Evidence that this is the case comes from animations of the kind originally created by Heider and Simmel (1944), and, much later, by ourselves (i.e., Abell, Happé, and Frith, 2000). When observers watch the triangles in these animations interacting with one another, they tell remarkably consistent stories about what they think is happening. In these narratives, the behavior of the triangles is described as being determined by mental states. For example, people typically describe what is happening in the sequence illustrated in figure 7.6 as a mother trying to coax her child to go out into the garden, with the child seeming reluctant to do so.

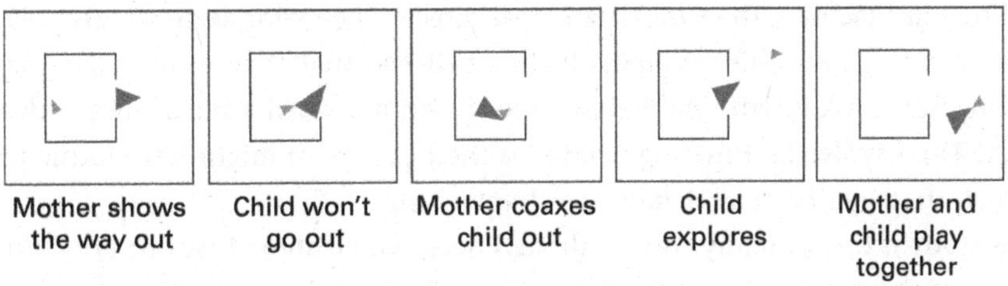

Figure 7.6
Can triangles play tricks? Still frames from a one-minute animation showing the most frequently used description (Abell et al., 2000). This figure shows an example from a set known as "Frith-Happé animations," which have been used widely as stimuli that trigger mentalizing.

The way that these animations are represented is very artificial. The interactions are seen from above, and the movement of the shapes is not biological in the sense of being constrained by bodily structure. The triangles have no bodies or facial expressions, and there is no sound. So, what is it about these animations that makes us take an intentional stance?

As first shown by Fulvia Castelli, Francesca Happé, and ourselves (Castelli et al., 2000), the contrast between randomly moving triangles and triangles perceived as intentionally moving was manifest in increased activity in the major components of the brain's mentalizing system. What provokes the attribution of hidden mental states in one set of animations and not another? We believe that these attributions are triggered when the world of agents opens up to the world of ideas. We shall discuss this and other studies of mentalizing in chapter 10.

But when we are interacting with an intentional agent, how do we tell when there are no movement cues? This turns out to be quite a tricky problem and was tackled by a fiendishly clever study by Jean Daunizeau and colleagues at the Paris Brain Institute (Devaine, Hollard, and Daunizeau, 2014a). They devised a competitive hide-and-seek game, to be played competitively with an opponent on the computer. The opponent was portrayed either as a one-armed bandit, which could win or lose, or as a person hiding behind a wall or a tree. However, both these opponents were fictional and had been programmed identically in both conditions to produce moves with varying degrees of sophistication. The game was played over an extended period, which allowed learning to occur. At lower levels of sophistication, the program simulated goal-directed behavior, and at higher levels, intentional behavior. The question was whether people would notice and automatically adapt to the difference.

As might be expected, the human participants used strategies that suited the way that the opponent was portrayed. If they believed that they were playing against a one-armed bandit machine, then they took goal-directed behavior as given, and played a simple "win-stay-lose-shift" strategy. If they believed that they were playing against a person, they took intentional behavior as given, and used a much more advanced strategy. This involved estimating what idea their opponent might have formed about their own strategy (for more details, see chapter 12).

This study is one of many we will discuss here, which show that our prior expectations about the nature of an agent are more important than our observations of the actual behavior. It may well be that because there are no unambiguous markers of intentional behavior, we depend on already knowing something about the agent. The difficulty of interpreting behavior is painfully clear in a criminal court, when a judgment must be made about whether an act was manslaughter (unintended) or murder (intended). In everyday life too, we often have to decide whether an outcome was intentional or accidental.

Three Levels of Agency

We can classify the prediction and control of the behavior of social agents using three levels of increasing sophistication, from self-propelled, to goal directed, and finally, to intentional. We suggest that there are different mechanisms operating in each, and we presume that they evolved in a hierarchy, likely driven by an accelerating competition. We propose that these mechanisms enable human agents to move freely among the world of objects, the world of agents, and the world of ideas, while most other animals mainly move between the world of objects and the world of agents.

The evolution of this hierarchy began with the emergence of self-propelled agents. In a world of plants, a self-propelled agent has no need to predict its movements. It can just go and eat them, and the poor plants—not having a brain—cannot run away.[3] Dealing with other moving agents requires greater cognitive complexity. The simplest self-propelled agents show fixed action patterns, which can be exploited by predators. Through evolution, predators can acquire complementary fixed action patterns, which enable them to catch specific kinds of prey. Goal-directed agents, however, have an advantage since they can learn about the fixed action patterns of simpler agents. This gives them the flexibility to deal with a wider range of agents.

But as we have seen in the last section, there is a higher level still. Intentional agents have an advantage over goal-directed agents since they not only represent the goals of these agents, but also guess *why* they have these goals. The assumption that agents have hidden intentions enables greater flexibility in understanding and predicting the behavior of others. Nevertheless, we acknowledge that the interpretation of intentions presents much ambiguity and is open to errors.

Triggers for Passing from One Level of Agency to the Next

As we suggested earlier, self-propelled motion is detected because it does not show conservation of momentum. It breaks a rule associated with the world of objects. This unexpected behavior suggests that we are dealing with an agent. We believe that, similarly, there might be a marker of the difference between goal-directed and intentional motion.

3. Plants have many other defenses at their disposal, such as toxins and thorns.

Goal-directed behavior is marked by rationality. The agent reaches his or her goal via the shortest or least effortful path. When the movement breaks these rules, can we infer that the agent is no longer associated with the world of agents? The agent belongs to a strange new world, where neither conservation of momentum nor the principle of least effort applies. This is the world of intentional agents—agents with ideas.

We propose that the interpretation of how people move in the world of ideas is analogous to the interpretation of how people talk. Relevance theory (Sperber and Wilson, 1986) guides us in this interpretation, and we will say more about this topic in chapter 16. Within this theory, an utterance must always be interpreted in terms of the intentions of the speaker—or else it is meaningless or nonsense. For example, the metaphorical statement "This room is a pigsty" is not meant to be an accurate description of the state of the world. So why does the speaker say it? Extra work is required from the listener to infer the intention behind the speaker's utterance. Perhaps the speaker has the intention to indicate that the room is so untidy that it flouts some commonly accepted norm. Perhaps the speaker intends the owner of the room to start tidying up immediately.

Following the same logic, if we see someone jumping unnecessarily high over a barrier, we spontaneously ask why they are doing this. One possibility is that this is an ostensive gesture indicating that this is more than just a movement. This is a deliberate signal, a meaningful communication, just like using the word "pigsty" in the previous example.

Cordula Vesper and Michael Richardson (2014) studied just such a situation. Two participants were instructed to quickly tap in synch on a sequence of targets. The critical point was that only one participant, the leader, knew which was the next target. The partners were separated by a screen, which was adjusted so that, on some trials, the participants could see each other's hands (see figure 7.7).

When the follower could see the hand movements of the leader, then the leader made exaggerated movements. These movements were irrational in the sense that they did not minimize the path needed to reach the target. But the exaggerated movements communicated the current target to the follower. We conclude from this finding that deliberate signaling of this kind is an indication that the movements were being generated by an intentional agent.

This interpretation is supported by a brain-imaging study from Antonia Hamilton's lab (Marsh et al., 2014b). In this study, participants saw rational hand movements in which a hand was moved over a barrier, and in another situation, irrational hand movements where the hand moved up over a nonexistent barrier (see panels A and B in figure 7.5). Observing rational movements activated the action observation (mirror) network,

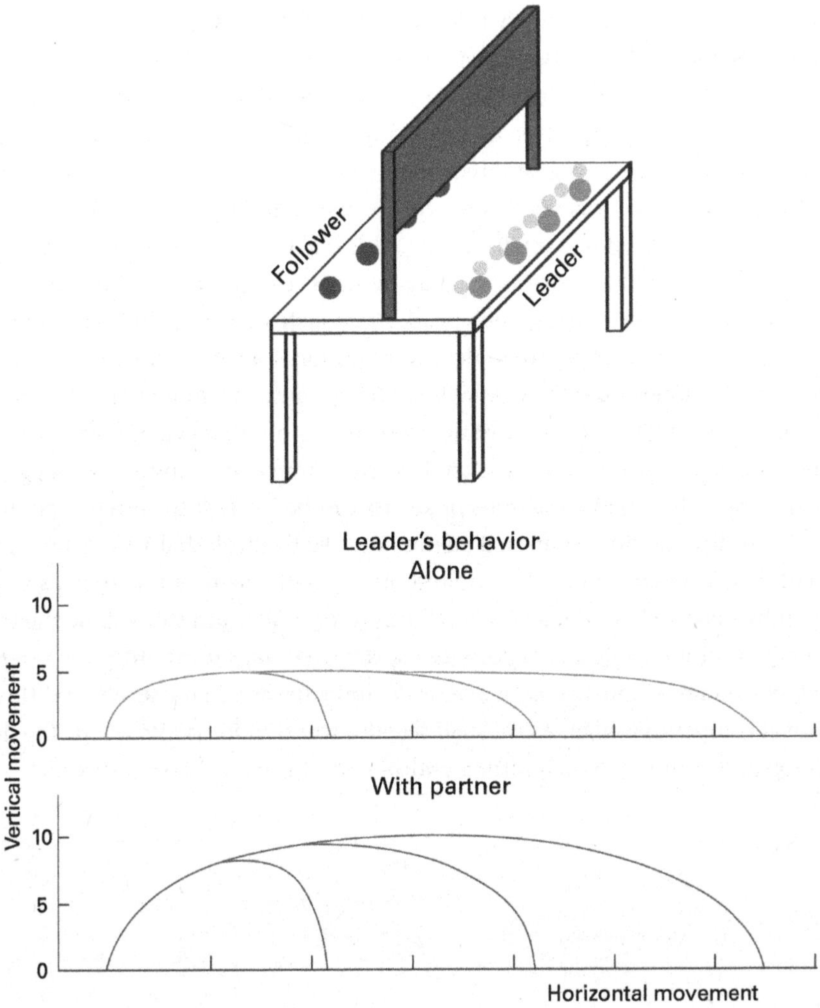

Figure 7.7
Actions as signals: the follower must move in synch with the leader. From the lights at the sides of the targets (T1 to T4), the leader knows the next target in the sequence, but the follower does not know it. When they work together, the leader makes exaggerated vertical movements. Redrawn from figures 1 and 2 in Vesper and Richardson (2014). Copyright 2014, with permission from Springer.

which is relevant to processing goal-directed actions. Observing irrational movements activated the mentalizing network, which is relevant to processing intentions.

These findings help us to explain how observers interpret the animated triangles mentioned earlier. The behavior of the triangles typically includes movements that seem irrational in terms of goal-directed behavior. For example, near the beginning of the sequence in the original Heider and Simmel animation, a large triangle rotates from side to side in front of a small triangle. This makes no sense in terms of moving to reach a goal, but it makes sense as a signal, perhaps communicating "no" to the small triangle, reminiscent of shaking one's head. Once such signaling has been inferred, the observer will assume that the triangles are intentional agents. This triggers processes that will seek to understand their behavior as being caused by hidden mental states (the intentional stance). The observer can now look for a story that explains the interaction.

From the experience of our own studies with animated triangles, we suggest that the merest hint that the interactions of agents can be made into stories is enough for observers to slip into an intentional stance. We like to think that this stance entails a different way of interpreting our own and other agents' behavior, perhaps a way that is only fully mastered by humans. We call this *mentalizing*, and this will be the topic of chapter 10. Can it explain our success as a species? Perhaps it can, to some extent. We can better explain and justify what agents do and what we do ourselves, and this gives us a novel competitive edge. As we shall see, the world of ideas also opens the door to deception and cunning, which form a critical part of the Machiavelli thread.

8 Us and Them

One of the most widespread sources of competition across species is between ingroups and outgroups. Beginning with our own family, most animals recognize the members of their group by familiarity and "like me" similarity. Humans automatically classify strangers as outgroup members if they look different, with racism being an infamous outcome of this process. Negative stereotypes of people who are dissimilar in appearance or behavior are easily formed and hard to overcome. But there are various means to mitigate them, such as getting to know individuals from outgroups and becoming friends with them.

We humans have a strong drive to be affiliated with our ingroup, and we greatly fear exclusion. If we are excluded, unconscious mimicry acts as a means of appeasement, which may allow us to gain readmission to the group. Our behavior is far more affected by the behavior of members of our own ingroup than it is by members of an outgroup. From a very early age, we prefer to learn from members of our ingroup and strive to be like them. But support for the ingroup goes hand-in-hand with hostility to outgroups. Outgroups are perceived as competitors and this competition enhances ingroup altruism, as well as outgroup hostility. We struggle to make different groups work together for the greater common good, and as a group, we work best when we are competing with an outgroup. The more we know about these processes, the better we will be able to overcome the negative consequences of "us versus them" distinctions.

* * *

Starting with Our Family

Only in the Garden of Eden can we find all animals peacefully living together. Within virtually all animals who live in groups, competition breeds disunity. War is more common than peace. But it is not everyone for himself or herself. Groups are formed spontaneously, and they protect individuals and furnish their identity.

What are these groups? The first to come to mind is the family. There are indeed ancient processes, inherited via evolution, that bind us to our families. For example, newborn mammals bond with their mothers. This model has been studied in detail in rodents, where it has been shown that bonding is supported by the hormone oxytocin (Francis, Champagne, and Meaney, 2000). Strong and secure affiliation to our close kin is not a luxury. It provides essential protection during the most vulnerable period of development.

But how do humans and other animals recognize members of their family? Similarity and familiarity are the best markers we have, even if they are not always reliable (Bressan and Kramer, 2015). The preference for similar others is known as "homophily." It is ubiquitous throughout the animal kingdom and emerges as a preference under a variety of conditions (Fu et al., 2012). In humans, we feel its ripples in gender and race discrimination. The term "neophobia" refers to fear of unfamiliar foods, objects, and places, while "xenophobia" refers to unfamiliar agents. Where there is xenophobia, chauvinism is not far behind.

Research into rodents and birds has demonstrated that they can recognize their kin by a multitude of sensory signals (Holmes and Sherman, 1983). Kin also tend to have frequent interactions with each other, which strengthens familiarity. In humans, learning who your kin are may well depend on these same mechanisms, and there is evidence of such learning. For instance, infants under one year old expect their kin to look similar and to speak the same language. Infants are spontaneously wary of strangers (e.g., Sroufe, 1977). They also assume that they can ask their kin for help, but not others (e.g., Buttelmann et al., 2013; de Klerk et al., 2019).

A child's first ingroup, the family, is soon extended to others who live nearby, speak the same dialect, and sing the same songs. Nine- to twelve-month-olds preferred familiar songs, as assessed by a number of indicators, including time spent listening, presence of positive affect, rhythmic movements, and autonomic relaxation (Kragness, Johnson, and Cirelli, 2021). There was robust generalization across singers, demonstrating that songs are a strong sign of group membership and identity, even beyond the family.

Here is yet another tricky question: how do human infants even recognize a group as a group? Lin Bian and Renée Baillargeon (2022) approached this question with a series of ingenious experiments and concluded that even young children have an abstract concept of what a group is. They expect individuals in a group to be similar, but similarity is not necessarily defined by fixed attributes. For example, one-year-old infants regarded similar outfits as a mark of group membership, but not when the outfits served a purely instrumental purpose (e.g., cleaning). Two-year-old toddlers regarded the use of the same verbal labels as cues to group membership, but not when these labels conveyed only incidental rather than categorical information. These results suggest that when recognizing a group, it is context that matters.

Now here's the thing: Where there is an ingroup, there is also an outgroup. Indeed, the ingroup is defined by the outgroup. The outgroup consists of people who are different in appearance, language, song, behavioral norms, and customs, even though these differences are sometimes laughably small. A comprehensive review of the literature by Yarrow Dunham (2018) concludes that the outgroup can be recognized, simply by being different from us.

The evolutionary payoff of homophily remains an open question. In humans at least, it seems that cooperation receives a boost among similar individuals in social dilemma games (Di Stefano et al., 2015). But how does it work? Perhaps mere exposure to close conspecifics in a particular location is the critical first step. But then there is a fateful second step: the "like me" group is automatically assigned positive values, and the "not like me" group is assigned negative values.[1] We like one and hate the other. We learn from one and not from the other. We trust one and distrust the other.

First Impressions

First impressions are hard to suppress, and faces are a good place to start. We are attracted to looking at faces from birth (see chapter 2) and build up a store of representations of familiar faces. We can spot a familiar face very easily. The appearance of a face can be used as a compelling marker of what in folk psychology is thought of as "race." We have no difficulty distinguishing, for example, a Japanese face from a European face.

But race is a slippery category that does not have clear boundaries. So how do we come to draw boundaries so readily, and what does it say for our feelings of identity? This was explored in an experiment using faces that morphed along a continuum varying from typical European to typical Japanese (Webster et al., 2004).

Observers were asked to categorize faces like those in figure 8.1 as European or Japanese. The results showed that the cutoff depended on who was doing the categorizing. Japanese observers put the cutoff closer to the typical Japanese face, while Europeans put it closer to the typical European face. A similar effect has been observed for mixed race (black and white) faces (Lewis, 2016). White observers tended to classify mixed-race faces as black, while black observers tended to classify them as white.

In this study, the Japanese observers had mainly been exposed to Japanese faces, while the Europeans had mainly been exposed to European faces. Webster and colleagues (2004) found that Japanese observers who had lived in North America for at least a year were more similar in their categorizations to Europeans. Exposure over time

1. An exception is when the outgroup is of high status, when positive values are usually assigned. This could be due to the ability of such a group to offer protection and useful alliances.

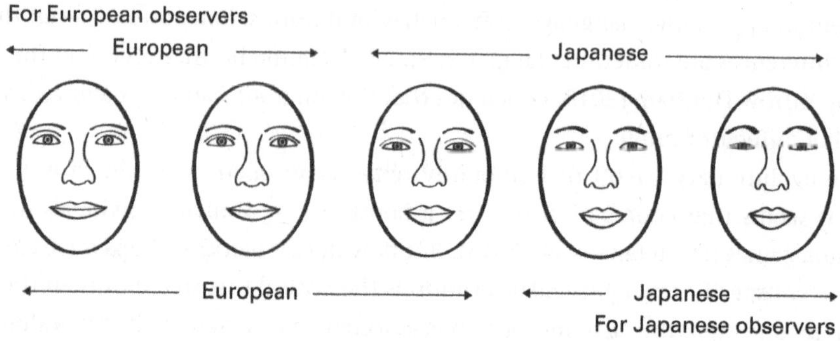

Figure 8.1
Categorizing faces as ingroup or outgroup: where you put the cutoff between ingroup and outgroup faces depends on your experience (Webster et al., 2004).

not only increases familiarity, but also reduces outgroup prejudice. Increased exposure to interracial individuals in one's local environment can also alter the way that this category is perceived (Freeman, Pauker, and Sanchez, 2016).

Reining in Prejudice

Are we merely surprised at seeing strange faces? Or is our ancient alarm system responding because we feel threatened by them? Elizabeth Phelps and her colleagues (2000) explored this question in terms of a notorious racial contrast. They presented white and black faces to white participants and found that outgroup faces elicited activity in the amygdala (see figure 4.3 in chapter 4), indicating an automatically occurring threat appraisal. The alarm system was indeed responding. The negative appraisal (amygdala activity) elicited by outgroup faces is very rapid and correlates with behavioral measures of implicit race prejudice (Cunningham et al., 2004).

Another negative consequence of the automatic detection of outgroups from a face is that it suppresses automatic empathy for pain. This was observed when comparing Chinese and European participants (Xu et al., 2009) who watched images of faces of either Chinese or European individuals either being touched by a cotton bud or touched by the needle of a hypodermic syringe. Participants typically did not show brain activity associated with empathy when they were observing images of an outgroup member (apparently) about to be hurt. This was equally true for Chinese observers of Europeans as for European observers of Chinese.

However, the empathic response could be reinstated after the outgroup members had become familiar as individuals (Sheng and Han, 2012). Familiarity can be accomplished quite simply by including outgroup members in a game as part of the ingroup team. Mere exposure to outgroup faces can also decrease dislike of such faces (Zebrowitz, White, and Wieneke, 2008).

A Question of Identity

In the middle of the twentieth century, Muzafer Sherif and Carolyn Sherif carried out some remarkably naturalistic, yet rigorously designed experiments in Connecticut and Oklahoma. They inivited a number of young boys from a middle class background to attend a summer camp (Sherif and Sherif, 1956). In several studies, two groups were randomly established and separated. They quickly stabilized as ingroups and adopted team names, such as Eagles and Rattlers, which they put on flags and T-shirts. Each group appropriated hideouts and swimming places of their own. During their stay, they embarked on typical summer camp activities while being carefully observed by the experimenters.

As hypothesized, intergroup stereotypes were quickly in evidence, as demonstrated by name calling, pranks, and outright aggression. Indeed, some violent attacks on each group's camping sites took place. When the groups were brought together again in joint activities to reestablish peace, antagonism and group allegiance proved hard to dispel. Fortunately, the experimenters hit on a highly effective means that eventually eased the intergroup tensions. They contrived an emergency by cutting off the water supply for all. This problem could be solved only by pooling the resources of both groups. This example suggests that working together toward an overarching aim can diffuse intergroup hostility.

These and other studies showed that entirely arbitrary differences are sufficient to give rise to ingroups and outgroups, and hostile intergroup behavior. Outgroup members need not be different from us in looks or language. For instance, people with different preferences for food might readily be treated as an outgroup (Hamlin et al., 2013). Just as in the classic summer camp studies, today's young people wearing the colors of different sports clubs can readily generate intergroup friction and can even come to blows.[2]

Ingroups and outgroups can be created simply providing two different-colored T-shirts to two random selections of participants. This alone is sufficient to create a sense of loyalty to one's current ingroup and fires up competition with a current outgroup (Dunham, Baron, and Carey, 2011). This is usually referred to as "minimal group design." Minimal group design has been used in numerous experiments to analyze not just intergroup

2. Readers of the *Harry Potter* books will recognize the fierce allegiance that the children have to the houses of Hogwarts once they are assigned to them.

conflict, but also the effects of belonging and/or exclusion of individual group members. These effects tell us something about self-identity. We cannot separate the value we feel by belonging to a particular group from how we value ourselves or wish to be valued by others.

The feelings are even stronger when political, ethnic, or religious affiliations are at stake. Everyone knows the intense feelings generated by international sports events. There is the joy that we feel on behalf of our own team when they win, and the misery when they lose. This speaks strongly to our self-identity as being bound up with group affiliation. Our self-identity hinges on where we belong and whom we affiliate with (see the review by Ellemers, Spears, and Doosje, 2002). An anecdote illustrates how even purely temporary grouping can affect the feeling of identity. During the filming of *Planet of the Apes* (1968), the (adult) cast remained in full costume during the breaks. At lunchtime, the chimpanzees sat with the chimpanzees, the orangutans with the orangutans, and so on (Nichols, 1998).

Once we have a marker of the difference between an ingroup and an outgroup, this difference will be exaggerated in our perception. For example, faces with typically African features are perceived as darker than they actually are. This biased perception occurs rapidly and without awareness of any conflict with actual luminance levels (Travers, Fairhurst, and Deroy, 2020). And focusing on such differences can have pernicious effects (Greenwald and Banaji, 1995).

However, our perception of ourselves is also malleable, and if we perceive ourselves as looking more like the outgroup, our prejudice is reduced. This change in perception can be created using the rubber hand phenomenon (Botvinick and Cohen, 1998). Participants in this paradigm cannot see their own arm, but instead see a rubber arm positioned not far from their actual arm. They then see the rubber arm being stroked while their own arm is also stroked. They feel stroking (on their own arm) and see stroking (on the rubber arm). When these visual and tactile experiences are well synchronized, the participants rapidly begin to have the illusion that they are *feeling* the stroking of the rubber arm. At this point, the participants experience the rubber arm as their own.

Maister and colleagues (2013) used this phenomenon to study the effects of self-perception on race prejudice. Here, the arm of a light-skinned participant was paired and synchronized with a dark-skinned rubber arm. They now perceived the dark-skinned arm as their own. Remarkably, this experience reduced implicit bias, as measured by the Implicit Association Test (IAT; Greenwald, McGhee, and Schwartz 1998). Indeed, the stronger the experience of ownership of the dark-skinned hand, the greater the reduction in implicit racial bias (see also Banakou, Hanumanthu, and Slater, 2016). This is an example of how strongly our people-detectors are interested in similarity to ourselves. It also shows that stereotypes can be overcome when we experience ourselves as similar. Sadly, these effects often prove less durable than one might hope.

Craving for Affiliation and Fear of Exclusion

There is no getting around it: the distinction between ingroups and outgroups acts like an apex predator in the social domain of our information-processing hierarchy. It dominates our feelings of identity and our relationship with others. A plausible reason is our deep-seated need for affiliation. It is hard to exaggerate the powerful and automatic processes that cause us to align ourselves with others. This is of great benefit to individuals, since each individual reaps numerous benefits simply from being with others (see chapter 5). These others are typically our ingroup. Our need to belong is well catered for by social media. These have made it easier than ever to form groups based on specialized interests, hobbies, health concerns, or grievances.

But beware. While the ingroup versus outgroup distinction seems to be a default setting of our brain's built-in implicit processing system, we don't have free access to the group that we wish to belong to. We are not in a summer camp, where we are randomly assigned to different groups for fun adventures and competitive sports! Just like luxury enclaves or clubs, some groups are impossible to join, especially high-status groups that pride themselves on being exclusive. However, there is a second factor that drives our almost-desperate desire to belong: We are deeply afraid of being expelled from our ingroup. We learn who we are from our ingroup, and to maintain this identity, we need to cultivate our affiliation with other members of our group and vigorously maintain our differences from an outgroup. Generally, members who deviate from the prevailing norms of their club should expect punishment. They are marked as traitors and may be thrown out into the cold (Ditrich and Sassenberg, 2016). The terror that we feel when we are excluded is the twin of our desire to be affiliated.

The effects of exclusion have been investigated through a substantial number of studies using the cyberball paradigm. Here, participants play a computer game where they interact with two virtual agents by throwing balls to each other. However, in one situation, these virtual agents do not throw any balls to the participant and share them only between themselves. This engenders the unpleasant feeling of being excluded from the team. But this is not all. People who have experienced such exclusion (ostracism) typically increase their nonconscious mimicry of ingroup members immediately afterward.

Take, for example, an experiment by Jessica Lakin, Tanya Chartrand, and Robert Arkin (2008). After experiencing ostracism, participants were given the opportunity to interact with a likable member from their own ingroup. This person was instructed to move his or her feet throughout the interaction. The previously ostracized participants tended to unconsciously copy these movements, and far more so than participants who had not been ostracized. Other experiments have showed that ostracism generates feelings of depression, and this is particularly evident during adolescence, when peer groups start to become more important (Andrews, Ahmed, and Blakemore, 2021). Increased mimicry of ingroup members appears to offer a way to recover from the hurtful effects of exclusion and to gain readmission to the group.

This effect is already observed in toddlers who show enhanced facial imitation if they feel excluded (de Klerk et al., 2020). Moreover, it occurs even if the exclusion is not personally experienced, but observed happening to others on video, and even when the protagonists are just animated shapes (Over and Carpenter, 2009). This finding has been confirmed and extended in other studies. For example, children aged five to six years old not only showed higher-fidelity imitation after being ostracized by their ingroup, but they also were made more anxious. In contrast, being excluded by an outgroup had no such effects (Watson-Jones et al., 2016).

Returning to what is probably our first ingroup, the family, we can draw on an experiment by Philip Brandner and colleagues from Eveline Crone's lab. They looked at brain activity in adolescents, who were able to earn monetary rewards either for themselves, for their parents, or for a stranger (Brandner et al., 2021). As expected, the brain's reward system (i.e., the nucleus accumbens) responded equally well to rewards for themselves and their parents but did not respond if the reward was assigned to a stranger.

We Prefer to Learn from the Ingroup

Chapters 2–6 of this book focused on cooperation, such as learning from others, being with others, acting with others, copying others, and sharing emotions with others, but we have skimmed over the question of who the others were. In fact, any general conclusions from these chapters apply only when the others in question are members of our ingroup. This is how basic this distinction is to our social nature and how much priority it enjoys when we process social information. Thus, we much prefer to learn from members of our ingroup than from outgroup members (Golkar, Castro, and Olsson, 2015). At the brain level, this preference is reflected in lateral prefrontal activity (Kang et al., 2021). The stronger the activity in this region, the more strongly participants weighted learning cues (action prediction errors) that they picked up from an ingroup demonstrator. However, they did not respond to such cues provided by an outgroup demonstrator.

Learning from existing ingroup members also allows learning whether an unfamiliar person is an ingroup member. Thomas, Saxe, and Spelke (2022) investigated how one-year-olds discover whether a newly encountred individual is a suitable social partner. To do this, they devised videos of interactions between adults and an unknown puppet while measuring looking preference. When their own parent showed affiliation toward one puppet, the children expected that the puppet would also engage with them, but they did not expect this when they watched the parent of another child interacting with the puppet.

It is our loss when we fail to learn from outgroup members. It is also our loss that the powerful and automatic processes by which we imitate and share our emotions are engaged only by the ingroup (Lin, Qu, and Telzer, 2018). Selective imitation starts early in life. For instance, in one study, infants aged eleven months were shown videos of models who either spoke the children's native language or an unfamiliar foreign language. The models also produced specific facial actions, such as mouth opening and eyebrow raising, which involved muscle activity that could be measured in the children themselves via electromyography (see chapter 4). The children showed more mimicry for members of their ingroup (de Klerk et al., 2019).

Eye gaze following, as discussed in chapter 2, can be seen as a type of mimicry that has strong implications for perceiving and sharing the goal of another person. Chinese infants aged seven months took part in a study where an adult model who was either from their ingroup or an outgroup, here defined as Chinese or African ethnicity (Xiao et al., 2018). As illustrated in figure 8.2, the adult model on the screen gazed toward one of the four corners to reveal an animal. When the gaze predicted the animal with 100 percent reliability, the infants followed the adults' gaze regardless of their ethnicity. If there was only a 25 percent (chance) reliability, they followed neither. However, with a 50 percent reliability, infants showed a strong preference to follow the ingroup member's gaze, not the gaze of the outgroup member.

How do we explain the preference for getting knowledge from ingroup members? It seems that all of us, including very young children, treat them as a source of more reliable knowledge. But why should the knowledge of the ingroup be better? One reason is that we most frequently interact with people from our ingroup. The more similar we become to these people, in terms of knowledge and norms of behavior, the easier these interactions will be. However, in addition to becoming more similar to the other people in our ingroup, we are more likely to be recognized as members and avoid being expelled. No wonder it is so important to assimilate to your ingroup.

Another reason for preferring to learn from members of our ingroup is trust. We believe that ingroup members are less likely to be trying to deceive us than outgroup

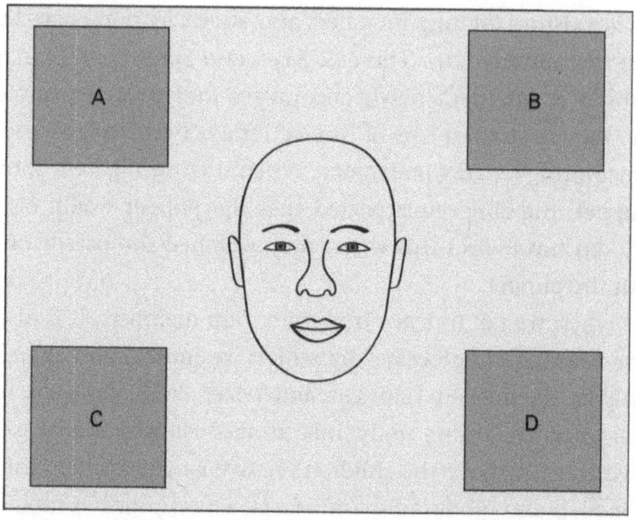

Figure 8.2
The person in the middle will gaze toward one of the boxes in the four corners. His gaze often reveals which box will contain an animal. To find the animal, infants prefer to follow the gaze of an ingroup member. Redrawn from figure 1 from Xiao et al. (2018). Copyright 2018, with permission from Wiley.

members. If they are found to deceive us, we are outraged and will take steps to punish them. In contrast, we are not surprised if members of an outgroup are trying to deceive us; we expect them to behave badly and not follow the rules. In a highly revealing as well as amusing study, Schmidt, Roakoczy, and Tomasello (2012) presented a puppet game to three-year-olds, where the puppets had either native (German) or foreign (French) accents. Within the carefully controlled design, the puppets sometimes transgressed norms. When this occurred, the children spontaneously protested and showed their disapproval ("You should not do this!"), but only for ingroup members, not outgroup members (Schmidt, Rakoczy, and Tomasello, 2012).

The emphasis on learning from the ingroup takes little account of individuals. Amazingly, while we may dislike some of the people in our ingroup, we still copy them. Wilks, Kirby, and Nielsen (2018) studied copying behavior in the context of observing how to open a puzzle box. Ingroups and outgroups were distinguished in a minimal way, using simple color coding of wristbands. The children did not simply like all members of their ingroup; they disliked those who behaved in an antisocial fashion. Indeed, they liked these antisocial ingroup members less than the prosocial members of an outgroup. Still, their dislike did not affect their tendency to copy ingroup behavior more closely than outgroup behavior. As we pointed out earlier, overimitation reveals that we are not copying in order to learn how to open a box most efficiently. We want to learn how to do it in the proper way—in other words, conforming to the rules and conventions of the ingroup. This will be appropriately demonstrated even by antisocial ingroup members.

This aspect of affiliation is just as evident in imitating the ingroup as it is in *not* imitating the outgroup. We are often very aware of this distinct aversion. By the age of five, children already consciously distance themselves from the behavior of outgroup members. In one study, having seen three outgroup members perform the same action with a novel toy, rather than imitating them, as they would with members of their own group, these children do something different with the toy (Oostenbroek and Over, 2015). This distancing from the outgroup is explicit, suggesting an incipient influence of the Machiavelli thread. In a study of delayed gratification, three- to five-year-old children practiced delaying more and valued it more when ingroup members delayed and the outgroup did not. This behavior was accompanied by remarks such as, "I'm in the green group; orange don't wait" (Doebel and Munakata, 2018). At this age, children also show considerable loyalty to their ingroup and will keep an ingroup secret in spite of being bribed to tell (Misch, Over, and Carpenter, 2016).

Love for Us and Hate for Them

The vast literature on the topic forces us to conclude that the ingroup versus outgroup classification is obligatory and happens so fast that it can act as an on/off switch for other automatic responses. It does not stop with perception; rather, it strongly influences our behavior toward love and care for the ingroup and often toward fear and loathing for the outgroup.

If this mechanism has a basis in evolution, there must be advantages. For instance, ingroup affiliation optimizes sharing of resources (Wilson et al., 2020), and cooperation within the ingroup is enhanced by competition with outgroups. We expect to work for the benefit of the group and to receive help from them when we need it. Rather than maximizing our own reward, we maximize the reward for the group. When adults from the same ingroup perform a foraging task together, they adapt to each other's goals (McClung et al., 2017). In contrast, people from different groups foraging together concentrate more on individual goals.

All the advantages of belonging to an ingroup translate into disadvantages for an outgroup. Ingroup favoritism is a persistent problem, even in societies that profess to uphold fairness and equality. Overall, as children get older, they behave more fairly, but this applies only to their ingroup (e.g., children from the same school). At seven to eight years old, they share even less with an outgroup (children from a different school) than younger children (Fehr, Bernhard, and Rockenbach, 2008), as illustrated in figure 8.3.

The idea that favoring the ingroup and disdaining the outgroup are opposite sides of the same coin is also suggested by the effects of oxytocin. This peptide is well known for its role in social bonding. It is supposed to increase our liking for others and has even been called a love potion. However, it may actually play a more divisive role. When oxytocin was administered to people, it was found to increase ingroup favoritism, but at the same time, it increased outgroup hostility (De Dreu et al., 2011).

Competition from an outgroup seems to increase cooperation within the ingroup. It also changes the behavior of some of our ingroup members. For example, as we shall

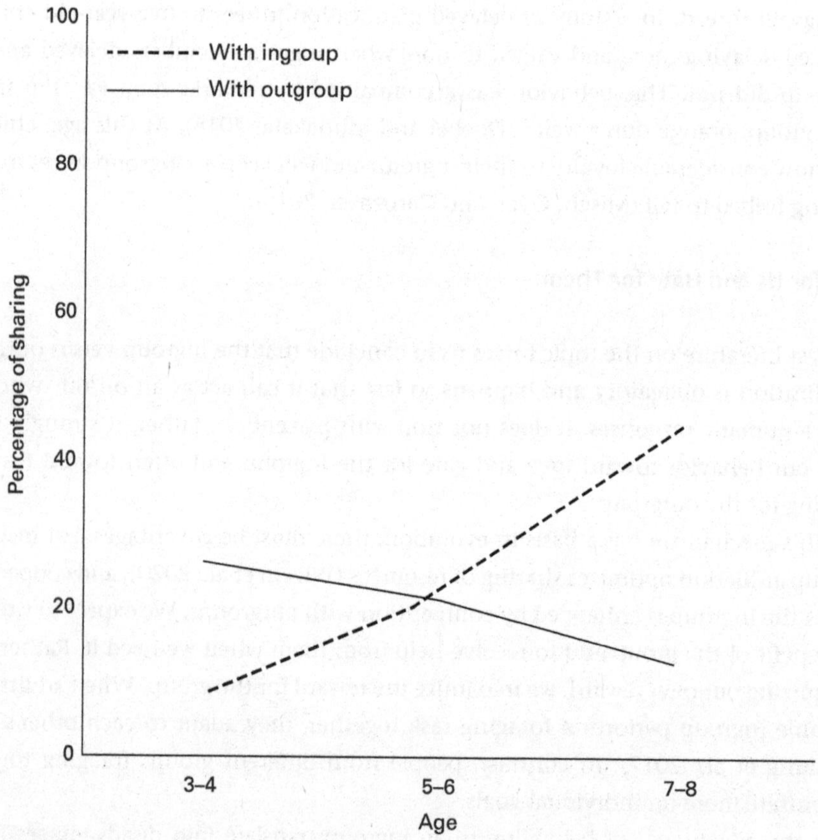

Figure 8.3
Ingroup favoritism increases with age: Children can keep all the money or share it equally with their partner. Sharing increases with age for ingroup partners, but decreases for outgroup partners. Redrawn with permission from figure 3b in Fehr et al. (2008), *Nature*. Copyright 2008.

see in chapter 9, males with wide faces, a feature associated with high testosterone, tend to be perceived as being less trustworthy and more aggressive. However, when there is competition with an outgroup, it is precisely these individuals who show greater self-sacrificing cooperation (Stirrat and Perrett, 2012). Countless tales celebrate the heroism of individuals when their group is under attack. However, in times of peace, these same people squabbled among themselves and only grudgingly contributed to the communal kitty.

Can We Escape from Automatic Hostility to the Outgroup?

There is compelling evidence that the distinction between *us* and *them* is not just a sideline of our social nature. It is a dominant theme. It determines whom we learn from, whom we laugh with, and whom we despise. The deep divide between *us* and *them* makes us liable to judge identical behavior very differently: positive for the ingroup and (mostly) negative for the outgroup (Schug et al., 2013).

There is an advantage in treating individuals of a group as being similar to each other, as it makes interactions less effortful. Our stereotypes simplify interactions and often drive the moment-to-moment decisions that we need to make in the course of social interactions. This can create a vicious circle of self-confirming biases. In the extreme, we might be treating outgroup members as merely goal-directed rather than thinking agents.

There is some evidence of this. Five- to six-year-old children applied more mental state terms to triangles that they believed were members of their ingroup than they did to those that were designated outgroup triangles (McLoughlin and Over, 2017). This led to the idea that encouraging children to take the perspective of an outgroup member and consider their internal mental states would increase prosocial behavior toward them. This idea was indeed confirmed (McLoughlin and Over, 2018).

It takes the Machiavelli thread and the capacity for reflection to make us question the ingroup versus outgroup distinction and to become aware of the value of diversity. Throughout history, major contributions to art and science have often come from strangers who might at first be met with some reluctance, or even suspicion. Great novels, for example, have been written by people such as Joseph Conrad and Vladimir Nabokov, for whom English was not their first language. As we shall see in later chapters, diversity offers a huge benefit to joint efforts to solve problems and build better models of the world.

Once we see ourselves as actors in the world of ideas, we automatically take an intentional stance toward other agents. We expect their behavior to be determined by their knowledge and beliefs. We do this readily for ingroup members. We treat them as being especially like us because we are convinced that their behavior is based on the same kind of hidden mental states as we experience ourselves. We assume that they will understand the behavior of other ingroup members in terms of intentions, desires, and beliefs, just as we do.

Thus, one obvious strategy to overcome our tendency to disparage the outgroup is to deliberately think about their mental states, with the proviso that we avoid a consistently negative interpretation. Getting to know outgroup members could result in making them part of the team.

So, why can't we continually enlarge our ingroup and all become one big happy family?

Unfortunately, there is a catch. When individuals compete within a group it is the uncooperative (selfish) individuals who do best, while when groups compete it is the groups with more cooperative individuals which do best (Sober and Wilson, 1998). If we all become one big group, it may not be a happy one. Selfish individuals can take over, and our lives will become very unfair. Speculatively, this might be the cause of family feuds that are notorious for their ferocity. If there are no competing outgroups, then the ingroup may cease to be a nice place to be.

How can we escape this dilemma? Most of the mechanisms involved in creating ingroups and outgroups are low-level, automatic processes that are restricted to the worlds of objects and agents. Once we enter the world of ideas, though, we can become aware of these processes and their unfortunate consequences. This is what is happening right now as you read this chapter. We can then take measures to overcome them. As we shall see in chapter 13, there are ways to modulate low-level processes via top-down control.

9 Reputation and Trust

Even within groups there is competition. There are always some people who tend to look out for themselves rather than promoting the interests of the group. To prevent and solve conflicts of this kind, a desire for a good reputation is key. A good reputation confers status and creates trust. Reputation is not only valued as a mark of good citizenship, but also plays a critical role in resolving the conflict between selfish and altruistic behavior and making groups more cohesive.

We intuitively infer our partners' trustworthiness by direct observation, but these impressions are often misleading. Learning about others' reputation through gossip turns out to be more reliable. Experiments with trust games have shown that when we can infer in advance from gossip that a partner is not to be trusted, we ignore their actual behavior. When their behavior during the game contradicts the prior information, the brain's response to their actual behavior is downregulated.

Experiments have explored the temptation to take advantage of the generosity of other group members while not contributing to the group oneself. Free riding is facilitated by anonymity. Hence, it is important that we can identify the individuals we are interacting with. Reputation management generates an arms race between sophisticated strategies for detecting free riders and strategies for avoiding detection. This competition has expanded with the rise of the internet. Here, the identification of individuals is particularly difficult, and solutions to the problem of trust remain to be developed.

* * *

Is It Possible to Get What You Want and Still Be Loved and Respected?

"How selfish soever man may be supposed, there are evidently some principles in his nature which interest him in the fortune of others," pondered the great Enlightenment philosopher and father of economics, Adam Smith (1759, 9). How can you get what is best for yourself and best for all the members of your group at the same time? How do we balance our need to cooperate with our need to compete? Throughout the ages, great thinkers and religious leaders have tried to suggest ways to solve this problem.

Smith suggested that it was possible to combine rational self-interest with competition, provided that this was monitored under a system of law. This combination, he believed, would be the way to achieve prosperity for all. But what would make this possible? The answer is—reputation! Smith proposed that human beings are born to crave a good reputation, and to crave it even more than money. The craving for a good reputation, sharpened by the fear of getting a bad reputation, would be sufficient to prevent self-interest from taking over. In this way, self-interest and group benefit could be held in balance.

Smith did not use the term "reputation," but instead the somewhat unusual term "regard": "Nature, when she formed man for society . . . taught him to feel pleasure in their favourable, and pain in their unfavourable regard . . . It is chiefly from this regard to the sentiments of mankind, that we pursue riches and avoid poverty" (Smith, 1759, 103). The term "regard" is instructive, as it implies that others are looking at you and observing what you are doing. It also implies that you must be identifiable for the reputation game to work. As we will see, both aspects are crucial to managing our reputation.

The power and attraction of reputation was recognized by philosophers in antiquity. Cicero wrote that "most people are generous in their gifts, not so much by natural inclination as by the lure of honour" (Cicero, 1913, 1.14.44). This statement is in line with ancient Greek philosophy and literature, which is pervaded by the belief that the quest for fame (i.e., the high regard in which you are held by others) is the primary motivation for heroic and generous actions. Outstanding individuals who proved themselves worthy of fame could become in a sense immortal. They might be honored with statues and their actions enshrined in epic poems.

Terms like "lure of honour" or "pleasure in favourable regard" might suggest some sophisticated and conscious ability. But this is not necessarily the case. There is a striking example to be found in fish.

Reputation Management among Fish and Humans

The cleaner wrasse (*Labroides dimidiatus*) are small reef fish that live in a delicate relationship with bigger fish, their clients, whom they clean from parasites and at the same time nibble their tissue (Grutter and Bshary, 2003). The client fish prefer cleaners that refrain from nibbling, and therefore, they spend more time next to less greedy cleaners (Bshary and Grutter, 2006). The cleaners, on their part, reduce their biting when prospective clients are watching (Pinto et al., 2011). Furthermore, immediately after a cleaner has taken a small bite, it will stroke the client with its fins, which the client seems to like (Bshary and Würth, 2001). It seems that this behavior acts as a placation, which enables the cleaner to regain its reputation.

How is a fish able to develop such sophisticated behavior? It is likely that cleaner wrasse have ample opportunities to learn via reward and punishment since they engage in more than 2,000 interactions per day with their clients (Grutter, 1996). Impressive as their behavior is, more complex cognitive mechanisms are not necessary to explain how they manage their reputation.

The situation is somewhat different for humans. We also engage in multitudes of interactions, and we learn about signals for trustworthiness in potential collaborators. But many of our interactions will be with people we have never met before and about whom we have no information. We cannot rely on reinforcement learning alone. We must use our more sophisticated information-processing capacities as well. We therefore can expect a variety of other mechanisms to be involved in the way that we manage reputation.

Craving a Good Reputation

There is ample evidence that a good reputation leads to rewards (Wedekind and Milinski, 2000), such as high status in your group and being a valued member of the community. High status will usually bring what we desire in terms of respect, admiration, and deference from others (Anderson, Hildreth, and Howland, 2015). You can almost guarantee to gain these rewards by performing plenty of altruistic acts (Nowak and Sigmund, 1998a). Conversely, a bad reputation leads to penalties (Fehr and Gächter, 2002).

There are at least two mechanisms involved. One provides a basis for selecting appropriate partners, while the other provides a basis for rewarding them. Individuals who have gained a reputation for cooperative behavior are more likely to be chosen as partners (Sylwester and Roberts, 2010), whereas those with a poor reputation are likely to be shunned and excluded (Panchanathan and Boyd, 2004). Similarly, individuals who have gained a reputation for cooperative behavior are likely to receive rewards, whereas those with a poor reputation may be punished (Fehr and Gächter, 2002; Wedekind and Milinski, 2000; Roberts et al., 2021).

It is obvious that being helpful and generous requires cost and effort. There may not be a chance of a direct reward, but your good deed may be rewarded in the future because it increases your reputation as a valuable member of the community. Thus, our craving for a good reputation plays a key role in maintaining social cohesion over the long term. According to Martin Nowak and Karl Sigmund (1998b), it even provides an evolutionary reason for altruism, an idea reviewed and refined by Gilbert Roberts and colleagues (2021). Interestingly, reciprocity provides strong motivation to be cooperative, even for individuals with a natural tendency to be uncooperative (Fehr, Goette, and Zehnder, 2009).

The Importance of Identifying Individuals

For successful cooperation in big communal projects, we should know what everyone is supposed to be doing. If some do not do their allotted task, group cooperation suffers. Selfish people can get away with uncooperative behavior and still benefit from the overall effort of the group, so long as they can remain hidden (Milinski, Semmann, and Krambeck, 2002). Anonymity makes selfish behavior easy to disguise. It has been shown that if donations are kept anonymous, charitable giving is reduced by about 25 percent (Alpizar, Carlsson, and Johansson-Stenman, 2008). No wonder charities like to publicize the identities of their donors. However, the picture is more complicated. Observers may judge helpful behavior as self-serving, and in this case, they are likely to withhold reputation benefits. This can lead to particularly generous donors preferring to remain anonymous (Raihani, 2014).

Why do people generally behave better when they can be identified? Clearly, they are motivated to acquire and maintain "a favourable regard," to use Smith's term. For instance, we give unfavorable regard to wealthy individuals who are found out to have evaded taxes by exploiting loopholes. Almost everyone will care about ruining their reputation if their dishonorable actions are uncovered. So they go to great lengths to portray themselves as having stayed within the law and as being philanthropists.

But why are we so set on acquiring and maintaining this "favourable regard"? Ultimately, it is because we need to trust each other so we can all gain the advantages of working together. But not everybody is trustworthy. The ability to predict the trustworthiness of a person we have not met before would be very valuable. Faces provide plenty of information, and, whether we like it or not, when we glance at a new face, we evaluate it for trustworthiness (Oosterhof and Todorov, 2008). And, on the whole, intuitive judgments of potential cooperators or competitors have a modest degree of accuracy (Bonnefon, Hopfensitz, and De Neys, 2017).

First Impressions

There is a high degree of consensus among people as to what a trustworthy face looks like (Rule et al., 2013: see figure 9.1). Exposure to a face for only about 100 milliseconds is sufficient to achieve such an evaluation (Todorov, Pakrashi, and Oosterhof, 2009). Furthermore, this quick appraisal of a face influences our behavior. In a trust game, people will initially invest more in a partner with a trustworthy face (Chang et al., 2010; see also figure 9.1). The consensus of what is a trustworthy face is shared even by six-year-olds (Cogsdill et al., 2014), and there is some evidence that infants aged only seven months avoid looking at untrustworthy faces (Jessen and Grossmann, 2016).

All this speaks for the presence of an automatic mechanism. How might it work? By morphing faces along anatomical dimensions, Alexander Todorov and his colleagues (Todorov et al., 2015) have identified the cues in facial appearance that make us spot various personality attributes, including submissiveness, dominance, trustworthiness, and competence.

Are quick appraisals of facial features in any sense valid? Do they actually reflect the trustworthiness of the person? This question was addressed in a study that used photos of people whose behavior in real life could be verified (Rule et al., 2013; see also Jaeger et al., 2022). In these experiments, it turned out that there was no valid relationship at all. Thus, some very trustworthy-looking individuals were known to have committed fraud (perhaps helped by having the perfect disguise), while untrustworthy-looking ones had led a blameless life.

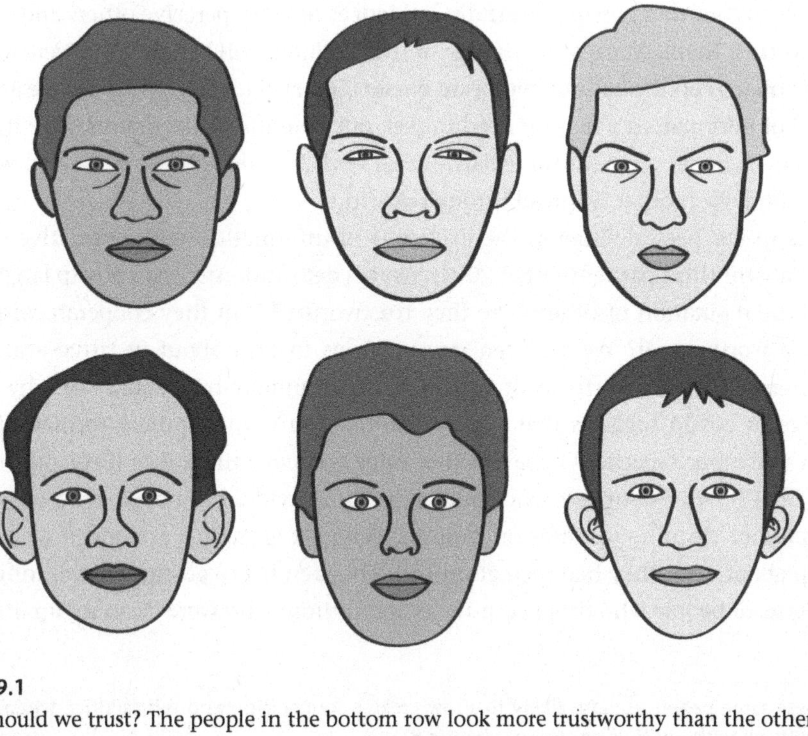

Figure 9.1
Who should we trust? The people in the bottom row look more trustworthy than the others, but are they really?

But perhaps we should not throw out our intuitions just yet. There are strong cultural conventions about what a trustworthy person looks like. On the whole, they serve us well. We let ourselves be affected by first impressions, but we can correct them. We might remember, even if only dimly, that they can mislead. We can be caught out, and this is precisely what suspenseful movies take advantage of. With time, we learn to expect that, in the movies at least, the submissive-looking blonde is most likely the cold-blooded murderer.

According to Stirrat and Perrett (2010), there is one marker that might act as a reliable warning sign. This is the facial (bizygomatic) width. It is a sexually dimorphic, testosterone-linked trait that predicts aggression in males. In this study, men with wide faces were more likely to exploit the trust of others and were also less likely to be trusted by others.[1] In another study (Lin et al., 2018), political corruption was detected using just a photo of a face, and again the determining factor was facial width.

The Power of Gossip

Faces are not everything. There is plentiful information about others in the form of gossip. It turns out that gossip powerfully influences how we perceive others and whether we can trust them. Already, by the age of five, children will behave more generously to other children on the basis of overheard gossip (Qin et al., 2020). Other people are a rich source of information when we need to seek out potential collaborators. Like it or not, everyone is under constant surveillance from both friends and enemies. And what we learn about each other is spread through gossip.

Gossip has been defined as the exchange of information with evaluative content about absent third parties (Foster, 2004). As this definition suggests, gossip largely concerns the reputation of others. Are they trustworthy? Will they cooperate with us to reach a worthy goal? We are even more curious to hear about negative evaluations of others, since this warns us of danger. A questionnaire-based study of why people indulge in gossip suggests that a major motivation is to acquire information about norm violation, especially behavior that might damage the group (Beersma and Van Kleef, 2012). For example, participants were shown videos of people who were either dropping or cleaning up litter on campus (Peters et al., 2017) and given a chance to gossip about what they had seen afterward. They tended to gossip twice as long about incidents of people who dropped litter as about those who were cleaning up litter.

1. These same people are also likely to show greater cooperation and self-sacrifice when there is competition with another group (see chapter 8).

It is surprising to learn from a diary-based study that gossip is the most frequently occurring speech event (Goldsmith and Baxter, 1996). Furthermore, it has been said that people "gossip with an appetite that rivals their interest in food and sex" (Wilson et al., 2000, 347). Whole media enterprises are based on this ravenous consumption, and we are indiscriminate in our appetite. We do not care too much about where the information comes from, and somehow we can factor in that it might be vastly exaggerated.[2] Our greed for gossip and stories parallels our craving for reputation. Both are intertwined to the extent that gossip can make or break somebody's reputation. In fact, it has been claimed that gossip is vital to all human societies and regulates such important attributes of group members as their status in the hierarchy (Dunbar, 2004).

The spread of information about other people and their reputation is not only a tool to regulate status, but also to promote social cohesion (Jolly and Chang, 2021). Gossip enables people to choose whom to cooperate with and whom to ostracize (Feinberg, Willer, and Schultz, 2014). Indeed, there is some evidence that gossip about an individual's reputation is more effective than punishing the person if we want to maintain cooperation within a group (Wu, Balliet, and van Lange, 2016).

However, gossip is often looked down on as unworthy of honest people who would rather check out the evidence for themselves. Gossip is feared, and for good reason, by anyone whose misdemeanors may be revealed. But it is also feared by innocent people, on the grounds that it is an ill-regulated instrument, so they can come under suspicion quite unjustly. In gossip, suspicion is enough to damage reputation, on the principle of "no smoke without fire", balancing the scales of justice toward "guilty."

The power of gossip is underlined by the steps we take to prevent its deliberate use and misuse. Laws against slander and libel exist as a protection, and sometimes gag orders are imposed on the media if someone has become the victim of malicious gossip. But it can be exceedingly costly for an innocent person to fight against gossip and redress a damaged reputation. Those in power will employ a variety of means to suppress gossip, especially when it is true. Dictators who try to put themselves above the law have been known to claim that what is said about them is "fake news," and some go so far as to kidnap or otherwise silence reporters, or shut down media altogether.

A series of studies in the lab showed that information about a gaming partner in the form of vignettes (gossip in other words) strongly affects people's behavior, over and above what they learn from direct interaction (Sommerfeld et al., 2007). Even when the players had access to firsthand information about the behavior of their partners

2. It is not a huge step from gossip to highly regarded and much-loved epic tales. They too contain a mix of fact and fiction about the deeds and misdeeds of individuals.

while interacting with them in a game, they relied more on the prior information. This study also revealed the disconcerting fact that gossip was effective even when it was provided by an uncooperative (and hence untrustworthy) person. It is rather alarming that someone with a poor reputation for cooperation can elicit strong prior beliefs about potential cooperators. This only highlights the potential for gossip to generate a reputation, good or bad, that is not aligned with reality.

The evolutionary biologist Manfred Milinski and colleagues, who have published seminal work on social cooperation and reciprocity in different species, suggest a possible solution to the problem of malicious gossip. When there are multiple *unrelated* sources of gossip available, the true stories will compete with the lies. If false accusations through gossip come from a single source or only a small number of sources, they can be singled out and discounted. For this reason, the power of malicious gossip can be diminished (Sommerfeld, Krambeck, and Milinski, 2008). This is why it has long been recognized that it is important to consult several independent referees when assessing papers for publication or choosing prospective employees. Unfortunately, such procedures are costly and are not implemented automatically. And on the internet, gossip may appear to come from several different sources when in fact it emerges from a single source using an army of bots.

Reputation in the Brain

In the lab, trust games have proved useful for investigating how people learn from their own experience, whether or not their partners are good cooperators (Berg, Dickhaut, and McCabe, 1995). In a typical trust game, the first mover can give up something of value (e.g., money) to a partner in the hope of gaining some benefit in the future. To give up money in this way requires some degree of trust in the partner. The second mover can behave selfishly and leave the first mover with a loss, or he or she can reciprocate. This makes for the classic dilemma: defect or cooperate.

Learning how to get a reward in this game requires only the simple model-free algorithm discussed in chapter 12. If our partners in the trust game give back more money than we expect, then our estimate of their trustworthiness goes up. If they give back less money, then their trustworthiness goes down. The magnitude of these prediction errors (the difference between expected and obtained returns) is correlated with activity in the dorsal striatum (King-Casas et al., 2005), a structure involved in learning about value more generally.[3]

3. Thus, cognitive neuroscience confirms that a high reputation is worth its weight in gold!

The added value of using these games in the lab is that they give a glimpse of what happens in the brain while we are interacting with a stranger, and what happens when we have prior information about this person. This was first explored using neuroimaging in Elizabeth Phelps's lab (Delgado, Frank, and Phelps, 2005) and later in Giorgio Coricelli's lab (Fouragnan et al., 2013). Prior to playing a trust game, participants were shown vignettes about their (fictional) game partners, highlighting their lifestyle and achievements in such a way as to paint a picture of either a good or poor moral character, and hence suggestive of whether they would be a good or poor collaborator.

These studies were consistent with data from the behavioral study mentioned earlier (Sommerfeld et al., 2007), showing that prior information about potential partners trumps learning about them through direct interaction. Remarkably, people continue to rely on prior information, even when their own experience should cast doubt on its accuracy. Why is this? Learning about others through what others tell you is an extremely efficient shortcut. It is so labor-saving that, at some level, you decide that you no longer need to bother to update data from subsequent experience—until, perhaps, there is a glaring discrepancy.

What happened in the brain? Giorgio Coricelli and his colleagues (Fouragnan et al. 2013) conducted a carefully controlled study to answer this question. Activity in the caudate nucleus (in the striatum) reflected updating estimates of trustworthiness when people did not behave as expected. But this happened only if no prior information was provided. When information about trustworthiness was provided, whether through gossip or vignettes, the actual experience of the partner's behavior during the game had little effect on activity in the caudate nucleus. Switching off attention to behavior seemed to depend on the medial prefrontal cortex (mPFC), which was activated in response to the presentation of the reputational prior, whether this was positive or negative.

There was also evidence of a role for the ventrolateral prefrontal cortex (vlPFC). This region was more strongly connected to the dorsal striatum when reputational priors (via gossip) were available. This region may play a role in modulating the function of lower-level cognitive processes, such as those involved in taking account of online prediction errors. We suggest that this modulation might involve lowering the precision of the prediction errors, so that they are given less weight (see chapter 12).

The effect of gossip on brain activity in these regions reflected reduced uncertainty about the behavior of the partner. In other words, the participants treated actual behavior as less important (or less reliable) once they had information that had already been processed by other social agents. The precision of their prior belief about the trustworthiness of the partner had increased.

Seen from a Bayesian perspective, the influence of prediction errors is weakened, while the prior expectation about trustworthiness becomes resistant to modification through learning. Our brain's decision to sideline prediction errors when it has prior information seems to be mediated by regions of the frontal cortex. This is what one might expect since this region is at the top of the information-processing and decision-making hierarchy.

Epistemic Vigilance

As we have seen in chapter 2, learning from others is hugely important over and above learning by ourselves. This is why gossip is so powerful. It demonstrates that preprocessed information is more valuable to us than information that we gather ourselves. But herein lies the danger. There are occasions when we should trust our own observation rather than what we get through hearsay. But this needs conscious reflection and requires the Machiavelli thread. Left to their own devices, our trust detectors are biased toward the opinion held by our group in the Zombie thread.

Let us suppose that on Twitter, a pronouncement is made by a Nobel prizewinner. This is likely to be treated as highly trustworthy because it comes from someone of authority, even if it concerns a field that is not that person's own. The pronouncement is retweeted avidly by those who feel that it confirms their own ideas and gathers even more trust as it continues to be spread by like-minded people. Looking for further evidence is not considered necessary. In this way, false theories thrive.

To counteract this kind of danger, we can draw on a formidable weapon—epistemic vigilance (Sperber et al., 2010). In everyday language, this means to always apply a healthy dose of skepticism and question the evidence that the information is based on. Skepticism is also one of the foundations of modern science. Science disdains trust in authority. It exhorts us to scrutinize the evidence from many sources, not just based on our own observations, but also those of other scientists.

Our ravenous appetite for gossip puts us at high risk for misinformation. Epistemic vigilance is a necessary countermeasure. According to Dan Sperber, it must have evolved together with our enhanced ability for communication. We will return to this notion in chapters 10 and 11, Here, we note that it is via the Machiavelli thread that human beings have become masters of the art of deception, and we can slide over to the dark side almost without noticing. Once we have learned about the cues that imply confidence and trustworthiness, we can fake them. This will be useful for our propensity to burnish our image. Also, because we know that these signals can be faked, we need to examine them more closely rather than relying on our gut responses. This is where epistemic vigilance is invaluable. Is the person making this irresistible offer a little too confident, a little too pushy? Is the offer a little too good to be true? Paradoxically, it is the criminals themselves who have to be the most vigilant. In his splendid book *Codes of the Underworld*, Diego Gambetta (2009) describes the arms race between the mafia and the police. The police want to infiltrate the mafia, but when the mafia suspect this, the signals needed to be accepted as a trustworthy member become more and more costly. Now, to become a member, you have to kill someone. (The police hopefully don't go that far.)

Betraying Trust

We take it for granted that some people are more selfish than others. But the games that we have been considering here are designed to throw light not on "bad" people, but on free riders.[4] Free riders are not a particular type of person. Anyone can become a free rider if the opportunity is there and the likelihood of being found out is low. For example, people might drop litter in a beautiful spot if there is no trash can and if they are unobserved. Anonymity is a very effective cloak that can tempt anyone to become a free rider.

Free riders come into the spotlight in a group version of the trust game known as the Public Goods game. Each member of the group can put some or all of their money into a common pot. The central banker then doubles the money in the pot and shares it among all the group members. If only one member of the group puts money into the common pot, then that member will suffer a loss while the others gain, since the money returned by the central bank is shared with the other members of the group who did not contribute. However, if enough members invest money, then everyone will gain. If the Public Goods game is played repeatedly and everyone invests some of his or her private goods, whether in the form of money or work, then the group as a whole will get steadily richer.

But some group members will realize that even if they do not put any money into the pot, they still will benefit from the investments of the other members. These individuals, just like tax dodgers, will get richer than the rest because they do not invest any of their own money. Once they start to emerge, more and more group members will cease to invest their money (or pay their taxes), since no one wishes to support those who are not investing. As a result, the group as a whole ceases to make gains. So-called fragile or failed states might be examples of the end result of such effects. Social cohesion has disappeared, and the country has splintered into groups competing with each other.

4. Originally, when it was introduced in the nineteenth century, the term referred to a person who rode on a train without having paid the fare.

How to Punish Free Riders

Policing with the threat of punishment is designed to prevent such calamities. However, most people dislike punishing others, so punishment has a cost attached. In some trust games, group members can punish free riders by imposing fines, but they pay a small sum for the privilege. This option is known as "altruistic punishment" (Fehr and Gächter, 2002).

Cooperation flourishes if altruistic punishment is possible and breaks down if it is ruled out. Still, altruistic punishment is only a partial solution of the problem because it leads to second-order free riders (Fehr, 2004). These are people who dodge their responsibility and rely on others to administer punishment. It is here where the group will benefit from having a few angry and aggressive individuals who are willing to apply sanctions (see chapter 8).

Punishment may sometimes consist of withdrawing help from free riders and not choosing them as cooperators for subsequent games (Rockenbach and Milinski, 2006). This makes their poor reputation public. Conversely, there are also third-party rewards for those who have gained a good reputation. In the lockdown period during the pandemic, almost everyone in Great Britain stood outside their homes to thank the staff of the National Health Service, applauding loudly on one day every week. We can recognize and reward trustworthy people, such as those who have been kind to strangers even when we were not the object of their kindness ourselves (Ule et al., 2009).

What would happen if we relaxed our social norms? When given a choice, players often say that they would prefer to be in a group where there is no punishment. Is this a good choice? This was tested in Bettina Rockenbach's lab (Gürerk, Irlenbusch, and Rockenbach, 2006), in a study where participants could choose whether to be in a group where free riders were punished or in a group where there were no such sanctions. To start with, many people chose the group without sanctions. In this group, cooperation rapidly collapsed. As a result, people started switching to the other group. By the end of the experiment, nearly everyone had switched to the group where free riders were punished.

The studies conducted in the lab showed that there is an irrepressible urge for justice. As soon as observers register the presence of uncooperative partners, they are primed to punish them. This is vividly illustrated in a study conducted by Tania Singer, Chris Frith, and colleagues (2004). Participants were paired with trained actors who behaved as either cooperating or noncooperating partners in a prisoner's dilemma game.

In one of the experimental conditions, the participants were told that their partners (the actors) were merely obeying instructions for their moves that they were reading

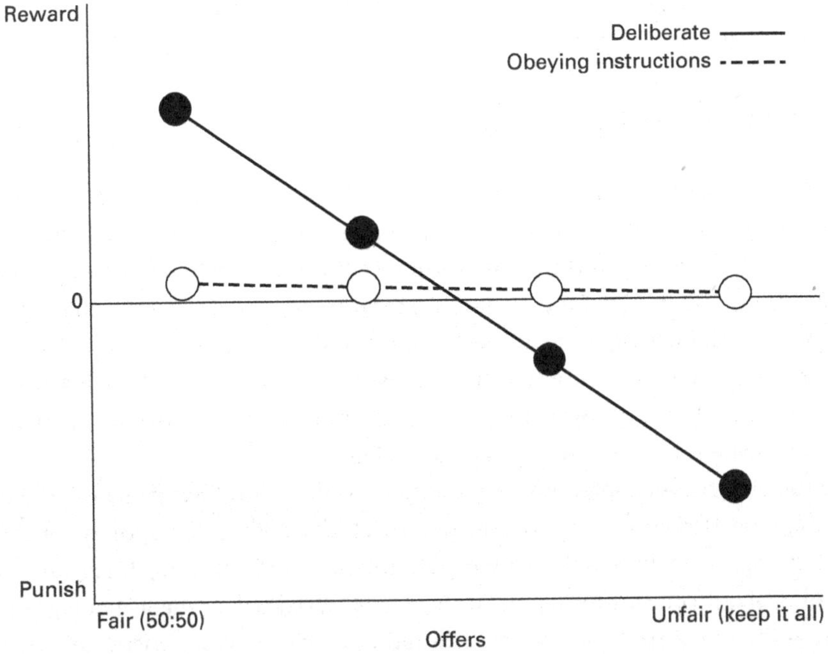

Figure 9.2
Punishing the uncooperative: people who persistently make unfair offers are punished, but only if they do it deliberately (Singer at al., 2006).

from a sheet. In another, the participants believed that the partners were deliberately being uncooperative. The participants disliked the uncooperative partners but chose to punish them only if they thought that their bad behavior was deliberate. This difference showed that the dislike had nothing to do with crude rewards from behavioral outcomes, since both conditions were rigged to yield exactly the same outcome. We can infer that the participants reacted to the intentions behind the actions and followed the moral maxim that only people who are responsible for their actions deserve punishment.

It seems that very little experience is needed to understand that transgressors should be punished, or at the very least not be helped by others. Ting, He, and Baillargeon (2019) found that one-year-olds showed surprise when a thief received help from a bystander. Furthermore, they showed no surprise when the thief was obstructed by another character.[5] Four-year-olds are well aware of free riders and dislike them, regardless

5. The result was obtained only when the victim of the thief was a member of the ingroup (see chapter 8).

of whether they are adversely affected themselves (Yang et al., 2018). Moreover, given the opportunity, they will punish them.

Boosting Your Reputation

We act differently when we are in the presence of others than when we are on our own. Consider a variation of the famous marshmallow test that was conducted in China (Ma et al., 2020). Just as in the classic version of this test, children are presented with a marshmallow and told that if they didn't eat it now, they would get two marshmallows later (see also chapter 14). Here, three- to four-year-old children were either left on their own or watched by someone else while they were waiting in front of the marshmallow. They were able to wait longer when they were watched by a classmate, and twice as long again when they were watched by a teacher.

Presumably these children wanted to impress others with their patience, and indeed self-discipline. However, the use of gossip to affect somebody's reputation is not observed before the age of about five (Engelmann, Herrmann, and Tomasello, 2016), and explicit reasoning about concern for reputation is not observed before the age of eight (Engelmann and Rapp, 2018). Clearly, this is a rather complex achievement, related to our ability to reason about intentions and beliefs following the Machiavelli thread.

Over time, societies have created systems to monitor good and bad reputations, chiefly through institutions. For example, we trust that doctors will treat us properly because they are registered with the General Medical Council. In this case, it is the institution that has gained a high reputation over time and, to preserve this reputation, it selects new members with great care. So a good way to boost your reputation is to become a member of an institution of high repute. Such organizations have strict entrance criteria and rules regulating professional behavior. They also apply sanctions for breaking the rules. Members who behave badly will be expelled. Institutions such as academies, universities, and guilds put in great effort to maintain their reputations and thus become part of reputation management.

Reputation Management

Reputation is too important to be left to gossip and incessant vigilance. Given the role of reputation in regulating the tension between self-interest and competition, it is plausible that it is fed by both the Zombie and Machiavelli threads. Our desire to boost our image when being watched may happen quite unconsciously, as in the case of the children who waited longer for their marshmallows. Deliberately spreading gossip to manipulate somebody else's reputation needs a lot of experience. When it comes to boosting our image, many of us have become very inventive and are able to use a range of deliberate strategies (e.g., securing endorsements) to signal our trustworthiness.[6]

It is not only the sharing of physical resources that enhances our reputation. We can also enhance our image by sharing ideas (Altay, Majima, and Mercier, 2020). However, it is not always easy to behave generously and altruistically or to come up with original ideas, especially if no one sees you or listens to you. This is when you have to remember that there are costs when you succumb to free riding and still want others to believe that you are a trustworthy partner (McNamara and Barta, 2020). You have to keep in mind that others may be spying on you, trying to catch you behaving improperly when you think you are unobserved (Milinski and Rockenbach, 2007). Only if you behave as if other members of your group are constantly watching you, are you playing a reasonably safe game (Whitfield, 2002).

If you have lost your reputation, how can you regain it? You can try to convince others of your change of heart by demonstrative behavior. You can generously reward people who have been kind to third parties. You can administer costly punishment to wrongdoers. However, fearing that you may revert to your old behavior, others will be very vigilant. The more you try to signal that you are a trustworthy person, the more suspicion you are likely to arouse. People may suspect that you are being hypocritical. Just remember that hypocrites are judged particularly harshly, not so much because they behave badly if they can get away with it, but because they send false signals about themselves (Jordan et al., 2017).

The problem of regulating and managing reputation has become vastly more complex in the age of the internet. We can now interact with ever larger numbers of people, most of whom we will never meet, and we can disguise our identities only too easily. This expansion provides greater opportunity for acquiring goods and information, but also much greater opportunity for theft and deception. How can we be sure that the goods we order will arrive and be satisfactory? How do we know which "facts" can be believed?

Mechanisms for the management of reputation on the internet are still being developed (Tennie, Frith, and Frith, 2010), but gossip and identifiability will continue to play key roles. We are asked to rate the suppliers of the goods that we buy. We are asked to rate the restaurants and hotels that we visit. These ratings provide multiple sources of gossip, creating reputations that will help others make their choices. At the same time, though, we are under pressure to give good ratings even if we are not entirely satisfied, perhaps

6. There are celebrity marketing experts who specialize in such activities.

because we need to use the same services again or we fear some backlash. Opportunists and cheats have plenty of opportunities to subvert the process. They may change their identities once their reputation has sunk too low, or they may create false identities to spread false information. For example, authors have used false identities to write reviews on Amazon, praising their own books and trashing those of their rivals (Lea and Taylor, 2010). We are very curious to see how reputation management will develop in these new circumstances.

10 Mentalizing: The Competitive Heart of Social Cognition

Mentalizing lies at the pinnacle of social cognition as it allows us to enter the minds of others. Through mentalizing, we can manipulate the minds of others to achieve our goals, thus boosting competition. Mentalizing allows us to understand the behavior of ourselves and others in terms of hidden mental states, such as intentions and beliefs. We use this knowledge automatically to predict what others will do. We distinguish two forms of mentalizing, implicit and explicit. Implicit mentalizing has been demonstrated not only in preverbal infants, but also in other animals. Explicit mentalizing, by contrast, has strong links to language, conscious reflection, and ostensive communication. It first emerges in children aged four to six years, and it has a protracted development up until adulthood. We briefly discuss the hypothesis that the specific social impairments associated with autism can be explained by a problem in implicit mentalizing. This hypothesis suggests that mentalizing relies on a dedicated neural system that is vulnerable to the hazards of brain development.

There are many neuroimaging studies using tasks contrasting the presence or absence of the requirement to mentalize. These studies have consistently converged on a circumscribed brain system that is activated by mentalizing. We suggest that it is a hierarchical system of three major hubs that are closely interconnected. We suggest that one of these hubs acts as a controller, initiating or inhibiting the process of mentalizing. Another hub acts as a connector, linking prior expectations and incoming sensory evidence, while a third hub is the navigator in social space. It has proved difficult to devise tasks that reliably distinguish between implicit and explicit mentalizing. But we suspect that explicit mentalizing involves an additional layer of information processing placed on top of an implicit mentalizing system.

* * *

A Special Sense

Think, know, believe, feel, guess, understand, attend, perceive, recognize, notice, ignore, prefer, want, wish, dislike, hope, expect, imagine, decide, deceive . . .

Have you ever wondered how easily we use a host of verbs to refer to wholly invisible internal states and actions? We use them as we constantly try to explain why somebody does something, and likewise to justify our own actions to ourselves and to others.

Here's an extravagant claim that we stand by: our brain has a special sense that allows us to perceive mental states, just as it has a sense that allows us to perceive goal-directed movements. The difference is that goals are out there, in the physical world, but mental states are inside one's head. This special sense gives us access to the mental world—the world of ideas. When fully developed, it lies at the very heart of the social life of humans and forms the foundation of our ability to communicate with each other. It is not a rarefied ability; it pervades everyday behavior. But it also has a dark side and acts as a powerful weapon in competition (see chapter 11). While the ability to mentalize allows us to predict what others are going to do next, it also lets us take advantage of others by inserting self-serving ideas and false beliefs into their minds.

It has taken experimental psychology quite a while to start to unpack what is behind human social behavior, and mentalizing ability has occupied a prominent place in this work. Nevertheless, as this book demonstrates, there is more to social cognition than mentalizing. This is still a new field of research, and hence it is surrounded by controversy. One part of mentalizing uses the Zombie thread. Another part uses the Machiavelli thread, which seems to typify it in its full-fledged form.

How did this area of research start? Mentalizing was first known under the curious term "Theory of Mind." This was the term used in the title of a paper by Premack and Woodruff (1978): "Does the Chimpanzee Have a Theory of Mind?" The concept resonated with ideas that came together from a variety of disciplines. Some came from philosophy (Dennett, 1987; Searle, 1995), some from evolutionary psychology and animal behavior (Byrne and Whiten, 1989; Woodruff and Premack, 1979), and others from developmental psychology (Wimmer and Perner, 1983; Leslie, 1987).

Why did the research take hold? We believe that it pinpointed something that we had all taken for granted and hence never thought about: the tendency to explain behavior as being caused by inner states of mind. While it is the essence of folk psychology, it has baffled experimental psychology. "Theory of Mind" was an awkward label, which was soon was shortened to "ToM." In weekly lab meetings at the MRC Cognitive Development Unit at University College London, we started using the term "mentalizing." This was because we grew tired of using awkward phrases such as "having a Theory of Mind," or "attributing mental states to self and other." The term was coined by John Morton, and it first appeared in print in Uta's book on autism (Frith, 1989).

We ourselves were in the midst of these developments. We found the concept captured by ToM or mentalizing to be extraordinarily useful to explain a variety of impairments in social interactions, and in particular those that were observed in autism (Frith, 1989) and in schizophrenia (Frith, 1992). We were also there at the beginning of the neuroimaging studies in the 1990s. Inevitably, this made us eager to discover the neural basis of mentalizing. But even today, our curiosity is not satisfied. Theories are still in flux, and our ideas keep changing.

Insights from False Beliefs

Remember our distinction between goal-directed and intentional agents from chapter 7? Simple goal-directed agents respond to their immediate environment (e.g., they avoid obstacles in order to reach their goals). In contrast, intentional agents respond to their beliefs about the environment. When their beliefs are true, then their behavior will match the environment, and they behave just like goal-directed agents: they move toward the location where the reward/goal is located. But when their beliefs about the location of the reward/goal are false, they do *not* move toward the location where it actually is. To understand this is beyond the capacity of simple goal-directed agents.

Understanding false beliefs was key to developing empirical studies of Theory of Mind. Heinz Wimmer and Josef Perner (1983) developed an elegantly simple task to find out at what age children were able to predict behavior on the basis of a false belief. This is known as the "Maxi task":[1]

> Maxi is given a bar of chocolate. Before he goes out to play, he puts his chocolate in a safe place; the blue cupboard. While Maxi is out, Maxi's mother moves the chocolate to the red cupboard. When Maxi comes back, where will he look for his chocolate?

Children before the age of four to six give the wrong answer ("in the red cupboard"), but older children find this task incredibly easy. They can readily explain why Maxi looks in the wrong place: "He didn't know the chocolate had been moved."

Insights from Make-Believe

Enter Alan Leslie, a member of the MRC Cognitive Development Unit. He had started to investigate how preschool children manage to understand pretense (make-believe) without confusing it with reality. They often playfully pretend that something is the case when it is not. For example, children can effortlessly pretend that a rigged-up basket resting on a cloth is a magnificent sailing ship negotiating a storm (figure 10.1).

Having recognized the deep relationship to false beliefs, Alan Leslie (1987) proposed a special mechanism, "decoupling," to explain how we might understand beliefs. Beliefs

---

1. The Sally-Ann task is a variation on the Maxi task and was devised by Simon Baron-Cohen, Alan Leslie, and Uta Frith (1985).

Figure 10.1
An example of pretend play—these children are pretending to be in a ship at sea.

are decoupled from the real state of affairs, and this is why they need not correctly represent reality. But far from this being a disadvantage, decoupling gives rise to behavior that is virtually unheard of in other animals: imaginative play, teaching, and caring for one's reputation but also deliberate deception of others. This is the world of ideas.

The decoupling mechanism explains that other people can have beliefs about reality that are different from our own, but we all share "the intentional stance" (Dennett, 1987). This is what allows us to better predict what other people are going to do next by taking account of their invisible mental states. This is a vast improvement over reasoning from visible, goal-directed behavior.

Insights from Autism

When Uta first met an autistic child at London's Maudsley Hospital, she was so fascinated by what she observed that she made autism the topic of her PhD. The thing that struck her particularly was that she could not engage the child in play, however much she tried. It was not until a decade later that Lorna Wing, a pioneer in autism research, presented evidence that make-believe play was virtually absent in young autistic children, while it did exist in other learning-disabled children of the same mental age (Wing et al., 1977).

Still, lack of pretend play came into focus only in combination with Alan Leslie's notion of decoupling. Could it be that the brains of these autistic children did not engage the decoupling mechanism? If so, then they should also find it difficult to understand mental states such as beliefs—an outrageous idea at the time. This idea was first put to the test using the Sally-Ann False Belief task by Simon Baron-Cohen, Alan Leslie, and Uta Frith (1985). Outrageous as the idea was, the experiment confirmed it. The finding proved robust, as shown in experiments using a whole range of mentalizing tasks (see Frith, 1989/2003).

The idea of a mentalizing problem had a strong attraction for theories of autism at the time.[2] It attracted attention because it could explain quite precisely and succinctly the typical social communication impairments of autism (Frith, Morton, and Leslie, 1991). While we take it as ground truth that behavior is driven by people's desires, beliefs and intentions, we proposed that this might not be the case for autistic individuals. If that is so, then they can't predict very well what others are going to do, which makes it difficult to negotiate the social world. At the same time, the theory allows for the fact that they can negotiate the physical world, where indeed they can show outstanding competence.

Dissociation between the physical and social worlds has been shown in experiments that compared tasks that require mentalizing and closely matched tasks that do not. For example, Francesca Happé (1994) showed that autistic children could understand the point of a story when it involved reasoning about physical causes, but they were nonplussed by a story that involved reasoning about mental states as causes of behavior.

In chapter 7, we mentioned Frith-Happé triangle animations. After watching these animations, both young autistic children and highly articulate autistic adults used fewer mental state terms to describe intentional scenarios than any other groups we tested (Abell, Happé, and Frith, 2000; Castelli et al., 2002). In the same chapter, we also mentioned a study involving the hide-and-seek game, in which people believed that they were interacting with either a robot or a person. When autistic adults played this game, they used the same strategy with both agents, taking no account of the difference (Forgeot d'Arc, Devaine, and Daunizeau, 2020).

This is not a chapter on autism, but we note some caveats. First, it gradually became clear that autistic individuals with sufficiently high levels of verbal ability are able to pass false belief tasks, albeit with a delay in age (Happé, 1995). This does indeed present

2. Autism at that time was diagnosed using highly restricted criteria, and hence the results discussed here do not apply to all individuals with autism spectrum conditions as currently diagnosed.

a problem for the original hypothesis. Second, the diagnosis of autism has changed over time. The criteria have been widened to the extent that there is now a highly heterogenous autism spectrum. Cases at the milder end of the spectrum are unlikely to have mentalizing problems.

A Shake-up of the Concept

It took time for nonverbal tests of mentalizing to become available so that younger children, as well as animals, could be tested for evidence of mentalizing ability. Likewise, it took time for more sensitive tests to become available for use with adolescents and adults. But once they had been invented, new territory was opened up for exploration.

The first cracks in the original concept of Theory of Mind appeared in a study by Wendy Clements and Josef Perner (1994). They showed that three-year-olds, who were unable to pass the Maxi False Belief task (they said that Maxi would look in the red cupboard, where the chocolate actually was), nevertheless had looked spontaneously at the "correct" (and now empty) location of the chocolate, where Maxi would believe it was. This finding pointed to two separate ways of approaching the task: one implicit and fast, which faithfully directed the gaze of the child; the other explicit, verbal, and slow, which led the child to give the wrong answer, probably due to the conflict between their own and the protagonist's points of view. The child knew where the chocolate now was but did not take account of Maxi's false belief that the chocolate was still in the blue cupboard.

A decade later, Kristine Onishi and Renée Baillargeon (2005) tackled the question with a live but wordless presentation of an adapted version of the False Belief task and measured eye gaze. Here, infants aged fifteen months were surprised and looked longer when the experimenter reached into a yellow box, where a toy had been hidden while she was not looking and couldn't know it was there. This showed that they expected her to reach into the green box, where she had originally placed the toy. Interestingly, infants of that age also detected violations in pretend scenarios (Onishi, Baillargeon, and Leslie, 2007). Furthermore, by eighteen months, they expected an agent's false belief to be corrected by relevant gestures or words. For instance, they expected the experimenter to say, "The ball is in the cup" when the agent wrongly thought the ball was in the box (Song et al., 2008).

Ágnes Kovács, Ernö Téglás, and Ansgar Endress (2010) were audacious enough to devise a task that might demonstrate mentalizing in babies only seven months old. They designed meticulously controlled video sequences, where a Smurf did or did not have a false belief about the presence of a ball (see figure 10.2). Adults were shown the same scenarios, but here reaction time was measured rather than eye gaze.

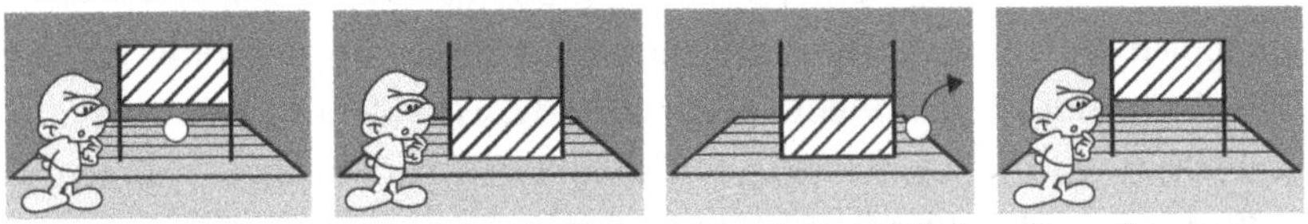

Figure 10.2
Inconsistent beliefs: at the end of this sequence, the Smurf believes that the ball is behind the barrier, while the observer believes that it has gone. The observer automatically takes account of the false belief of the Smurf and takes longer to report that the ball is not there (Kovacs, Téglás, Endress, 2010). Redrawn from figure 1 from Xiao et al. (2018). Copyright 2018, with permission from Wiley.

The surprising finding was that both infants and adults spontaneously tracked another agent's false belief.[3] This put a very different gloss on earlier findings where mentalizing ability was found to emerge only slowly and was not reliably found before the age of four to six years.

At around the same time, experimental (rather than anecdotal) evidence for mentalizing was found in other species besides humans. Rhesus monkeys prefer to steal grapes from a human who is not able to see the grapes rather than from one who can see them (Flombaum and Santos, 2005). Birds too were found to be able to take account of a conspecific's knowledge. Nicky Clayton and her colleagues (Clayton, Dally, and Emery, 2007) studied caching behavior (burying food for later use) in scrub-jays. They found that these birds would rebury their food in another place if another bird had been watching them the first time they buried it. By now, several other species have been added to the list (Krupenye and Call, 2019).

Other developments produced more sensitive mentalizing tests for use with adolescents and adults, which were difficult even for these older groups. Contrary to the original idea that the ability to mentalize was fully formed by age six, it became clear that the development of mentalizing continued to improve into adolescence and young adulthood. These changes were accompanied by systematic changes in brain activation (Blakemore et al., 2007).

3. There have been questions about the replicability of some of these demonstrations of implicit mentalizing in infants younger than eighteen months (Kulke et al., 2018), but there is good evidence for this automatic and inflexible form of mentalizing in adults (El Kaddouri et al., 2020).

All these studies, with infants on the one hand and adolescents and adults on the other, as well as studies with other animals, shook up our ideas about mentalizing. As we grappled with understanding the new findings, two opposing explanations were offered. One idea is that mental-state tracking, as evident in infants and in some other animals, is not true mentalizing, but should perhaps be called "submentalizing" (Heyes, 2014). The other idea is that, on the contrary, mentalizing has an old evolutionary basis and is manifest in several animal species, including humans, where it is manifest in their first year of life.

The evolutionary account appealed to us, but it demanded an explanation of why Maxi/Sally-Ann False Belief tasks are not passed by children below the age of four. Perhaps the explanation is quite simple. According to Leslie, these tasks are tricky because they demand inhibitory control; to pass the task, it is necessary to inhibit one's own point of view (Wang, Hemmer, and Leslie, 2019). Such inhibitory control is rarely in evidence before the age of four to six.

Two Forms of Mentalizing

We were more attracted by a third idea—namely, that there is not just one form of mentalizing, but two. One is the implicit tracking of other minds, while the other is the explicit attribution of mental states (Apperly and Butterfill, 2009; Edwards and Low, 2017). We speculate that the implicit form is part of the Zombie thread. It is likely one of nature's starting kits, emerging even in some nonhuman animals. The explicit form on the other hand, is part of the Machiavelli thread. It can be seen as a special human adaptation, dependent on cultural learning (Heyes and Frith, 2014), hence its late appearance and continuous development until adulthood.

Josef Perner and Johannes Roessler (2012) also endorsed the idea of two forms of mentalizing: Implicit sensitivity to false beliefs simply involves an expectation of what will happen. In contrast, explicit false belief attribution requires explicit consideration of the agent's perspective and reasons for action.

The distinction between implicit and explicit mentalizing reoriented our ideas about autism. Would it allow us to reconcile the findings of their early mentalizing failure with their delayed success? Perhaps autistic individuals lacked implicit mentalizing, but not explicit mentalizing. Perhaps they could acquire an explicit Theory of Mind through being taught and having repeated experiences of behavioral norms.[4] For instance, they might rehearse individual scenarios ("If Sally was out, she did not

4. This presupposes a sufficient level of language ability.

see what happened and she can't know where the marble is"), and in this way, they can learn to reason explicitly about other people's mental states. This would allow them to succeed on lab tests where sufficient time for explicit reasoning is allowed, but they would still struggle in real-life situations, as it is hard to reason on the fly.

Uta with Atsushi Senju, Victoria Southgate, and Sarah White tested this hypothesis (Senju et al., 2009). Here, autistic adults were selected as participants if they had performed at ceiling on a whole range of explicit mentalizing tests. The implicit test measured anticipatory eye movements in a version of the Sally-Ann task developed for use in the baby lab (Southgate, Senju, and Csibra, 2007). Just like the infants tested by Southgate and her colleagues, ordinary adults in our experiment looked in anticipation to where a teddy bear (falsely) believed the ball was. But this was not so for the autistic adults, exactly as the hypothesis predicted.

Of course, one experiment is never enough. Later studies (Eigsti and Irvine, 2021; White, 2013) showed similar results. However, it would be highly desirable to develop new measures to study implicit mentalizing (Kulke and Hinrichs, 2021; Kampis et al., 2021). Also, more work is needed to test the idea that it is possible to acquire skill in explicit mentalizing without the backup of implicit mentalizing. In typical development, the two forms of mentalizing would be expected to build on each other and work together.

Assuming that there are two forms of mentalizing, a metaphor that works for us is our ability to use both an automatic global positioning system (GPS) when we navigate space and a map that requires us to consciously reason about where we are and how to get to our destination. Map reading, like explicit mentalizing, has to be learned, whereas the GPS is intuitive and hardly needs any learning. Perhaps the brain is equipped with a GPS that continuously tracks where we are in the world of agents. Instead of (or as well as) tracking the geographical position of other agents and their goals, it tracks their mental states. This then enables us to predict what they are going to do on a moment-to-moment basis. All this would be part of the Zombie thread.

In contrast, explicit mentalizing functions like a map, in that it represents the mental rather than the physical world, providing high-level behavior rules that guide us through such tricky ground as moral responsibility and reputation management. This makes it part of the Machiavelli thread. As with other cognitive processes that can be linked to Kahneman's system 1 and system 2, the explicit version is not necessarily superior to the implicit one. Both have their advantages and disadvantages, and they may well be used in parallel, with increasing skill from about the age of six years onward.

When we keep track of others' mental states in the world of agents, we are, in effect, taking their perspective on the world. Taking the perspective of others is often considered to be more complex than taking one's own, egocentric perspective. Victoria Southgate (2020) has made the attractive suggestion that vicarious representations, which she refers to as "alter-centric," may dominate during early infancy. This means that there is no conflict between own and other agents' perspectives. She explains the early appearance of implicit mentalizing as the default adoption of another person's perspective. This makes sense when children are completely dependent and have to learn about the world vicariously through others.

As children get older, the conflict between own and others' perspectives emerges, and this leads to them typically failing false belief tasks such as the Maxi/Sally-Ann task. From about the age of four to six, metacognitive processing becomes possible, and with it explicit mentalizing. This is when children are beginning to have the level of language to reason about differences in perspective. This involves recursion[5] (I think that you believe that . . .), which we will discuss in more detail in chapter 14.

The conflict between the viewpoints of the observer and the agent remains quite hard to resolve, even at later ages in more complex tasks. For example, in the Director task, what you can see and what the director can see is sometimes the same and sometimes different, as indicated by the presence or absence of a screening device. In this task, even adults occasionally adopt an egocentric perspective and fail to take into account the director's perspective (Dumontheil, Apperly, and Blakemore, 2010). This failure has been explained by lingering egocentric biases among adults (Epley, Morewedge, and Keysar, 2004).

Explicit Mentalizing—a Tool to Step up Competition

What does our consciously controlled, explicit Theory of mind do for us over and above the automatic tracking of others' mental states? GPS systems are useful because they are fast and automatic, but they are also inflexible and can lead us astray. For example, we automatically and inflexibly follow someone's eye gaze, even when this is consistently misleading (Bayliss and Tipper, 2006), as discussed in chapter 2. This is when explicit mentalizing can save the day. But perhaps its most valuable contribution is that it enables us to infer that somebody is deliberately trying to mislead us.

Explicit mentalizing is effortful because it requires an extra layer in the information-processing hierarchy, which serves to consciously monitor our own and others' mental states (Shea et al., 2014). This monitoring facility enables us to either deliberately hide or deliberately reveal our own desires, intentions, and beliefs. It also makes us vigilant about the signals sent by others. Can we trust in their supposedly good intentions? It

5. Recursion: see *Recursion*

pays to be vigilant to avoid being taken in by deception, and takes a lot of learning. Eventually, we can hold in mind that a thief can lie about the place where the stolen goods are hidden, a trader can overpraise an object to increase its value, and a trickster can pretend to be a celebrity.[6]

How do we acquire and refine explicit mentalizing? It seems reasonable to think that its appearance is intimately connected with the development of explicit metacognition (Heyes et al., 2020; also see chapter 14). It also depends on a reservoir of knowledge of social situations, which goes beyond our own direct experience and is dependent on language. Cultural learning is key, and this is not unlike learning your native language, at first from our own family (Tompkins et al., 2018), then from our peers, and then from others we are interacting with, in ever-widening circles. Over the years, we will have internalized all those norms that govern the behavior of members of our group.

The cultural influences on explicit mentalizing are revealed in the way that children, and indeed adults, explain and predict behavior. Depending on their culture, they may use references to magical, to religiously inspired, or to scientific concepts. They may even deliberately avoid talking about mental states when predicting behavior (Curtin et al., 2020; Robbins and Rumsey, 2008). In line with their experience and personal preferences, they may show particular attribution biases. The most influential and sophisticated example of cultural teaching may well still be the guidebook for princes written by Nicolo Macchiavelli five centuries ago (Machiavelli, 2008/1532).

The Brain's Mentalizing System

It still amazes us that theory of mind/mentalizing, a concept that had not been anticipated by social psychologists, has been taken up with enthusiasm by cognitive neuroscientists. In hindsight, it was outrageous to expect neuroimaging to reveal the brain's mentalizing system, at a time when positron emission tomography (PET) scanners had only just been invented and magnetic resonance imaging (MRI) scanners were only a glimmer on the horizon. Given the many uncertainties of these then very novel techniques, we were extremely lucky to be at the right place at the right time. It was not just the new technology and the new possibilities of statistical mapping of brain images, but also the application of experimental designs to make brain imaging functional, not merely structural.

Thus, just after the first functional imaging laboratory in the United Kingdom was established, we were able with our colleagues to identify changes in activity in certain

6. We will return to this dark side of mentalizing in chapter 11.

brain regions while volunteers were asked questions that made them think about mental states in one condition and about physical states in another condition, with other factors kept constant. For our first attempt, we used Francesca Happé's "Strange Stories" (Happé, 1994). This early PET study revealed the components which together form the brain's mentalizing system (Fletcher et al., 1995). We were gratified that the same basic components (see figure 10.3) were observed in subsequent studies that used functional magnetic resonance imaging (fMRI) (see, e.g., Schurz et al., 2021).

We do not yet understand how these brain regions interact to share their workload, but some researchers are already zooming in at the single-neuron level. Indeed, it seems likely that there are individual neurons that respond preferentially to social stimuli (Lockwood et al., 2020). It is even likely that there are neurons that code for intentional behavior (Yoshida et al., 2011). The existence of such dedicated neurons has been demonstrated in a study using single-cell recordings in the dorsal medial prefrontal cortex of eleven patients undergoing surgery. The patients were solving either false belief problems or physical cause-and-effect problems presented in story format (Jamali et al., 2021). This enabled the identification of populations of neurons that represent the contents of an agent's belief across different scenarios. This is an astonishing finding. However, we are far from knowing what actual mechanisms allow neurons to represent mental states, one's own and those of others.

One promising approach might be to study neural activity in different species while they observe goal-directed and intentional actions. The idea of an evolutionary basis of the intentional stance received a boost by a study by Julia Sliwa and Winrich Freiwald (2017), which showed that brain regions similar to the human mentalizing system were activated when macaque monkeys watched videos of social interactions between two other monkeys.

What Happens during Development?

By now, there have been numerous brain-imaging studies that target mentalizing in volunteers of different ages. Fortunately, there is a comprehensive review by Hilary Richardson and Rebecca Saxe (2020). These authors concluded that despite increasing competence in solving mentalizing tasks, the mentalizing network appears to be "preferentially engaged by similar stimuli in adults, 3-year-old children, and 7-month-old infants." They also stated that "these results point to a very early developmental origin of a cortical network for ToM" (Richardson and Saxe, 2020, 470).

While the mentalizing system of the brain is in place at a very early age, an egocentric bias (Epley, Morewedge, and Keysar, 2004) may add conflict at later ages. Also, it takes time to perfect the understanding of complex recursive phrases, such as "He thinks that she thinks that he knows." Still, seven- to eight-year-olds are aware of white lies and the existence of double bluffs. Mentalizing performance continues to improve over adolescence, while dramatic structural changes are taking place, in particular a marked decrease in gray matter volume (Mills et al., 2014).

There are also functional changes linked directly to mentalizing. In a study carried out by Sarah-Jayne Blakemore with Chris Frith and colleagues, participants had to answer questions about intentionally caused events and events involving physical causality (Blakemore et al., 2007). While responding to intentionally caused events, adolescents showed greater activation of the anterior regions (i.e., part of the medial prefrontal cortext (PFC)) than adults. On the other hand, adults activated posterior regions (i.e., part of the right superior temporal sulcus (STS)) more than adolescents. Speculatively, this change suggests that practice succeeds in turning a previously effortful process into an automatic one. Presumably, in adulthood, explicit mentalizing is so well practiced that it becomes an automatic skill. This would be consistent with the idea that extensive learning frees up the anterior frontal regions, known to be required for conscious control during novel tasks (Jueptner et al., 1997).

The Three Hubs of the Brain's Mentalizing System

There have probably been more brain-imaging studies in which participants were asked to make inferences about mental states than of any of the other processes discussed in this book. In most of these, participants are presented with social scenarios and invited to think about them, and in some tasks, they were also asked to empathize with the protagonist. Stories about false beliefs or deception are frequently involved, which can be presented either verbally or visually in videos, cartoons, or animations. There are as yet relatively few studies where participants are directly involved in a social interaction, such as by playing an online game with a partner.[7] There is an urgent need for more studies in "second-person interaction," as it has become known (Redcay and Schilbach, 2019).

A multitude of brain-imaging studies in the field of social cognition have recently been assessed by Matthias Schurz, Philipp Kanske, and other colleagues in a giant meta-analysis (Schurz et al., 2021). This has provided evidence for a clear distinction in patterns of brain activation obtained with mentalizing tasks (cognitive cluster) and others obtained with tasks based on observing pain and emotions (affective cluster). The cognitive cluster, engaged by mentalizing, extends over a somewhat broadly defined, but still circumscribed, set of brain regions (see figure 10.3). These included anterior midline structures (medial prefrontal and anterior cingulate cortex, mPFC/ACC), posterior midline structures (posterior cingulate cortex and the adjacent parietal cortex, PCC/

7. These interactive studies allow computational modeling of processes involved in mentalizing (see more about this in chapter 12).

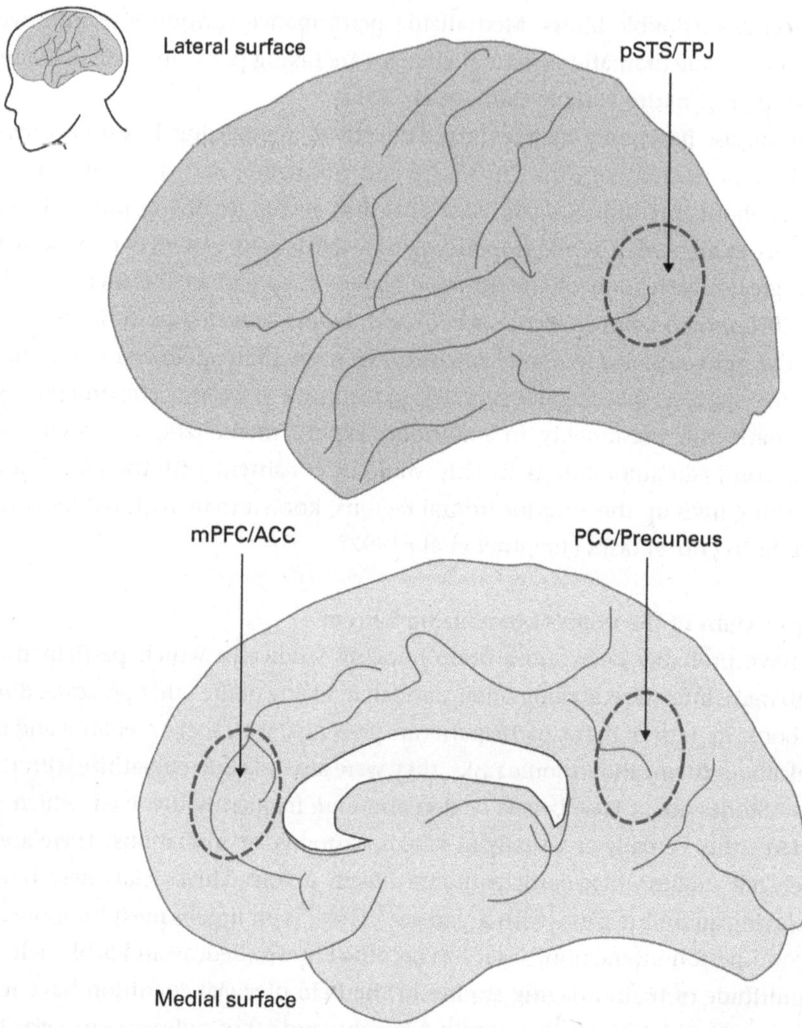

Figure 10.3
Mentalizing in the brain: the location of the three principal hubs in the brain's mentalizing system; controller (mPFC/ACC), connector (pSTS/TPJ), and navigator (PCC/precuneus).

Precuneus) and bilateral temporoparietal areas (posterior superior temporal sulcus/ temporo-parietal junction (pSTS/TPJ).

One of our major assumptions in this book is that the brain is a hierarchical information-processing system (see chapter 12), and this applies to mentalizing as well. Matthias Schurz and colleagues make a similar assumption. In terms of brain connectivity, their hierarchy extends from primary sensory and motor areas (unimodal) at the lower level to increasingly abstract (transmodal) areas at the higher level. The mentalizing cluster is located close to the transmodal end of the gradient, supporting its functional interpretation in terms of abstract, stimulus-independent thinking. How might the three hubs fit into this hierarchy, and what are their individual roles when they work together during mentalizing tasks? Somewhat speculatively, we have labeled them "connector," "navigator," and "controller."

The Connector (pSTS/TPJ)

We suggest that the pSTS/TPJ region is the gateway from the physical world into the mentalizing system. In particular, this region is concerned with the question, "Is the agent I am interacting with doing what I expected?"

The pSTS is a region that is activated by biological motion (Grossman et al., 2000) and has an important role in action observation (Urgen and Saygin, 2020). In the macaque monkey, this region is sensitive to violations of expectation in social settings (Roumazeilles et al., 2021). Here, pSTS is adjacent to the temporo-parietal junction (TPJ), which has been repeatedly implicated in studies of mentalizing. While mPFC/ ACC (the controller) occupies the top level in our hierarchy, we believe that pSTS/TPJ (the connector) occupies an intermediate level, perfectly positioned to connect top-down expectations set by the controller/director, with the incoming evidence provided by the senses. In this position, it can compare the behavior that is expected on the basis of beliefs with the behavior that actually occurs. Thus, pSTS/TPJ (the connector) might be concerned with prediction errors that emerge from the difference between the prior expectations derived from the (top-down) mentalizing mode and the (bottom-up) perception of the incoming stimuli.

Several studies suggest that pSTS/TPJ is involved in prediction. And this is the case when we are monitoring our own actions, as well as the actions of other people. When the outcome of our action is not what we expect, greater activity is seen in the TPJ (Miele et al., 2011). When we are monitoring others, greater activity is seen when people move their eyes in an unexpected direction (away from, rather than toward, a target stimulus; Pelphrey et al., 2003), as well as when people appear from behind a barrier at an unexpected time (Saxe et al., 2004).

The incoming evidence that is used to create these prediction errors is not restricted to direct visual perception of what other agents are doing. For example, in a seminal study by Alan Hampton, Peter Bossaerts, and John O'Doherty (2008), greater activity in pSTS/TPJ occurred when the partner in the game made an unexpected choice. In this case, the partner's choice was indicated by a symbolic representation. Furthermore, in many of the studies in which pSTS/TPJ is activated, the behavior of people is described in words (e.g., Saxe and Kanwisher, 2003).

The idea that pSTS/TPJ plays a specific role in prediction is also supported by a study from Bardi, Gheza, and Brass (2017). Here, transcranial magnetic stimulation (TMS) was used to disrupt the TPJ (on the right side of the brain), while participants watched videos involving an agent with a belief about the location of an object (the task developed by Kovács et al., 2010). Performance of this task requires a representation of the belief of the agent (true or false) and a prediction of what is going to happen. TMS did not disrupt the representation of the agent's belief (the top-down signal), but it did interfere with the predictions.

Evidence for the position of pSTS/TPJ in the hierarchy can be obtained from studies of brain connectivity. For example, Hauke Hillebrandt, Karl Friston, and Sarah-Jayne Blakemore (2014) used fMRI data from the human connectome project. The data was gained during spontaneous mentalizing elicited by the Frith-Happé set of animated triangles. When these movements elicited the perception of mental states, there was an increase in connectivity between V5, a region concerned with low level motion detection, and pSTS. This is an example of bottom-up signals being passed from the sensory regions of the brain into the mentalizing system.

Using the same task, Moessnang et al. (2017) also observed increased coupling between pSTS/TPJ and visual areas, and between pSTS/TPJ and mPFC/ACC, during mentalizing. They suggest that mPFC/ACC is involved in belief representation, while pSTS/TPJ is involved in perception-based processing of social information.

The evidence from all these studies converges to confirm that pSTS/TPJ receives the sensory input and therefore is an entry node into the mentalizing system. Neurons in this region receive bottom-up evidence concerning people's behavior and make sense of this behavior by checking how well it matches expectations (top-down) about the kind of behavior that the actors should display. We speculate that this connector region supports the automatic tracking process that underlies implicit mentalizing.

The Navigator (PCC/Precuneus)

We suggest that the PCC/precuneus region is concerned with navigation through the social world. It is concerned with the question, "Where am I in relation to others in this social space?"

Here, we are being even more speculative since the implication of PCC/precuneus (retrosplenial cortex) activity in mentalizing has rarely been considered. But just because it is rather obscure does not mean that it is unimportant. Even in our first PET study (Fletcher et al., 1995), the precuneus was robustly activated during mentalizing. Its precise function outside mentalizing is also unclear since it is engaged by many tasks, including spatial navigation and scene processing, episodic memory retrieval, mental imagery, and self-referential processing (Chrastil, 2018). We suspect that its role in spatial navigation, for which there is extensive evidence, may provide a clue for its role in mentalizing (hence our label "navigator").

In their review of the role of this region in spatial navigation, Mitchell et al. (2018) suggested that it has access to the same spatial information represented in either an egocentric or an allocentric representation, and therefore is needed to switch between these representations. During navigation, this would enable us to establish and maintain our bearings in the scene (Hartley et al., 2003). As we have noted, the bringing together of perspectives is likely to be needed for implicit mentalizing. As with TPJ/pSTS, this region does not require direct evidence concerning spatial scenes. It is also engaged by words describing spatial scenes (Auger and Maguire, 2018; Vukovic and Shtyrov, 2017).

Vogeley et al. (2004) found that this region was activated when people had to take somebody else's viewpoint into account (visual perspective taking). A comparison of brain-imaging studies involving either visual perspective taking or mentalizing showed that this region has a role in third-person perspective taking, which is a common feature of both kinds of task (Arora, Schurz, and Perner, 2017). For example, in a false belief task, it is necessary to take into account the protagonist's false perspective on the world, which is different from the observer's own perspective. This conflict of viewpoints has often been highlighted as the main obstacle to solving the Maxi/Sally-Ann task. The conflict does not occur with an altercentric representation (Southgate, 2020) and is resolved by taking on an allocentric representation or We-mode, in this case applied to the social world rather than the physical world.

A suggestion made by Matthew Schafer and Daniela Schiller (2018) is of interest in this context. They propose that navigation in the social world involves the same neural mechanisms as navigation in space. This idea had been explored earlier in a study in which participants had to navigate through a social space, interacting with people (avatars) who varied along the dimensions of affiliation and power (Tavares et al., 2015). Here, activity in the PCC/precuneus tracked the social distance between the participant and the avatar they were interacting with (how similar am I to this agent in terms of affiliation and power?). A more recent study used social networks derived from social media (Facebook) and showed that social network distance was encoded in this brain region (Peer et al., 2021). A greater social distance would require a greater shift in point of view to take account of the mental state of the partner.

This account would suggest that this region is situated at an intermediate level in the hierarchy of the mentalizing system, maintaining a point of view rather than responding to incoming evidence. This region might also represent a staging post where implicit and explicit mentalizing processes join together. But this is speculation.

The Controller (mPFC/ACC)

We suggest that the mPFC/ACC region comes into play when we engage the mentalizing mode (or the intentional stance, as discussed by Dennett, 1987). It is concerned with questions such as "Am I dealing with an intentional agent?" and "How sophisticated is this agent?"

The mPFC/ACC region plays important roles in many control tasks, but also a special role in mentalizing. As already mentioned, this was recently confirmed in a study of single-neuron activity in humans, all in the PFC, responding when challenged to think about intentional social interactions (Jamali et al., 2021).

But what is this role? We suggest that the mPFC sits at the top of the hierarchy of information processing and is the source of top-down signals connecting to other hubs. In our Bayesian framework, we believe that these top-down influences provide information about priors. There are always some prior expectations about the situation in which people find themselves, which are just as important as the exposure to the situation itself. We believe that these prior expectations and their influence are part of our conscious experience, and thus components of explicit mentalizing.

The controller also acts as the interface to the world of ideas and can thus convey information from other people and from our culture more generally. This would explain the effect of context and instructions. For example, if the experimenter (see chapters 7 and 12), tells somebody before they play a hide-and-seek game, "You are interacting with a robot," the player will have the expectation (the prior) that the behavior of the other player will be either random or goal-directed. In contrast, if the experimenter says, "You are interacting with another volunteer," the player will expect the behavior to be intentional.

We call the state created by such prior expectations the *mentalizing mode* by analogy with the retrieval mode in studies of episodic memory (Lepage et al., 2000). This mode is a cognitive state that determines how incoming stimuli will be processed. In terms of brain function, this will be revealed as tonic activity maintained throughout a period in which people are engaged in mentalizing. There are several studies in which this pattern of activity was observed in mPFC/ACC. In these studies (Rilling et al.,

2004; Gallagher et al., 2002), activity in the mPFC/ACC was seen when participants believed that they were interacting with a person, even though there was no difference in behavior from the condition in which they believed their partner was a computer.

That mPFC/ACC operates at this high level (as a controller or director) is also indicated by the observation that activity is less affected by the actions of the agent being observed as by beliefs about the nature of that agent. An elegant study by Wheatley, Milleville, and Martin (2007) found evidence for the effect of context on brain activity, which resulted in a switch between mirror and mentalizing systems. Here, people observed an agent (a conical shape) making identical movements (a figure eight). The brain activity elicited depended on whether they thought the agent was an object or a person. These beliefs were induced purely by a background image. Thus, a snowy landscape suggested that the image was a skater, while toys on a playground suggested a spinning top.

A landmark study from Chris Miall's group (Stanley, Gowen, and Miall, 2010) went one step further. Participants viewed a series of animated point-light figures, which were derived from motion-capture recordings of a person moving. In one situation, they were told that the moving images were from recordings of a human, and in another, they were told that they were computer generated. When they had the prior expectation for human movement, they immediately engaged mPFC, and this was even more the case than with the prior expectation for computer-generated movement. In contrast, when they simply observed humanlike movements, without instructions, they did not engage this region more than when observing robotlike movements.

These strong context effects that arise from independent external cues support the reality of a world of ideas. These effects tell us that it is a mental state, a belief, and not physical reality that determined how the images were processed. We presume that the controller hub sets the scene, inserting the prior expectation that the incoming information is (or is not) caused by an intentional agent.

Returning to our example of interacting triangles, viewers spontaneously believe (but would explicitly deny if asked) that these shapes are intentional agents, and therefore they would take an intentional stance to interpret their movements. This could occur spontaneously because the triangles do not behave like rational goal-directed agents. Instead, they behave irrationally (e.g., by turning on their axis and moving toward and away from each other). As we argued in chapter 7, this irrational behavior triggers a switch in the observer's stance from goal-directed to intentional agency.

The suggestion that activity in mPFC/ACC represents a tonic state (the mentalizing mode) fits with the observation that there is considerable overlap between components of the mentalizing system and the so-called default mode network. The default mode network is active when people are immersed in their own thoughts. There is substantial evidence that this cognitive mode is reciprocally suppressed by another cognitive mode, the "task-positive mode," so called because it is active while performing particular tasks.

What are these tasks? Tony Jack and his colleagues (2013) have shown that they concern the world of objects. Thus, it is reasoning about the casual/mechanical properties of inanimate objects that engage the task-positive network. In contrast, the default network is engaged by problems that are concerned with the world of ideas (i.e., reasoning about one's own and others' mental states). In Jack's study, a wide range of tasks was used to characterize the two modes. Here, activity associated with mentalizing was most pronounced in the mPFC/ACC and PCC/precuneus. A similar pattern was observed by Michael Gilead, Nira Liberman, and Anat Maril (2014). They asked their participants to consider either *why* (mental causation) or *how* (physical causation) an action was performed. The comparison of why versus how also revealed activity in the mPFC/ACC and PCC/precuneus.

How Might the Hubs Work Together?

Of course, we cannot stop at simply contemplating the role of the individual hubs, as they necessarily interact in the hierarchical system. The importance of connectivity within this network is pinpointed by evidence from autistic participants in functional imaging studies of mentalizing. So far, this evidence points to weaker connectivity between these hubs, while the hubs themselves are present and active (e.g., Castelli et al., 2002). Weaker connection may well result in less precise activation in the various hubs of the hierarchy, but what this means is hard to interpret at present. In a study by Charlotte Grosse-Wiesmann and colleagues (2017), young children who did not yet succeed at false belief tasks showed weaker connectivity in the white-matter tracts of the mentalizing system, compared to those who did succeed.

To better understand the nature of the hierarchy and the interactions between the components of the mentalizing system, a computational model is needed. Such a model should predict behavior and also enable better specification of the role of the regions involved, as with model-based fMRI (e.g., O'Doherty, Hampton, and Kim, 2007). In addition, a full neuropsychological, neuroanatomical, and evolutionary approach is needed to discover and understand the neural mechanisms of mentalizing and their origins.[8] A start has already been made in work on social cognition in primates, which

8. Mentalizing in neuropsychological and schizophrenic patients has often been found to be impaired, but this may be part of a more general impairment in self-awareness, involving the metacognitive processes at the top of the information-processing hierarchy (see chapter 14).

we do not review here. Instead, we recommend the comprehensive review by Marco Wittmann, Patricia Lockwood, and Matthew Rushworth (2018).

Do Brain-Imaging Studies Distinguish between Implicit and Explicit Mentalizing?

This question has not yet been answered satisfactorily. Unfortunately, we can never be sure which of these two forms of mentalizing is used by the participants during scanning. For all we know, both might often be engaged simultaneously. Telling participants to do nothing and just watch animated triangles does not guarantee that only implicit mentalizing will be engaged. While they are watching the movies, they may well start to think about the intentions that fit the movements of the triangles. They might even start to speculate on the experimenter's intentions in showing them these animations.

One solution to this problem is to eliminate participants who report afterward that they *did* think about mental states. Another option is to use a distracter task designed to fully occupy the conscious system with the aim that no spare capacity is left to engage explicit mentalizing. Using both these options, Naughtin and colleagues (2017) studied implicit processing of false beliefs and observed a substantial overlap with brain activity during explicit processing. However, this overlap was restricted to the connector and navigator hubs. The controller hub in the PFC was not activated by mentalizing when participants had to do the distracter task. Presumably, it was fully engaged in that task.

This result suggests that implicit mentalizing is achieved by the lower levels of the processing hierarchy, while explicit mentalizing requires a higher level. The explicit system works on the same incoming information, but it can engage the mentalizing mode on the basis of cues provided in the context or given by instructions. This system's main function may be to reason about the hidden intentions underlying behavior (see also Van Overwalle and Vandekerckhove, 2013).

Our hierarchical account has direct parallels with accounts of the differences in brain activity between conscious (explicit) and unconscious (implicit) processing in visual perception. So long as perceptual processing remains at an unconscious level, activity is restricted to visual-processing regions (e.g., the fusiform face area for faces and the visual word-form area for words). When these stimuli become conscious, additional activity is seen in the intraparietal sulcus (IPS) and dorsolateral prefrontal cortex (DLPFC) (faces, Beck et al., 2001; words, Dehaene and Cohen, 2001). By analogy with the processes leading to visual consciousness, we might expect mPFC/ACC and PCC/precuneus in the mentalizing system to take the roles of the DLPFC and IPS in the visual system. If so, this provides a solid basis to our speculations about the role of the three major components of the system, the controller (mPFC/ACC), the connector (pSTS/TPJ), and the navigator (PCC/precuneus).

We look at the period between 1980 and 2000 as the most exciting chapter of our working lives thus far. It laid the foundation for our passionate interest in social cognition and ultimately led us to write this book. Autism provided a constant source of inspiration to think about different processes being brought into play in the social and in the physical world. Brain imaging provided another source of inspiration—the possibility of finding evidence of mentalizing in the activity of different brain regions. We are very aware that we are only at the beginning of a new field of research and so far, we have taken only baby steps.

11 The Dark Side

Selfishness reigns on the dark side of social interactions. In humans, there is a spectrum of selfishness, varying from extreme altruists to extreme egoists. However, rather than trying to eradicate the dark side, we should acknowledge that competition is part and parcel of life, and variation is important for the survival of the group.

Competitive, selfish behavior cannot be characterized as an ever-dominant gut feeling. Cooperative, altruistic behavior can be equally intuitive in initiating generous acts. Likewise, our deliberate, considered responses can be selfish as often as altruistic. Surprisingly perhaps, mentalizing, our human ability to take account of the mental states of others, is not necessarily conducive to altruism or empathy, nor is it always helpful when cooperating. Instead it can be a distinct advantage in competitive situations. For example, mentalizing is required when we deceive others by implanting false beliefs. Vigilance is needed to combat such deception. While prosocial tendencies emerge early in development, they are not applied when interacting with outgroups. Top-down control, via reason, can be used to overcome this behavior, but unfortunately, it can also be used to justify bad behavior and demonize outgroups.

* * *

Selfishness and Altruism

As much as we can delight in all that is good about social interactions, this is not the whole story. In chapter 8, we discussed some of the harmful effects of the group distinctions that have run like a gash through the history of conflicts. In chapter 9, we discussed the ease with which we can turn into free riders when the opportunity arises, and in chapter 10, we did not gloss over the fact that mentalizing ability has given us powerful weapons for competition (particularly deception). Here, we take a closer look at the tension between competition and cooperation and between selfishness and altruism. Why is altruism not more prevalent? Why is inequality so widespread in many human societies, with consequent injustices for individuals who are at the bottom of the hierarchy?

According to William Hamilton (1964), the emergence of altruistic behavior from the competitive forces underlying evolution depends upon kin selection. It is the genes that must survive, not the individual that carries the genes. Thus, altruistic actions tend to benefit our own kin. Increasing relatedness results in increased cooperation and decreased competition. This process is illustrated by J. B. S. Haldane's off-the-cuff remark (*New Scientist,* August 8, 1974): "I'd lay down my life for two brothers or eight cousins."

So what are we to make of the awe-inspiring people who save complete strangers by sacrificing themselves? In a small corner of the City of London, there is a unique Memorial to Heroic Self-Sacrifice (see figure 11.1). It is very moving and suggests that we humans can feel deeply connected with others, even if they are not family members. Granted, these are exceptions. If we take an honest look at ourselves, we can

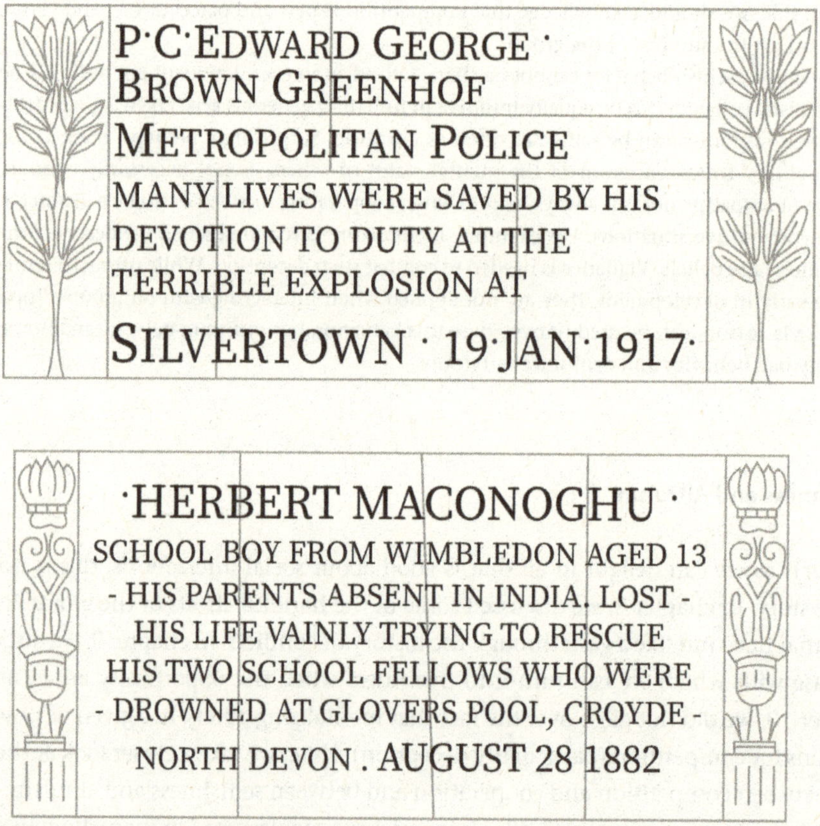

Figure 11.1
Inscriptions in Postman's Park in the City of London, honoring some of the heroes who died saving others.

admit that we will behave selfishly if we think we can get away with it. At the same time, whether or not we act selfishly ourselves, we prefer not to cooperate with a selfish person, as we could not rely on her or him to share any gains with us or help us.

But there is more than one way of being selfish: Paul van Lange has identified three main types of people: "prosocials," "individualists," and "competitives" (e.g., van Lange et al., 1997). Prosocials value fairness and are prepared to lose a little to maintain equality (also referred to as "aversion to iniquity"); individualists take whatever they can and keep it for themselves; and competitives aim to get more than others are getting, rather than having as much as possible. This classification is established by asking people to choose between three possible splits for sharing money with one other, usually anonymous, person (see figure 11.2).

Masahiko Haruno and Chris Frith (2010) used this method to identify people as prosocial or individualistic and then scanned them while they observed various sharing options (e.g., you get 110 yen while your partner gets 60). Prosocials disliked unfair shares, while individualists were indifferent to them. These unfair shares elicited activity in the amygdala (shown in figure 4.3 in chapter 4) for prosocial participants but not for individualists. This emotional response to unfairness seen in prosocial people seems to be rapid and intuitive since it is not affected by a distracting memory task (Haruno, Kimura, and Frith, 2014).

As we already hinted, there can be much larger variations in selfishness and altruism than these categories capture. For example, there are people who volunteer to donate organs to complete strangers. On the other hand, there are psychopaths who might decide to kill somebody for an organ if they needed it. Both groups have been studied by Abigail Marsh and her colleagues. Unlike the organ donors, psychopaths care about themselves and do not care about people in distress. The two groups show opposite

	Self (£)	Other (£)
A	100	20
B	110	60
C	100	100

Figure 11.2
The three main types of sharing preferences.

Which of these splits (A, B, or C) would you choose? Selfish people (individualists) simply choose the one giving the highest value for the self (B). Prosocial people maximize the amount for the self and the other (C), even though they themselves could receive more. Competitive people will choose A to achieve the greatest difference between self and other, even though they could receive more. When students were tested, about 37 percent were individualists, 52 percent prosocial, and 11 percent competitive (van Lange et al., 2011).

responsiveness to fearful facial expressions, associated with structural and functional differences in the amygdala (Marsh et al., 2014a) (see figure 11.3).

This work prompted Marsh to propose a continuum that reflects variations in neural and hormonal systems that underpin the urge to care for others (Marsh, 2019). A study from her lab delved deeper into the basis of this urge (Brethel-Haurwitz et al., 2018) and looked at empathy for pain (see chapter 4). Here, extreme altruists showed greater overlap in their neural responses (in the anterior insula) to distress in self and others compared to typical adults, who acted as control participants. Extreme altruists really did feel the pain of others.

Perhaps these are people who have a very permeable self-to-other boundary. It is tempting to think that their ingroup includes not only their kin, but possibly all of humanity. One might argue that compassion, as practiced by devout Buddhist monks, goes even further by extending the ingroup to all living creatures. Clearly, most of us never reach this beatific state of universal caring.

Given that there is a continuum, we do not have to rely on extreme groups to find examples of when altruism trumps selfishness. Patricia Lockwood and colleagues (2016) tapped into prosocial tendencies by tracking how quickly individuals learn

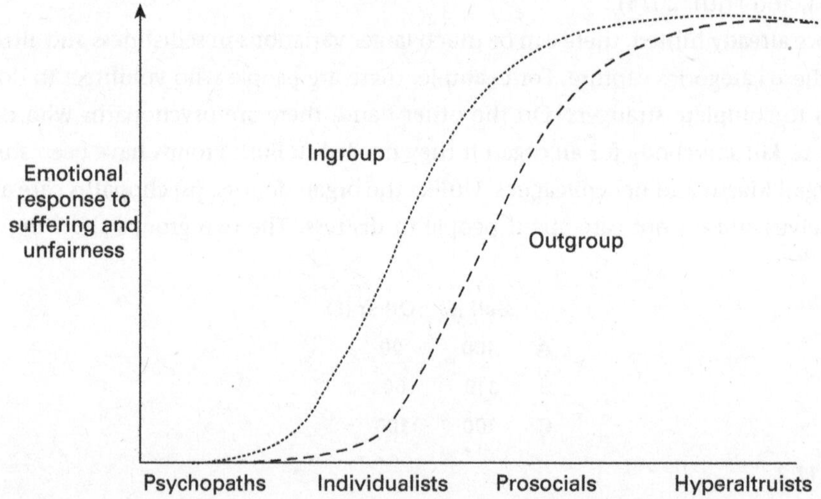

Figure 11.3
A continuum of altruism. There are individual differences in altruism: Psychopaths show no emotional response to the suffering of others. Hyperaltruists respond to suffering in everyone. Prosocials respond more than individualists, but both respond less to the suffering of an outgroup member.

to benefit others as opposed to benefiting themselves. People with stronger prosocial tendencies learned more quickly to benefit others. In another study from the same lab (Lengersdorff et al., 2020), prosocial individuals learned more quickly how to avoid causing pain to another person than how to avoid pain to themselves, all the while being unaware that they were doing this. From these and other studies (e.g., Crockett et al., 2014), it seems that human beings can overcome egocentrism without having to engage in top-down, conscious control processes, especially when this means avoiding harm to another person.

Social Hierarchies and Signals of Dominance

Many of us have a dream of a society based on equality. But history tells us that this may be achieved only in exceptional cases, in small groups, and perhaps only temporarily. In contrast, throughout the animal kingdom, there is plentiful evidence for conflict and competition between different strata of hierarchically organized societies. Life is tough, but perhaps even more so at the lower levels. Higher status in the hierarchy usually means better access to resources and a better life. Signals of high status, such as big antlers in deer, are on display in many social species. They are important to recognize when choosing a mate and measuring up rivals. They advertise fitness, and their benefit is that they help to reduce aggression. We may see a majestic stag in our mind's eye, but fish too can infer social rank by observation alone and decide whom not to attack (Grosenick, Clement, and Fernald, 2007).

Take the African cichlid fish, which thrive in Lake Victoria, where they live in small groups. Dominant males hold territory through chasing competitors and displaying their prowess toward the females and subordinate males. Their status signal is color. Males can either be yellow or blue, and the dominant males are yellow. They are more aggressive than the blue males and also have lower cortisol levels. Blue males are lurking at the edges, seemingly hoping to become dominant one day. When this day arrives, they change color, rapidly turning from blue to yellow and from meek to aggressive. This means they can increase their chances for reproduction. However, this example also shows that dominance comes at a price. When dominant, these fish suffer increased stress-related oxidative damage, which renders them less healthy, and hence less competitive. Then they turn blue and move back to the periphery. Once recovered, the fish may turn yellow again and rise in dominance. This cycle, with its change in color, is underpinned by the melanocortin system (Dijkstra et al., 2017).

The importance of dominance makes it likely that detectors evolve in the brains of animals that live in hierarchically organized groups, including humans. Countless movies are based on the observation that people have a facility to infer somebody's social status from subtle cues almost instantly. This facility is not surprising, as we need to take care whom we approach, avoid, and seek out as powerful allies. The signals that we use for assigning status are more subtle than the yellow color of the cichlids. Useful signals come from many sources, such as clothes and possessions, tellingly referred to as "status symbols," but, the face is probably the most important source (see chapter 9).

Alex Todorov and his colleagues used the morphing of faces along anatomical dimensions to identify the cues that enable us to spot submissiveness and dominance, as well as other traits such as aggression and competence (e.g., Todorov et al., 2015). These attributes are already recognized as being well above chance by three- to four-year-olds when presented with pairs of faces and asked simple questions such as, "Which of these is very strong?" By the age of five to six years, children do as well as adults, with a consensus of around 80 percent (Cogsdill et al., 2014).[1] Interestingly, with the question "Which of these is very mean?" consensus was between 87 and 95 percent, exceeding that of adults (76 percent).

These studies suggest that stable facial appearance, rather than changeable facial expression, is the basis for making intuitive judgments of dominance and selfishness. We expect some people to be consistently more dominant than others. Oliver Mascaro and Gergely Csibra (2012) studied infants who were fifteen months old and measured how long they looked at animated shapes. They designed scenarios in which these animated shapes competed for the same goal. The infants did indeed expect the shape that won one contest to win other contests as well. They treated dominance as a stable trait.

As children get older, they may learn that cues to dominance are not always valid. As we have seen in chapter 9, just because everyone agrees that certain kinds of faces look more trustworthy than others, it doesn't follow that such people are actually more trustworthy. When we judge others, we need to exert deliberate top-down control to check these first impressions, and if necessary reject them.

The Problem with Top-Down Control

We can readily imagine a primitive form of selfishness that is present in most animals and is largely driven by an automatic cost-benefit analysis. This analysis is a kind of gut response that determines whether we cooperate with others or instead do what is flippantly known as "looking out for Number One."

1. They did equally well for trustworthiness (nice) and competence (smart).

As mentioned in chapter 6, there is a widely held belief that our basic urges are selfish and that, to overcome such urges, we must exert self-control. But this belief is not wholly correct. For many of us, the first impulse is to cooperate. On the other hand, it is correct that through slow and deliberate reflection on what we are doing, we can overturn our impulses. But this top-down control too can go either way. Depending on the context, even the most selfish person might choose to cooperate. Even the nicest person might occasionally choose to be selfish.

Higher human reason can also be used to manipulate others into overcoming their basic altruism and empathy for fellow creatures. This is the strategy employed by leaders who want to demonize some outgroup for their own advantage. They provide compelling justifications for callous, antisocial behavior, and at the same time, they raise the specter of ostracism if this behavior is not adopted. This is perhaps the darkest part of the Machiavelli thread.

The idea of individual differences on a spectrum of altruism implies that each of us has a threshold above which an unequal distribution of resources starts to feel unfair (Roberts, Teoh, and Hutcherson, 2019). This threshold is lower for prosocials than it is for individualists and competitives. But it can be altered by top-down processes. Slogans such as "Greed is good" or "To each according to his needs" can cause us to adjust our fairness thresholds, up as well as down.

Our ability to reflect on the causes of our behavior empowers us to alter our behavior. The same processes can be used to think about the behavior of others, powered by our mentalizing ability (see chapter 10). We noted that if we can establish the causes of other people's behavior, then we can predict what they are going to do next. But from there, it is only a small step to ruining their plans to overtake you in the race to the top.

The Dark Arts of Mentalizing

Being good at mind reading tends to be seen as a precious accomplishment of human beings. Being able to predict what their partners are going to do by tracking intentions and beliefs might be a source of altruism and general prosocial motivation for people. It should be a great boon for people wanting to cooperate. But is it? In fact, mentalizing can be a distinct *disadvantage* in cooperative situations. On the other hand, it gives humans a huge advantage in competitive situations.[2]

Marie Devaine, Guillaume Hollard, and Jean Daunizeau (2014b) explored these seemingly paradoxical effects by setting up an army of artificial agents to play against

2. This is where the Machiavelli thread lives up to being named after the historic plotter of intrigues and evil plans.

each other in two types of games—a cooperative (or coordination) game and a competitive game. The cooperative game was modeled on the battle of the sexes, discussed in chapter 6 (A and B want to be together, but they prefer different activities). The competitive game was modeled on hide-and-seek (A doesn't want B to find him; B wants to succeed in the search). Both games involve recursion (Crawford, 2013). For example, factoring in what the other player prefers, is level 1 recursion, factoring in what one player thinks the other player prefers, is level 2, and doing just what you prefer, would be level 0.

As set out in figure 11.4, at the beginning of the simulation, there were five equal groups of agents whose level of sophistication varied from recursion level $k=0$ to $k=4$. There is a cost associated with high levels of recursion since they are cognitively demanding. To be successful, you need to achieve the best outcome with the least effort. This was a tournament played many times, and over time, the unsuccessful agents were eliminated. Which agents prevailed as the winners?

With the *cooperative* game, the tournament ended amicably. It stabilized with a mixture of one-third of the agents operating at the $k=1$ level and two-thirds at the $k=2$ level. However, there was no advantage for the more sophisticated (mentalizing) agents, $k=3$ and $k=4$. They were eliminated. Only if one of the cooperating agents uses a lower level of sophistication will coordination be achieved: I think you prefer football

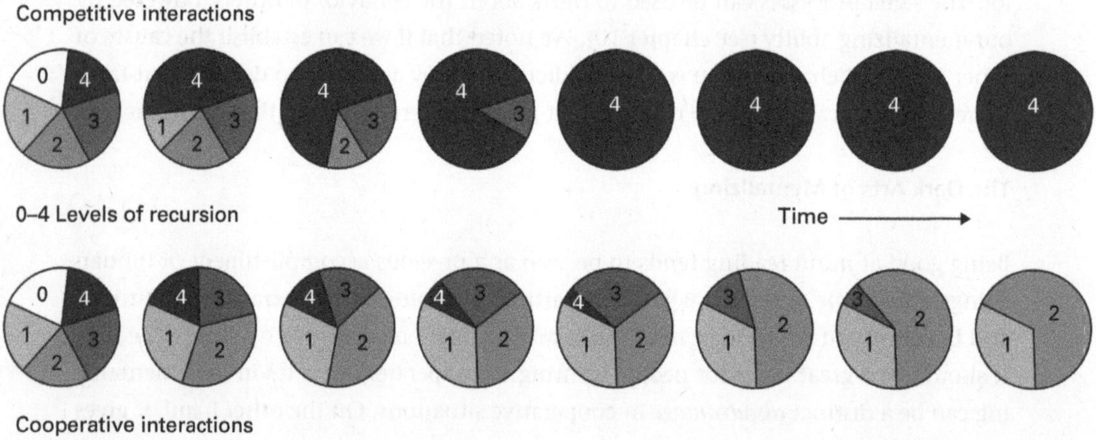

Figure 11.4
Evolution of mentalizing in competitive and cooperative interactions. For competitive interactions (e.g., hide-and-seek), the less sophisticated agents were eliminated. For cooperative interactions (e.g., battle of the sexes), mixed intermediate levels prevailed (Devaine, Hollard, and Daunizeau, 2014b).

($k=1$), and you think I think you prefer football ($k=2$): in this case, we will both go to the football stadium and can be together.

In contrast, with the *competitive* hide-and-seek game, all the agents using lower levels of sophistication were eliminated. Only those with $k=4$ capacities prevailed. In other words, the more sophisticated player wins. We conclude that in a competitive world, recursive thinking pays. And recursive thinking is inherent in our mentalizing capacity. This means that your mentalizing should be ramped up when you compete with an opponent who is also capable of mentalizing. But remember that higher levels of recursion are costly as they require hard work.

The tournament among artificial agents is only a feeble version of what happens when human beings compete with each other. In the world of ideas, perhaps the highest level of sophistication is reached when competition is disguised. The person who is going to lose the competition doesn't even notice that it is happening. We could list plenty of examples of such behavior: A reviewer can damn our book with faint praise.[3] A competitive friend tells you, "My garden is a wasteland" or "I haven't practiced the piano for months," and you can bet that this is just meant to lull you into a false sense of superiority. These ploys are sure signs of a competitive strategy that tries to take the wind out of the sails of escalation in—make no mistake—a fight for supremacy in the spirit of Machiavelli.

The games played by the artificial agents give weight to our claim that strategies that aid cooperation, such as copying, coordination, alignment, and affiliation, are independent of mentalizing and do not involve recursive thinking. They also throw doubt on the idea that mentalizing would make them work better. In fact, it could make them work less well through overthinking.

The competitive arms race that already flourishes in the world of agents has ramped up its weaponry in the world of ideas, which is why we are inclined to classify mentalizing among the dark arts. You can't get darker than Satan, and we can go back to the Garden of Eden to learn of the affinity between Satan and mentalizing. A sophisticated agent, the serpent,[4] tempts Eve to taste the fruit of knowledge. It does this by attributing to Eve the desire for knowledge and by inserting doubt into her mind about the prohibition to eat the fruit. The serpent promises that "you will be like God, knowing good and evil." Satan's temptation is intimately linked to the mental world, and with morality in particular.

3. We would notice this actually, not being strangers to reviewing.

4. In the King James version of the Bible, the serpent was described as "more subtle" than the other creatures, while in the New English Bible, the word "crafty" is used.

Deception—the Trojan Horse and the Invisibility Cloak

Putting aside microorganisms that cause disease, our only major competitors are other humans. We are constantly vying with each other for physical resources, such as land and precious objects, and psychological resources, such as reputation and dominance. Physical force is an effective means to gain supremacy, but deception often proves superior. Cunning and disguise are mightier than fists and swords.

The classic false belief task (see chapter 10) has long been considered the acid test of mentalizing. It involves deception simply because an action is hidden from sight: Sally is not there when Anne removes the marble from her basket. However, the interpretation of deceptive behavior is far from straightforward. Did Anne deliberately remove the marble so Sally would not find it? Behavior is always ambiguous and can have very different causes. If we accuse people of lying, we need to prove that they have the intention to deceive. In turn, they may try to convince us that their intentions were pure. There are many nuances in the realm of deception. It is no wonder that claims of cheating, lying, fraud, and duplicity are keeping lawyers busy.

Deceptive behavior can be observed in many animal species, but it usually does not imply an intentional form of action and does not involve mentalizing. An example is fixed camouflage, such as the pattern on the wings of a butterfly that suggests huge eyes and deters predators. An example of even more sophisticated deception without intention is seen in female fireflies of the genus *Photuris*. They hunt other species of firefly. When they detect the male flash of their prey, they respond with the sexual flash of the prey's female (which is different from their own). When the male approaches, the mimic pounces and devours him (Lloyd, 1986).

Even if this is an example of deception as "a false communication that tends to benefit the communicator" (Bond and Robinson, 1988, 295), a fixed action pattern has evolved. It is not a case of deception as "an act that is intended to foster in another person a belief or understanding which the deceiver considers false" (Zuckerman, DePaulo, and Rosenthal, 1981, 3). Even if we look to nonhuman primates, there is no convincing evidence that they can understand and manipulate others' beliefs to achieve advantages through deception (Hall and Brosnan, 2017). There is little doubt, then, that deliberate deception is a weapon that is unique to humans—always exempting the cunning fox of ancient folklore.

Plain Lies and Hidden Manipulation

The kind of deception where verbal communication is used to generate false beliefs is typically called "lying." It is arguably one of the most powerful tools in human competition, and strategic lying seems to emerge at the same time as explicit mentalizing (Hsu and Cheung, 2013). It is a honed skill, but it needs to be practiced in the dark. In the open, it is very much discouraged. Received wisdom holds that "a good man does not lie" (Fried, 1978, 54). The implication is that lying does harm.[5] This firmly links the study of lying with the study of morality. However, even people who strongly value and care about morality will sometimes take advantage of the opportunity to deceive (Gino, 2015).

Lies work only because we expect that most communications are true. This may be the case in the world of agents, where we generate our predictions by observing the behavior and largely involuntary cues emitted by our competitors. However, in the world of ideas, we are interacting with agents that send signals deliberately and consciously. Here, we are on much weaker grounds if we expect the truth. How did it come to this?

As Sissela Bok (1978) points out, we all benefit enormously by living in a world in which the practice of truth-telling is widespread. According to the linguist philosopher H. P. Grice (1989), as interpreted by Dan Sperber and Deirdre Wilson (1995), there are tacit rules that govern how we use language. One of these is that an utterance should be truthful. This simplifies communication. We don't need to engage tortuous thought processes with every utterance if we can implicitly assume that it is truthful. All the important things that we want to do in life are made possible by pervasive, if preliminary, trust. It is in this context that people believe lies.

The liar capitalizes on the fact that most people tell the truth, in the same way that the economic free rider capitalizes on the fact that most people subscribe to the common good. This is where deception serves as a Trojan horse. Indeed, lies can be used to prevent the detection of free riding. Luke McNally and Andrew Jackson (2013) call this "tactical deception" and use a game theory framework to show that such behavior frequently emerges in the context of widespread cooperation. Tactical deception makes it more difficult for conditional cooperators to detect cheats. This allows tactical deceivers to elicit cooperation at lower cost while simple cheats are recognized and discriminated against.

There are some softer options, such as persuasion, that serve to manipulate others. We tend to believe that certain forms of advertising manipulate our desires without us becoming aware of it. This can prompt us to buy products that we did not know we wanted. One-upmanship is often used to appeal to our competitive nature. When we become aware of the manipulation, such as fake positive ratings of sellers or service providers, we react with outrage.

Fake news is another example of manipulation of beliefs. It takes the form of an accusation of deception in the face of facts, if these facts are damaging to one's own side.

5. There are circumstances in which lying is harmless or even good, as in the case of white lies.

This is a form of lying that assumes a virtuous stance and does not engender feelings of guilt. To complicate matters, people may well sincerely believe in the truth of the misinformation that they spread. Robert Trivers (2000) suggested that self-deception is a useful mechanism to achieve better deception of others.

Countermeasures—How to Detect Deception

As convincing as the wiles of deception are, there are means of protection. With experience, we learn that some utterances may be untruthful and we must be on our guard, especially if we are interacting with a competitor. We learn that there are free riders who disguise their behavior. In response to the appearance of these tactical deceivers, there will be pressure to evolve the ability to detect liars.

It is supremely important for us to know whether someone is trustworthy or not (see chapter 9), but it is very difficult to make this judgment. We get little information from what people look like and how they behave. Classic stories and movies, like *Star Wars,* in contrast, give us plenty of such information so we can quickly side with the heroes and against the villains. But these expectations are subverted in many detective novels and films, where the most satisfying plot leaves the villain hiding in plain sight.[6]

As we saw in chapter 6, intentional behavior is quite difficult to detect. Thus it is not very surprising to learn that in general, people are not very good at detecting deception. A meta-analysis of studies involving several thousand people found that people achieved an average of 54 percent correct judgments of lies and truths, only just above chance (Bond and DePaulo, 2006). One reason for this difficulty is our automatic tendency to copy the actions of others. This happens in face-to-face conversation, and it makes social interactions smoother (Chartrand and Bargh, 1999), but it also makes it more difficult to detect if our partner is being devious and deceitful.

Marielle Stel, Eric van Dijk, E., and Einav Olivier (2009) found that if people were instructed *not* to mimic, they were better at detecting when their partner was lying. Perhaps this makes it less surprising that when we interact with someone from an outgroup, we automatically suppress mimicry. We are suspicious of them by default. The same happens when we know that our partner has different goals from us (Kavanagh et al., 2011). In both situations, we have a prior expectation that deception might occur. If we expect that we may be deceived, we cease to take forthcoming information on trust.[7]

---

6. Another alternative involves using casting against type to surprise us. For instance, in *Once Upon a Time in the West,* the psychopathic villain emerges from the mist to reveal Henry Fonda, the classic good-guy actor.
7. We adopt a stance of epistemic vigilance (Sperber et al., 2010); see chapter 9.

Suspicion evolved alongside the emergence of tactical deception. This means that we need to spend some cognitive effort in every novel situation, as we calibrate our degree of suspicion based on what we know about the people we are interacting with. We must evaluate their arguments, and we must gauge the plausibility of their claims. Han Solo is a supreme example of someone who is always in a state of near-paranoid suspicion, and thus immune to evil agents who try to convince him of their good intentions.

These sophisticated cognitive abilities only emerge gradually during development. Three-year-olds show a remarkable degree of trust and will accept the testimony of others even when it directly conflicts with their own experience (Jaswal et al., 2010). They are particularly affected by verbal statements, especially when they can see the person making the statement. Children find it hard to treat assertions as false before the age of five. This difficulty does not seem to result from any problem with their understanding of the concept of falsity, but rather from their being too trusting (Mascaro and Morin, 2015). This does not apply to children who are brought up in toxic and hostile environments, who may well come to assume that adults are not to be trusted, ever (Kidd, Palmeri, and Aslin, 2013).

Before the age of four years, children start to suspect claims made by informants who are uncertain (e.g., Sabbagh and Baldwin, 2001), who lack relevant episodic knowledge (e.g., Robinson, Champion, and Mitchell, 1999), who have made inaccurate claims in the past (e.g., Koenig and Harris, 2005), or who are malevolent (Mascaro and Sperber, 2009). At around four, children begin to understand what it means when they are told that an informant is a liar. At age five or six, children begin to recognize the role of intentions and motivations in lying (Mascaro and Sperber, 2009; Mills, 2013). This enables them to recognize the kinds of social situations in which lying is likely to occur. There are interesting parallels here between this developmental progression for the detection of deception and the emergence of explicit mentalizing, which marks our entrance into the world of ideas. As we have seen in chapter 10, it is not before the age of four to six that children are able to verbally reason about false beliefs (Wellman, Cross, and Watson, 2001).

Morality for Us but Not for Them

So far, we have been discussing the potential victims of liars, but what about the liars themselves? Given all the advantages of cooperation and honesty, why is there so much "ordinary unethical behaviour" (Gino, 2015)? A laboratory experiment used a Big Robber game, where participants could obtain money by taking the earnings of groups of sixteen victims at once, without fear of punishment (Alós-Ferrer, García-Segarra, and Ritschel, 2021). It turned out that almost nobody declined to rob these large groups, and more than half the participants took as much as possible. Yet the same participants behaved prosocially in standard trust games where only one other person was involved. Clearly, human beings can be both prosocial, especially with small groups, and antisocial, especially with large groups.

Michael Tomasello and Amrisha Vaish (2013) reviewed work on the evolutionary and ontogenetic origins of human cooperation and morality. They proposed a first step where sympathy is shown to particular others, especially close kin, and this is seen in great apes, as well as in human infants during the first year of life. The second step leads to an agent-neutral morality, in which ingroup-wide social norms are followed and enforced. This applies only to humans, but a full understanding of fairness and the role of merit is not expected until four to six years of age. We might conclude then, that by the second year of life, the basic foundations for prosocial behavior are in place. Where does it all go wrong?

We addressed a possible reason in chapter 8. Between the ages of four and eight years, children indeed become more egalitarian. They prefer to share resources (e.g., sweets) equally between themselves and another child. However, this applies only when the other child is a member of the ingroup. When the other child is an outgroup member (e.g., from a different school), the tendency to share decreases with age (see figure 8.3 in chapter 8). The development of good behavior in relation to the ingroup goes hand-in-hand with the development of bad behavior toward the outgroup (Fehr, Bernhard, and Rockenbach, 2008). This also provides an explanation for the predominant stealing behavior in the Big Robber game. In the case of large groups, it is easier to designate others as an outgroup, while this is difficult with only one other player.

While the basic foundations for prosocial behavior may already be in place quite early in life, the basic foundations for the distinction between us and them is there even earlier. Even infants as young as nine months old like people who treat similar others well and mistreat dissimilar others (Hamlin et al., 2013). When we interact with people who are not like us, we can readily go over to the dark side.

The competitive nature of our relationship with outgroups is revealed by the pleasure that we take in seeing members of an outgroup come to harm (intergroup Schadenfreude). Baseball fans feel pleasure when their rival team fails to score, even when these rivals are not playing against their own team (Cikara, Botvinick, and Fiske, 2011). Soccer fans are less willing to help a supporter of a rival team (Hein et al., 2010; see Cikara, 2015, for a review). Still, we justify our nasty treatment of the outgroup by pointing out that its members habitually behave immorally, so attacking is the best defense. After all, the default assumption is that outgroups will compete with us and hence try to deceive us.

A rather disconcerting observation about the dark side was highlighted by Nicola Raihani and Eleanor Power (2021) when reviewing work on helping behavior and generosity. They identified the existence of a surprising tendency to punish altruistic others and to disparage good deeds. What can be the reason for this seemingly perverse behavior? There are in fact many reasons, such as suspicion of ulterior motives, dislike of benefiting a hated outgroup, and not wanting to be seen as mean by those who give generously. As Raihani and Power point out, it is not surprising that prosocial individuals frequently prefer to remain anonymous. They trade the protection of anonymity against the benefit of gaining an increase in reputation (see chapter 9).

There Are Many Shades between Light and Dark

Should we abandon the belief that there is goodness in humans? Of course not. Nicola Raihani (2021) gives many eloquent examples in her book on the evolution of cooperative behavior. While violent aggression toward outgroups is a common feature of most animals, from ants (Batchelor and Briffa, 2011) to chimpanzees (Mitani, Watts, and Amsler, 2010), humans have more flexibility in their social behavior. They are more able to exert top-down inhibitory control on violent impulses toward potentially dangerous strangers. Anne Pisor and Martin Surbeck (2019) suggested that it is the unusual degree of tolerance toward individuals from outgroups, observed particularly in times of plenty, that has given humans an evolutionary advantage over other primates. This flexibility may well have facilitated the creation of hugely important cultural norms, such as "Thou should not kill."

Today, we are much more aware of the suffering of distant strangers, and we care when we get news reports of disasters or wars. Teams of helpers are sent to places where an earthquake has struck. Vaccines are sent to places to prevent epidemic illnesses. According to Steven Pinker (2011), there has, for a long time, been a steady decrease in violence in human societies. These changes are sometimes linked to the Enlightenment and the idea that our base urges can and should be overcome by higher reason. But as noted earlier, our base urges are determined just as much by altruism as by competition. While we are sometimes selfish, we are also motivated to help others, to avoid harming others, and, more generally, to increase the happiness of our group. At the same time, although we can exert top-down control to inhibit bad behavior, top-down control through reason does not always lead to the greater good.

Reason can be used to develop clever arguments that justify selfish behavior. One example is the idea that the poor benefit from tax cuts to the rich. Another, the idea that high-income earners, via their donations, can do more good than charity workers

(MacAskill, 2015). We have also become very good at inventing new outgroups of people who deserve to suffer (football fans of the rival team, or hedge fund managers). The dark side has not gone away in the world of ideas. It has just become more sophisticated.

As we keep arguing, individual variation is critically important for the success of the group. And this applies even to people whom we fear have been won over by the dark side, such as those with a high degree of selfishness, aggression, and lack of empathy. In times of intergroup conflict, members who lack aggression (i.e., doves) can be a disadvantage for the group, and aggressive members can be an advantage. For example, they may administer sanctions on wrongdoers that others shrink from.

Just as there can be problems for the group if there are too many doves, there can be problems if there are too many people with high degrees of empathy. For example, having to cut into the body of another human will be hard for someone who is easily distressed by the pain of others, even if doing so actually does good. So it is not surprising that surgeons are reported to have lower empathy scores than other medical specialists, especially compared to psychiatrists, who tend to score high on measures of empathy (Hojat et al., 2002).

III Computation—A Hierarchical System of Prediction and Action

12 Modeling the Social World: The Computational Approach

Describing mechanisms of social cognition is dangerously close to hand-waving. Therefore, this chapter attempts to look under the hood of how we learn and process information at the various levels of the information-processing hierarchy. We consider processes that are important for the worlds of objects and agents, as well as for the world of ideas.

At the most basic level of the hierarchy, research has long identified several kinds of learning, which apply to both social and nonsocial events. *Model-free learning* is pervasive throughout the animal kingdom, including humans. This enables us to learn about the world directly, as well as from observing other agents. It is readily accommodated within the framework of predictive processing.

Model-based learning enables us to estimate causes, including causes of behavior. We humans tacitly assume that agents do things for a reason. As well as improving the prediction of the behavior of agents, a model of the world frees agents from being enslaved by their present environment. It gives rise to the primary form of consciousness referred to as "sentience." It also enables counterfactual thinking (e.g., what would the outcome have been if we had chosen another option). This is critical for determining the causes of agents' behavior.

A further step up in the computational hierarchy, and something possibly unique to humans, leads to *metacognition*. This gives rise to self-awareness and goes well beyond simply exploring the causes of actions. Metacognition enables us to think about the models that we ourselves and other agents are using to identify causes and discover the hidden states of the world—not just states of the physical world but the mental world. The three systems are intimately linked in a single complex unit, and a Bayesian hierarchical framework provides a productive starting point for discussing the lower- and the higher-order computations and the interactions between them. The interaction between the individual and culture is found at the top of this hierarchy, closing the loop between the brain and the social environment. This will be a major theme of the following chapters.

* * *

J. B. S. Haldane (1964) wrote that "a mathematical theory may be regarded as a kind of scaffolding within which a reasonably secure theory expressible in words may be built up. . . . [W]ithout such a scaffolding, verbal arguments are insecure" (250). Clifford Truesdell (1966) made the same point more forcefully. "There is nothing that can be said by mathematical symbols and relations which cannot also be said by words. The converse, however, is false. Much that can be and is said by words cannot be put into equations—because it is nonsense" (35).

In this chapter, we will describe some of the mathematical theories underlying the words that we use when talking about social cognition, taking a look at the chapters so far, as well as the ones to come. (Do not fear—the few equations that we present will be in the footnotes.)

In this book, we have structured this book in terms of three worlds: the world of objects, the world of agents, and the world of ideas, and this is how this chapter is structured. So far, we have delved mostly into the world of objects and agents. Objects are physical things that exist in the world: rocks, tree, animals, humans, and so on. Agents are a special kind of object that can act on the world. These include humans and other animals, as well as robots. Ideas, in contrast, are part of our mental world. These three worlds differ in complexity, with the world of objects the least complex and the world of ideas the most complex. Nervous systems have evolved to cope with this increasing complexity, but the simpler systems are not thrown away (Cisek, 2019).

We propose that in the human brain, there is a hierarchy of systems with three levels, each associated with one of the three worlds. We will outline the computations that occur at different levels of this hierarchy, as far as they have been worked out, paying particular attention to their roles in social cognition. There are, of course, constant interactions between the levels to constitute the Umwelt[1] of an organism. They form a series of loops, with the highest being where individuals interact with the outside world of ideas: the culture in which they are embedded.

When we first studied psychology, there were no computers to be found in psychology departments. There were no word processors. Our theses were written on typewriters. There was no internet. We went to the library to read the latest journals. We collected reprints of papers and used postcards to request reprints from our colleagues. Now, we take computers and the internet for granted, and it is easy to forget the extraordinary changes brought about by these developments.

It was the advent of computers that ushered in the discipline of cognitive neuroscience and dramatically increased our ability to model behavior in mathematical terms. In parallel, there was the equally remarkable development of artificial intelligence (AI). Further, ever since the classic paper by Warren McCulloch and Walter Pitts (1943), there has been a constant exchange of ideas between brain science and AI (e.g., Hassabis et al., 2017).

1. According to the visionary biologist Jakob von Uexküll (1864–1944), who coined the term, the Umwelt is what a particular organism can perceive and act upon. Perception and action are connected by feedback loops, anticipating cybernetics.

When AI was first being developed, it was anticipated that some tasks, like perception, would be easy to solve, while others, like playing chess, would be difficult. This was because this is how it seems to us. We perceive the world effortlessly. Playing chess, on the other hand, requires much effortful thinking and years of practice. Now we know that we had it all wrong. Computers can play chess, and even Go, a far more complex game, better than the best human players. But computers are still not so good at abilities we take for granted, like seeing the world about us and moving about in it effortlessly.

The mistake was to equate difficulty and complexity based on our subjective experience of mental effort. In fact, many (if not most) of our abilities depend on complex processes carried out in that vast realm of the cognitive unconscious (Kihlstrom, 1987). We are entirely unaware of them. For example, the processes by which we understand speech are very complex, but we rarely think about them, and even if we do, we cannot penetrate them. A related mistake concerns the nature of learning. We tend to equate learning with what happens at school. There, we learn from a teacher who gives us rules and tells us when we are right or wrong. But much learning occurs without any teacher (Barlow, 1989). It is this unsupervised—indeed mindless—learning that occurs at the bottom of the brain's hierarchy. Even the modest nervous system of a lowly worm manages this kind of learning from experience and thus can adapt to its Umwelt.

The World of Objects

Learning by Mere Exposure

Even the world of objects, which we have situated at the bottom level of our hierarchy, is very complicated and involves huge amounts of information. Most of the information bombarding us is of no use to us and is ignored. Much of this filtering has already been done by evolution. Our eyes and ears can detect only a very limited range of light and sound waves. In other animals, the filters have been set slightly differently: bees can see ultraviolet light, and dogs can detect very-high-frequency sounds. But not all this filtering is fixed by evolution. From birth, and even before, filters are adjusted and new filters developed. Some of these new filters become stabilized; others are continually adjusted.

Our Umwelt is repetitive and structured, and brains are constructed to detect and take advantage of this structure. This depends on learning by mere exposure. We acquire a feel for how frequent, how probable, or how surprising an experience is. Such learning occurs throughout our life span and in every aspect of our experience.

Learning through mere exposure is often called "statistical learning" (Aslin, 2017). It is a great way to extract and condense useful information from our Umwelt while still retaining the signals that matter to us. By taking account of the most relevant

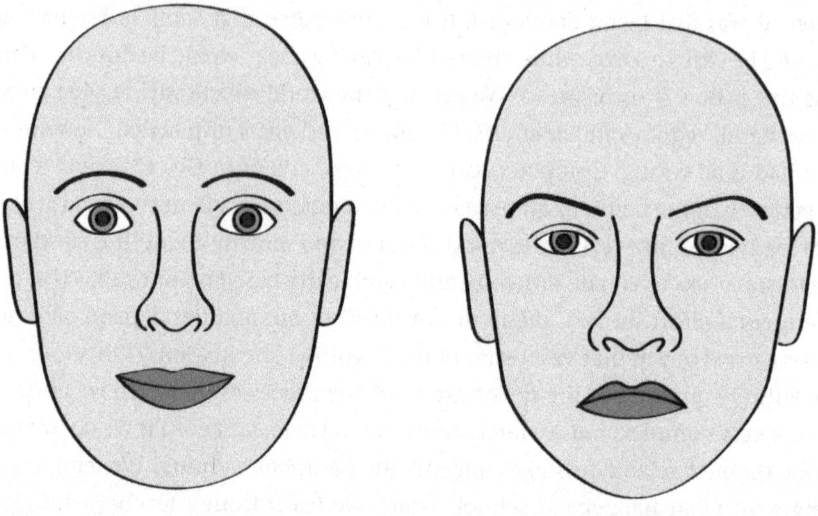

Figure 12.1
A trustworthy face (left) and an untrustworthy face (right).

dimensions of the Umwelt, we are able to form much more efficient representations and also discover hidden relationships. This is what statistics is all about: for example, representing lots of data by a mean and a variance. This condensed information allows us to represent a relationship by a correlation, and indeed, to represent a whole set of correlations by just a few principal dimensions. The human face is an example of a very complex object. It varies along about fifty shape dimensions (height of eyebrows, width of chin, and so on). But 80 percent of this variation can be explained by just two abstract perceptual dimensions: trustworthiness and dominance (Oosterhof and Todorov, 2008) (see figure 12.1).[2]

The importance of mere exposure is obvious in the learning of language. The basic units of spoken language are the phonemes from which words are constructed. Languages differ widely in the number and nature of their phonemes. At birth, children have the ability to perceive all possible phonemes, but after a few months, mere exposure to one particular language reduces the ability to perceive phonemes that are not in that language (Werker et al., 1981). The filter becomes stabilized. The advantage of this stabilized filter is that babies now perceive speech in terms of a relatively small number of phoneme categories. The disadvantage is that they will no longer be able to hear some

2. However, as we saw in chapter 9, a person with a face that looks untrustworthy need not actually be untrustworthy.

phonemes in other languages. For example, /l/ and /r/ form two categories in English, but only one in Japanese. As a result, Japanese speakers find this distinction difficult to hear.

Learning by mere exposure has a pervasive effect on cognition in many domains and modalities (Sherman, Graves, and Turk-Browne, 2020). For example, our perception of faces is continually adjusted on the basis of the types of faces to which we are currently being exposed (Webster et al., 2004; also see chapter 8). As a result, we find it difficult to distinguish between faces from groups to which we have little exposure.

Learning by mere exposure applies not only to the way we perceive the world, but also to the way we act in the world. Actions that are constantly repeated, become habits. The extent to which an action becomes a habit is simply determined by how often in the past that action has been taken in that context. If the action occurs, the habit strength for that action increases. If the action does not occur, the habit strength decreases.[3] The change in habit strength is not affected by the outcome of the action (Miller, Shenhav, and Ludvig, 2019). Through this process, even complex sequences can become habits, such as tying your shoelaces or playing a scale on the piano.

Habits are advantageous because they enable us to perform actions quickly and without having to think about them. But they can also be disadvantageous. If you have been brought up in the United Kingdom, you automatically first look to your right when you start to cross a road. This behavior can be quite dangerous when you visit other countries where people drive on the other side of the road.

At the end of this chapter, we will describe some of the ways in which we can inhibit such habitual behavior, but it is not easy, as it needs top-down control. When we had to practice social distancing during the pandemic, we found it tough to inhibit the habit of shaking hands, and of embracing our loved ones.

Learning by mere exposure, whether this relates to perception or action, is of great importance for social cognition, but this is because most of our experience is in the social domain. There is nothing specifically social about this process. Learning by mere exposure is a consequence of Hebbian processes in the brain—"neurons that fire together wire together" (Hebb, 1949; Oja, 1982).

Association Learning

Learning by mere exposure changes the way we perceive the world and the way we act upon the world. This makes it easier for us to interact with the world. But we would gain even more advantages if we could predict what was going to happen next. A

3. $H_{t+1} = H_t + \alpha_H(a_t - H_t)$. H is the habit strength. If the action occurs, $a_t = 1$. If it does not occur, $a_t = 0$; and α_H determines the rate at which the habit strength changes.

delicious smell suggests that food will soon arrive. If we press the button, we expect the elevator to come. Such predictions need not be made in any conscious or deliberate way. Instead, they are made possible by mindless association learning. Association learning is a computational workhorse exploited by all animals and applied in every domain whether or not there is a social component.

This kind of learning comes in different forms. "Classical conditioning" or "Pavlovian conditioning" is the term used to describe the process by which reflexive, autonomic responses come to be associated with certain stimuli (known as "conditioned stimuli"). In his pioneering work in this field, Pavlov (1927) observed that his dogs would begin to drool as soon as they saw the food and before they had begun to eat. Then, just before he let the dogs see the food, Pavlov sounded a buzzer, which had no prior association with food. Soon the dogs were drooling as soon as they heard the buzzer.[4] The same phenomena can be seen with responses to threat, if a neutral stimulus, such as a red square, is repeatedly presented before the administration of a mild electric shock, then animals, including humans, will show a threat response when they see the red square (Maren, 2001).

Another kind of learning, known as "operant conditioning" or "instrumental conditioning," involves an animal or human performing an action that, if it produces a positive outcome or reward, will probably be repeated in the future, whereas if the outcome is disadvantageous (i.e., punishment), it will probably not be repeated. For example, pigeons will learn to peck at a particular disk in order to get food (Skinner, 1938).

Prediction and Prediction Errors

But what do we mean by learning? We used to think about learning in terms of rules. For example, an animal might learn that the food in a maze was at the end of the left arm rather than the right. We imagined it had to learn the rule: *Go left*. We also tended to think that learning was about the past, a record of what had happened. Now we see learning as much more about the future. I am using my past experience to predict what is likely to happen next. And most of the time, I am making this decision in the face of uncertainty. There is no fixed rule. The food may usually be on the left, but not always. In association learning, we are predicting the likely outcome of our action. For example, will our action lead to food? An action that leads to food is of more value to us than an action that does not.

We now know a great deal about the neural basis of this kind of prediction. Activity increases in dopamine neurons in the midbrain (see figure 12.2) when the outcome of

4. Pavlov used a buzzer instead of a bell because it was a stimulus that can be more easily controlled, but the bell has taken hold as part of the legend.

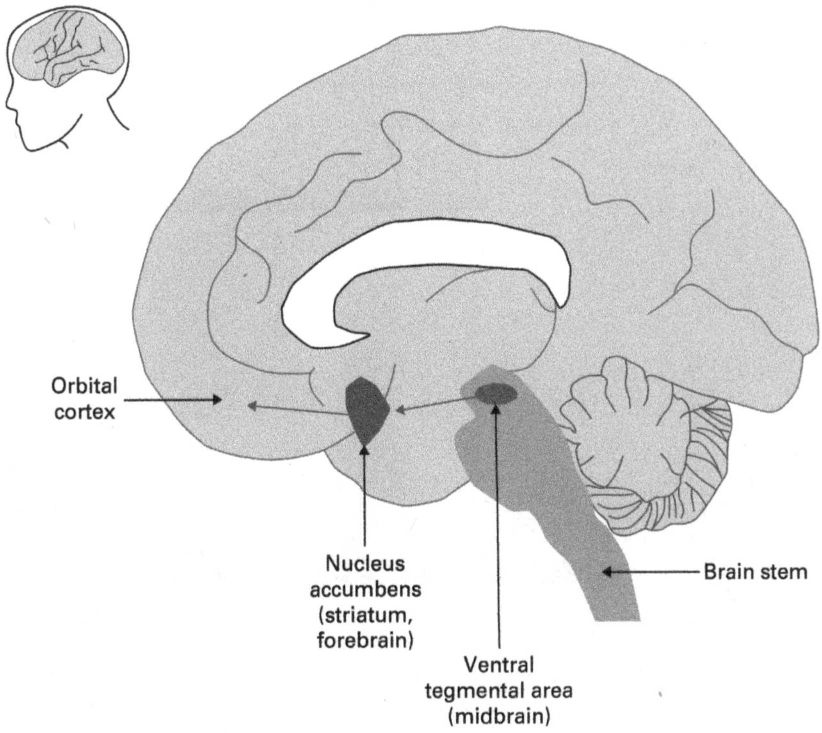

Figure 12.2
The reward pathway in the brain: This pathway connects dopamine neurons in the ventral tegmental area in the midbrain to the nucleus accumbens in the forebrain. Activity in these subcortical areas reflects reward prediction errors. Activity in orbital cortex relates to the more abstract concept of value.

our action is better than we expected and decreases when the outcome is worse than we expected[5].

This change in dopaminergic activity can be thought of as a reward prediction error (Schultz, Dayan, and Montague, 1997). This error tells us whether the outcome is better (more food) or worse (less food) than we expected (see figure 12.3).

Prediction errors help us to learn. When we choose an action, we are predicting what the outcome is likely to be. If the outcome is better than we expected, then we will be more likely to repeat that action. If it leads to punishment, then we are less likely to

5. Dopamine is only involved in learning via reward. Learning to avoid pain probably involves serotonin, 5-HT.

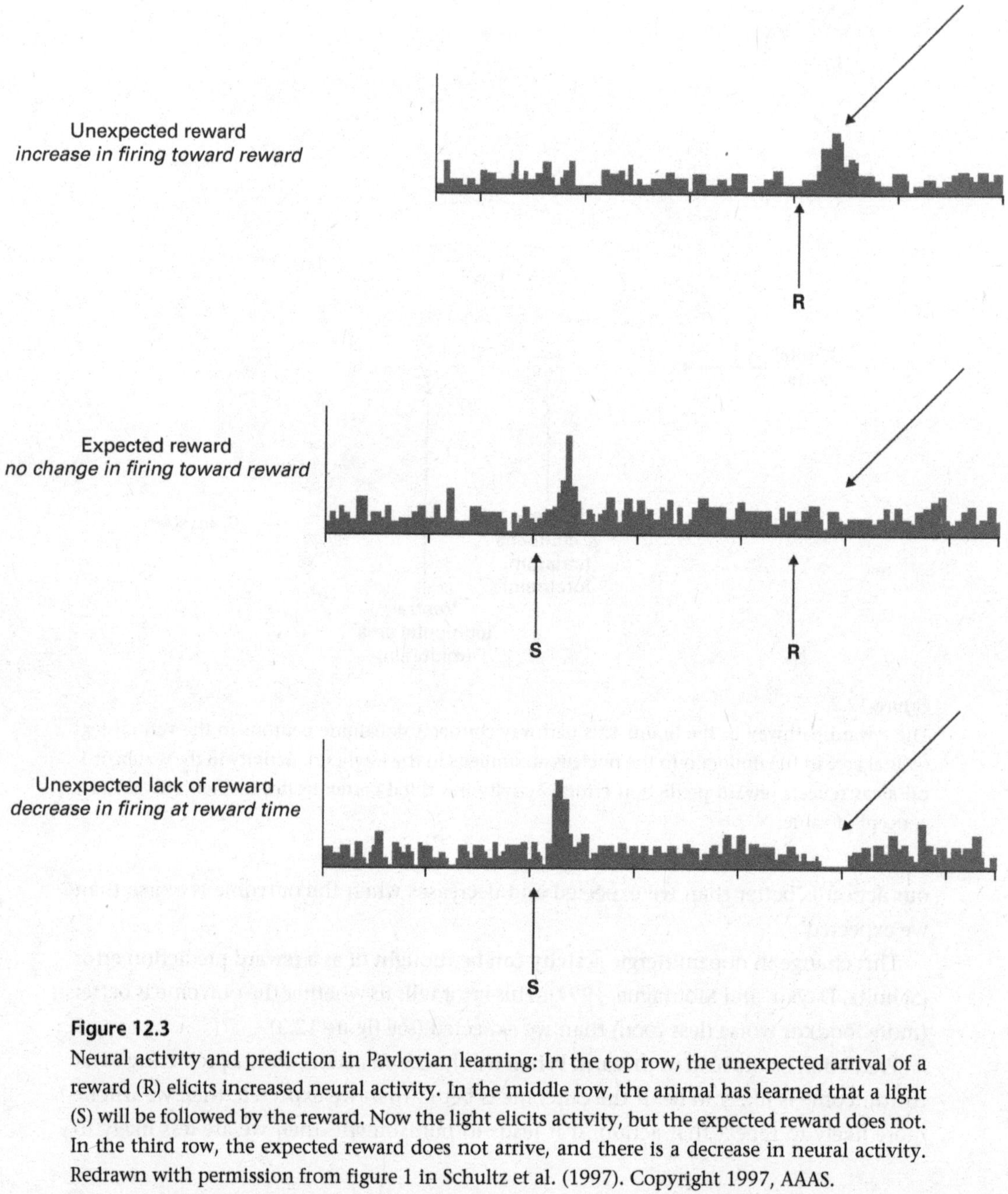

Figure 12.3
Neural activity and prediction in Pavlovian learning: In the top row, the unexpected arrival of a reward (R) elicits increased neural activity. In the middle row, the animal has learned that a light (S) will be followed by the reward. Now the light elicits activity, but the expected reward does not. In the third row, the expected reward does not arrive, and there is a decrease in neural activity. Redrawn with permission from figure 1 in Schultz et al. (1997). Copyright 1997, AAAS.

repeat it. What we are learning here is the *value* of the action; the better the outcome, the higher the value of that action. The process can be captured in a simple equation[6] (Sutton and Barto, 1998), which shows that the value of an action is updated after we discover the outcome of the action. If the outcome was better than we expected, then the value of the action increases, and we are more likely to perform that action again.

Learning the Value of Actions and Objects

Value is a very valuable concept! We can apply it to objects as well as actions and learn which objects, or people, are nice and which are nasty and should be avoided. Value is a continuous variable, so it distinguishes not only between good and bad, but also between good and better and bad and worse. This is "subjective value," the value for us at a particular moment in time. And we can attach such value to any kind of action or object. Hence, value provides a common currency (Levy and Glimcher, 2012). Value allows us to compare apples and oranges or choose between going for a walk or sitting in front of a computer. When I make these choices, I am basing them on predictions. I might predict that if I go for a walk, I will feel better, while if I carry on trying to write, I will feel worse.

We can also learn a lot about the value of actions and objects simply by observing the behavior of others. In chapter 2, we saw that many animals learn about the value of actions by observing the behavior of others (Huber et al., 2009). If we see someone approaching an object, or even just looking at an object (Bayliss and Tipper, 2006), then the value of approaching that object ourselves increases for us. This learning through observation is another example of association learning, and similar physiological mechanisms are involved. When another rat or human is observed being rewarded, the midbrain area (see figure 12.2) is activated in the observer (Kashtelyan et al., 2014; Mobbs et al., 2009). This is the same region that is activated when the self is being rewarded directly.

Explore or Exploit?

So there is a mechanism through which we can keep track of the value of actions and objects. But when we need to act, how do we make our choice? If one action (or object) has a much higher value than all the others, then the decision is easy. We simply choose the action with the highest value. In this case, we are *exploiting* the knowledge that we already have about the world. But what if there is little difference in the values

6. $V_{t+1}=V_t+\alpha(\Delta_t)$. **V** is the value associated with an action, where Δ_t is the prediction error and α is the learning rate. If the prediction error is positive, the value goes up. If it is negative, the value goes down.

of the actions? In this case, we don't have enough knowledge to make a decision. Rather than exploiting the knowledge that we already have, we need to *explore* the world a bit more. Perhaps some of the lower-value actions are better than we realized (see Cohen, McClure, and Yu, 2007 for a review).

Deciding whether to exploit what we know or explore the world further is a problem faced by all animals. For example, it is important for all animals to know where to find food and all the other resources essential for life (Stephens, Brown, and Ydenberg, 2008). Once a source of food has been found, this knowledge can be exploited. Whenever you need food, you now know where to go. You are choosing the action with the highest value. However, this knowledge will never be quite enough. Perhaps there is better food somewhere else. And worse, maybe the food supply that you know about has run out. In these cases, exploiting the knowledge that you have will not help. You need to explore the world to acquire new knowledge. You don't choose the action with the highest value. You may even choose to do something that you haven't done before. You are searching for knowledge, not value (Friston et al., 2015).

A simple mechanism for determining whether to explore or exploit is provided by the softmax function[7] (Sutton and Barto, 1998). This function includes a parameter, τ, labeled *temperature*. If τ is low, then the action with the highest expected value is very likely to be chosen. In this case, we are exploiting the knowledge that we already have. But if we always do this, we become stuck in a rut and never discover anything new. In contrast, if τ is very high, then we will essentially choose our action at random. This is a simple and effective way of discovering something new.

Whether we exploit or explore depends on the situation. When our current knowledge is inadequate, we will need to explore. But some of us have a greater tendency to exploit and some to explore. These individual differences have an important role in social cognition. Group decision-making is enhanced if there is a mixture of explorers and exploiters (see chapter 15). For example, most honeybees exploit known sources of food, but they rely on a small number of scouts who explore to find new sources of food (Seeley, 1983). Similar individual differences are also seen in humans, and the neural basis of these differences seems to be conserved across many species (Liang et al., 2012).

Prediction Errors Are Not Everything—Priors Set the Scene

There is no doubt about the importance of prediction errors for learning the value of actions or objects. These errors are the differences between expected outcomes and the

7. $P_a = \exp(V_a/\tau)/\Sigma_i(\exp(V_i/\tau))$ P_a is the probability of choosing action a, V_a is the value of action a, and τ determines whether to explore or exploit.

outcomes that we actually observe. In Bayesian terms, the expected outcome is the prior belief about what will happen after the action, while the observed outcome is the evidence of what actually happened. The Bayes equation tells us how much we should change our prior belief, given this new evidence.[8]

But how do we know what to expect in the first place? Where does our prior belief come from? According to the Bayesian view, which action we choose will depend on some already existing belief or expectation (known as a "prior"). A simple assumption is that our priors are generated by past experience. Yes, but how far in the past? Some priors might have been based on experience of creatures living millions of years ago. Such priors, and it is not necessarily obvious which, have been built into our brains by evolution: the past experience of our ancestors. As babies, we have built-in priors for some basic social stimuli, such as faces and voices (Johnson et al., 1991). These innate priors are analogous to the factory settings built into our computers (Scholl, 2005). They form part of a starter kit for learning about agents (see chapter 2). Many but not all of these settings can be modified by experience. Innate priors in humans seem especially open to modification. Throughout our lifetime, many of our priors are being constantly adjusted.

But what if there is no relevant past experience? What is our estimate of the value of an object going to be when we see it for the first time? There are some possible solutions to this problem. For example, new objects might be given a starting value of zero (i.e., neither positive nor negative). But learning will occur faster if we can somehow choose a more appropriate prior. Most novel objects and situations will have a resemblance to something that we have encountered before, and the prior for that object or situation can be applied. For example, an unknown face is given the same evaluation as a known face that it resembles (Gawronski and Quinn, 2013).

But we can do better. As we keep emphasizing, we learn about new objects and actions by observing the behavior of others.[9] We assign a high value to objects that we see others approaching or actions that we see others performing. We are taking over their priors. We will also attend to signals from others that are designed to tell us the values of objects and actions. These might be the smile from Mother indicating that an object can be approached (Feinman et al., 1992), or, once we enter the world of ideas, a verbal statement such as "Spinach is good for you" (Lovett, 2005). Through such learning, we

8. The Reverend Thomas Bayes was an obscure, eighteenth-century nonconformist clergyman whose key insight about probability was not published in his lifetime. Since the 1950s, his equation has become increasingly influential in statistics and beyond. This is his equation: $P(A|B) = (P(B|A) * P(A))/P(B)$. This tells us how much to update hypothesis A, given the new evidence B.

9. This can happen even from observing the traces left behind by the behaviors of others.

end up sharing the priors of others. This is one of the undeniably social mechanisms through which culture can emerge.

Let us return to the expression "prior belief." This phrase should not mislead you. When Bayesians use the term "belief," they are not talking about conscious beliefs. They are talking about an estimate of the probability that an outcome will occur.[10] The higher the probability, the stronger the belief. These computations, as well as the beliefs and predictions associated with them, occur at a deep level of brain processing. Conscious awareness is not involved. Both Pavlovian and instrumental conditioning can occur in humans without awareness (Olsson and Phelps, 2004; Pessiglione et al., 2007). Both kinds of learning have been observed in many animals, including sea slugs (Brembs et al., 2002). Conscious reflection and a complex nervous system are not required to apply these processes to learning about basic values of agents and actions. However, by altering priors, culture can have a top-down effect on this most basic kind of learning.

The World of Agents

Guided by the Causes of Behavior: Model-Based Learning

We are moving up in the information-processing hierarchy. Here, we come to a new and different form of learning, which we believe is associated with the eventual emergence of consciousness (Frith, 2002). It is very different from association learning. There, we gradually form a link between a particular action and a particular outcome, but what we have learned is simply a correlation. It does not tell us *why* this action leads to this outcome (Pearl and Mackenzie, 2018). Through this kind of learning, animals (particularly including humans) can acquire superstitious behaviors, in which actions that have no role in causing the outcome are persistently performed (e.g., knocking on wood) (Foster and Kokko, 2009).

Association learning is called "model-free" because it does not involve the model of the world that we need if we are to make inferences about causes. This limitation is particularly apparent when we enter the world of agents. Agents, we assume, do things for a reason: They choose goals and actions to maximize the rewards they expect to obtain (Jara-Ettinger et al., 2016). By observing the behavior of others, we are strongly drawn to ask the question, "Why are they doing that?" Only a hard-nosed behaviorist can resist this question.

Unlike such a behaviorist, we are curious about other people's preferences and knowledge. We can ask them directly, but this is often not possible, and they might refuse to

10. Philosophers should note that this kind of belief is not in the form of a proposition.

tell us anyway. However, by observing their behavior, we can learn about what individuals like and even what they know (Ereira et al., 2018). Such knowledge is very important when we are dealing with other agents, whether we intend to cooperate or compete with them.

To make such inferences about other people, we need *model-based* learning (Decker et al., 2016). We have to have a model—a hypothesis—about the causes of their behavior. It is already a hypothesis that when dealing with agents, their preferences and knowledge are the causes of their behavior. This hypothesis may not always be correct. Perhaps the behavior that we are observing is mindless and random. What we observe is not an agent, but a leaf blowing in the wind (Schultz and Bülthoff, 2013).

A Problem for Learning from Others: Value for Them versus Value for Me

If we see someone avoiding marmite, does that behavior indicate that marmite is nasty? What if some people approach marmite while others avoid it? A model-free learning agent would simply average all these approaches and avoidances and come up with some intermediate answer for the value of marmite: moderate.

A model-based agent can't help asking why, and it automatically latches onto the idea that different people can have different preferences. This idea is so basic that it is already appreciated by eighteen-month-old children (Egyed, Kiraly, and Gergely, 2013). They know that different people have different preferences, but they also know that if a person deliberately attracts their attention using an ostensive gesture (say by calling their name), they are signaling a more general feature of the object and wish to get joint attention. In this case, showing distaste for marmite will lead the child to assume that marmite is nasty.

This example involves a tool that is typical of human teaching and depends on the ability to use ostensive communication. If this tool is not used, but the child merely observes another person avoiding marmite, then the child is not learning that marmite is nasty. The child implicitly recognizes that this behavior is caused by personal preference. This is a person who dislikes marmite. In this case, it does not follow that marmite is of low value for the child. Learning about the preferences of others can be very useful (Robalino and Robson, 2012). It is learning about agents as opposed to learning about objects. It is crucial to the basis of trade, which is all about selling a commodity that I don't want to somebody who does.

Models of the World and Counterfactuals

Having a model of the world opens up many new possibilities. Once a system is in place for this kind of learning, very little more computational power is needed to entertain more than one model at the same time. For example, we can have a model of what the world is like now and what it was like in the past. This is the basis of autobiographical memory. We can have models of what the world looks like to us and what it looks like to someone else from a different viewpoint. This is the basis of our ability to mentalize (chapter 10).

In these examples, the models are based on the real world—what it was like in the past and how it would be seen from a different viewpoint. But here is a new possibility that is opened up by model-based learning: We can also develop models of worlds that don't exist. For example, we can think about the future.[11] We are able to think, *what if?* This is an example of counterfactual thinking.

Counterfactual thinking is extremely useful for elucidating causes (Lewis, 1973; Pearl and Mackenzie, 2018). It is not sufficient to say, Ben always drinks when he is thirsty, and hence thirst causes him to drink. If it is thirst that causes him to drink, then it must also be the case that, if he is not thirsty, he does not drink. And what if he drinks when he is not thirsty? We will need to find another cause.

The advantages of model-based, counterfactual thinking over model-free, association learning can be seen in a very simple paradigm called "reversal learning." An example is the hide-and-seek game, where an opponent either hides behind a tree or behind a wall. At first, the opponent usually hides behind the wall. For a simple, model-free learner, the value of looking behind the wall will become much greater than the value of looking behind the tree, so the learner will mostly look behind the wall. Then, suddenly and without warning, the opponent switches to hiding behind the tree most of the time. Our model-free learner will continue looking behind the wall until the value of this action falls below the value of looking behind the tree. Such an agent does not know (or need to know) that anything changed. It simply relearns the task afresh each time a reversal occurs.

Contrast this with an agent who can use counterfactual thinking. This agent has a model of the task. There are only two actions, and they are mutually exclusive. Now, the agent can ask, "What if I had looked behind the tree?" The opponent must have been there since he wasn't behind the wall (Heilbron and Meyniel, 2018). This means that the learner can increase the value of looking behind the wall even though he did not choose that action (known as a "fictive reward"; Lohrenz et al., 2007), and learning the reversal can occur more quickly. Using the rock-paper-scissors game, Lee, McGreevy, and Barraclough (2005) have shown that the behavior of rhesus monkeys is determined by these fictive rewards. It is not just humans who can entertain counterfactuals.

11. It is now widely acknowledged that the same brain systems are involved in remembering the past and imagining the future (as shown by Mullally and Maguire, 2013).

Models of the World and Hidden States: When Agents Change Their Minds

Counterfactual thinking in the hide-and-seek game is based on a model of the task. We believe the opponent must be behind either the wall or the tree. So if the reward wasn't behind the wall, it must have been behind the tree. This allows us to learn more quickly. But humans and many other animals can deal with reversals even more efficiently.

If the reversal occurs many times in succession, then animals will switch to the alternative action more quickly after each reversal (Harlow, 1949). The animal has learned that in this task, the opponent has two relatively stable states: hiding behind the wall or hiding behind the tree. So, after making only a few errors, they know that there has been a change in the state of the opponent (Soltani and Izquierdo, 2019), As social creatures, we are keenly interested in spotting significant changes in the states of others. Model-based learners in the reversal task don't have to relearn the values of the alternative actions. They know that they must now simply switch the values. This is an example of learning that allows us to keep track of the hidden states of the opponent.

As we saw in chapters 7 and 10, it is very useful to be able to make inferences about the hidden states of our competitors. Is the lion hungry or not? Can the alpha male see where the food is? A model-based learning agent perceives the world in terms of the hidden causes of the behavior of other agents. We see living creatures as moving purposefully, to attain food and other rewards and to avoid danger.

A basic tenet of our information-processing framework is that our perception of the world is not based on raw sensation, but rather on their interpretation, and this increasingly involves beliefs about the world. These beliefs are our model of the world, and, without any model, there can be no perception. This model may well kickstart subjective experience—sentience. If we accept increasingly rapid reversal learning as a marker of model-based learning, then sentience is by no means uniquely human (Frith, 2021; Birch, 2020). Such learning has been observed in many species, including rats, pigeons, and goldfish (Shettleworth, 2010).

Keeping track of hidden states does not require a special kind of computation. We can update our belief about a hidden state via prediction errors, in the same way that we update our belief about the value of an action. If the lion is hungry, then we predict that he will chase the antelope. If this is not what happens, we have a prediction error. It becomes more likely that he is not hungry. The same computation is being applied, but in different brain areas with different inputs and outputs. As we already have noted, prediction errors for model-free agents are associated with activity in the midbrain (see figure 12.2). In contrast, prediction errors, when we track the behavior of others, are associated with activity in the temporo-parietal junction (TPJ; see chapter 10; also see Hampton, Bossaerts, and O'Doherty, 2008; and Behrens et al., 2008).

The World of Ideas

Modeling the Modeler—Meta-Learning

Model-based learning can deal—in a limited way—with the behavioral world of agents, be they animals, robots, or conscious human beings. But a major step up is needed to enter the mental world of ideas, the world beloved and continually explored by us humans. To do this, we must go beyond estimating the hidden states that determine the behavior of agents. We need to compute what model other agents are using for them to estimate *our* hidden states. We must go beyond thinking about doing and start thinking about thinking. What are people thinking? Why do they think in this way? Here, we can ask why the agent abruptly changes his location from the wall to the tree. We can even ask why we are talking about this game. This is when we move over to the Machiavelli thread, and it becomes really interesting to think about social cognition because, at this level of the computational hierarchy, there is self-consciousness and we can talk to each other about social cognition. We can even write a book about it!

Although the computations are more complicated,[12] a predictive processing framework remains a very useful starting point. We have prior expectations about what model an agent might use, and we can update these expectations based on the behavior of the agent with whom we are interacting. Take for example, the hide-and-seek game used in Jean Daunizeau's lab (Devaine, Hollard, and Daunizeau, 2014a), as described in chapter 7. When people believed that they were playing with a one-armed-bandit, their expectation was that this machine will be mindless. It may have a biased habit or be a rather slow reinforcement learner. Such an agent can be beaten with cognitively undemanding strategies like *win-stay lose-shift* (equivalent to very rapid, model-free learning). On the other hand, when people believed that they were interacting with another person, then they expected this person to be trying to predict what they were going to do next and used a more sophisticated strategy.

In fact, in both cases, they were playing against computer algorithms of varying degrees of sophistication. If they thought that they were playing against a person, they could achieve a draw against even the most advanced algorithm. But if they thought that they were playing against a one-armed-bandit, their prior expectation about the behavior of such an agent was so strong that they consistently lost against the most advanced algorithm. They failed to switch to the successful strategy, which they could readily use when they believed they were playing against a person. This is an interesting example of the tenacity of prior beliefs in the mental world. They are affected

12. Too lengthy to put in footnotes.

rather weakly, or perhaps not at all, by actual experience. This is the case when we use gossip to predict what another agent is going to do (see chapter 9).

Metacognition: Thinking about Thinking

Thinking about the models that people use to guide their behavior is an example of metacognition: thinking about thinking. When we apply metacognition to the thinking of others, we call it *mentalizing* (see chapter 10). However, metacognition is more typically concerned with reflecting on our own thinking. One example is when we think about our own decisions. Am I making a good decision? How likely is it that I have made the right decision? Here, we are assessing our confidence (see chapter 13).

How do we compute our confidence in our decisions? Steve Fleming and Nathaniel Daw (201) make the distinction between first- and second-order computations. For example, if we have to decide between two actions, we should choose the one with the highest value. Our confidence in this decision could be based on the difference between these two values. When the difference is large, we are more confident in our decision. This is an example of a first-order computation. Both our choice and our confidence are based on the same estimates of the values of actions. We are concerned *only with states of the world*, such as the reliability of the evidence and the difficulty of the task. These determine both our estimates of the values of actions and our confidence in those estimates.

In contrast, a second-order computation of confidence involves estimating *the state of the decision-maker.* This estimate is based on the responses chosen by the decision-maker and by the relationship (so far) between his choices and his successes. The estimate may reveal how much attention the decision-maker has been paying (Davidson, Macdonald, and Yeung, 2021). One important advantage of this second-order computation is that it can be used to make inferences about the confidence of others, by observing their decision-making behavior (Patel, Fleming and Kilner, 2012).

How Good Is Your Model?

This breaking of the direct link between decision-making and confidence helps to account for phenomena such as our ability to detect errors in our impending responses (the "whoops" effect; Rabbitt, 1966a). Here, the confidence-rating mechanism, which detects the upcoming error, seems to be better informed than the decision-making-mechanism, which made the error. And this ability to reflect on cognition applies not only to our decision-making, but also to our models of the world. At the first-order level, the system is generating models of the world. At the second-order level, the system is reflecting on these first-order models and can estimate how good these models are (generative adversarial model Gershman, 2019).

This second-order or meta-level of computation, through which we reflect upon our understanding of the world, is a typical example of higher-order thought (Lau and Rosenthal, 2011). This is a form of consciousness, self-conscioueness, that goes beyond sentience. We don't just have subjective experience, we reflect upon it. The most important development enabled by this level of meta-consciousness is that we can discuss our understanding of the world with others and together generate better models. This is fundamental to the mental world of ideas and the basis of cumulative culture.

A Hierarchy of Control

We have discussed these various computational levels in the order in which they evolved, starting with simple learned habits and finishing with meta-level cognition. The new forms do not replace the old but are linked in a hierarchy of control (Friston, Kilner, and Harrison, 2006) (see figure 12.4). We can explain all the processes we are considering here via mechanisms that are built up from predictive coding. This applies to actions as well as all kinds of perception (Rao and Ballard, 1999). These mechanisms are not specific to social cognition.

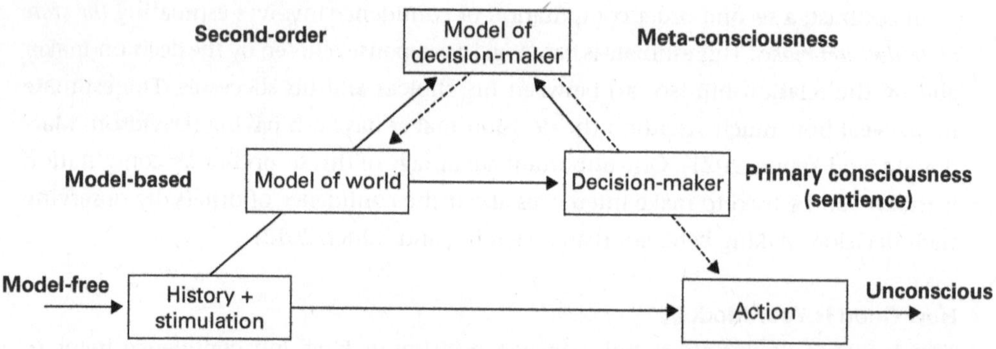

Figure 12.4
A hierarchy of control: essentially the same mechanisms are in play at each level of the hierarchy. Each level has bottom-up signals coming from the lower level and top-down signals coming from the higher level. And the bottom-up and the top-down signals are combined in the same way. That which is below is like that which is above, and that which is above is like that which is below.[13]

---

13. Isaac Newton's translation from the Emerald Tablet of Hermes Trismegistus (~1670)

Constraints on the Interpretation of Sensory Signals

In Bayesian terms, bottom-up signals provide the evidence (e.g., sensory signals, prediction errors), while top-down signals provide the priors for the lower levels, which constrain the interpretation of the evidence at those levels. We can see such a hierarchy in action when we consider our ability to read the intentions of others from their movements.

If an agent has the intention to pick up a glass and drink from it, then this intention will determine the movements that he makes. However, for an observer, it is harder to go in the reverse direction and infer his intentions from his movements. In this reverse direction, the mapping is inherently ambiguous. The same picking-up movement could be caused by many different intentions. The agent might want to pour out the contents and put the glass in the dishwasher, or he might want to throw its contents in the face of an irritating person (Jacob and Jeannerod, 2005). A solely bottom-up computation in which intentions are inferred from observing movements usually will not work.

The problem of ambiguity can be resolved with a hierarchical system (see figure 12.5). At the highest level, the context (is the party over? Is an irritating person

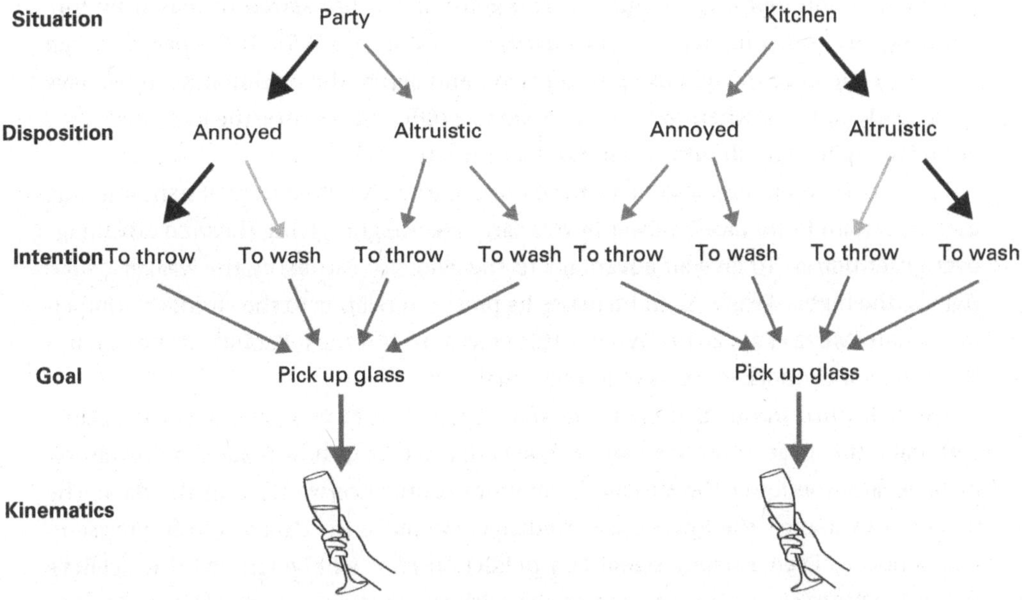

Figure 12.5
Inferring hidden states from movements: constraints from the top of the hierarchy constrain our interpretations of the intentions of the person whom we are observing. A person at the party who is annoyed is more likely to pick up the glass and throw it than a person who is having fun.

present?) constrains the intention. At the middle level, the intention is constrained by the situation and, in turn, constrains the goal (e.g., picking up the glass). At the bottom level, the goal constrains the movement kinematics (Kilner, Friston, and Frith, 2007).

As an observer, I will have prior expectations about someone's intentions, given the context (where they are and who they are). A person in the kitchen is more likely to have the goal of putting the glass in the dishwasher. From this expectation, I can use my own motor system (including mirror neurons, as discussed in chapter 3), to predict the movements that I am likely to see. But my prediction of the behavior might be wrong. He is not taking the half-empty glass to the dishwasher, but advances with it toward X. My prior belief about this agent's intention must be wrong. Their intention is to attack X! In a very short time, the hierarchical system has resolved all inconsistencies between the various levels, and the intention has been correctly read.

Biasing the Horse Race

At each level of the hierarchy, there are many competing processes. Within each level, the competition is resolved in a winner-takes-all manner. The loudest sound attracts our attention. The strongest habit determines our movement. The most valuable outcome determines our choice. But this competition can be skewed or biased by top-down signals from a higher level (Desimone and Duncan, 1995). At the party, we can choose to attend to the quietly spoken person and ignore the loudmouths. If we have been nearly run over when crossing the road in a different country, the increased value of looking right can inhibit the habit of looking left.

How does this biasing work? The mechanism is a bit like the system of handicapping used in certain horse races. Slower horses carry less weight, giving them an advantage over faster horses. To give an advantage to the evidence carried by the weaker neural signals, the higher-level system increases its precision relative to the stronger, competing signals (Mirza et al., 2019). What is this precision? It is the reliability of the signal—the degree of confidence we have in the signal.

As we learned in our first course in statistics, there are two parameters associated with data: the mean and the variance. The variance tells us how reliable our measurement is, and the lower the variance, the more confidence we have in the data. The higher the variance, the lower our confidence. We judge the data as too imprecise to take seriously. Even a strong signal of a prediction error will be ignored if we believe that it is imprecise. In this way, tampering with the precision of signals from the top down automatically achieves the desired adjustment of the competition.

If we believe that the precision of our evidence is high, then we will attend to it and be more likely to change our belief—our model of the world. If the precision is low, then we will stick with our prior belief. Precision determines the balance between evidence and beliefs (Clark, 2013; Yon and Frith, 2021). By biasing our beliefs about the precision of the bottom-up signals, we can control the competition between the lower-level processes. This biasing is itself informed by our model of the world. There is a downside to this mechanism: it increases the danger that we override evidence that we can't fit into our model. This may prevent us from generating better models.

How to Tell When Prediction Errors Are Important—and When They Are Not

We can illustrate the biasing process with the hide-and-seek game that we have already used to distinguish model-free and model-based learning. If our opponent consistently goes behind the tree most of the time (say 85 percent), then occasionally we will choose the wrong hiding place. But these errors don't tell us anything. They are just random. We rightly treat these errors as unreliable. They are not signals, but noise. It is an excellent idea *not* to pay attention to them. But if our opponent suddenly switches his preference and now mostly hides behind the wall, then the errors become important. They indicate that there has been a change in the state of our opponent (Soltani and Izquierdo, 2019), and as social creatures, we are keenly interested in spotting significant changes in other agents' behavior.

For unguided (model-free) learning, all errors are equally important. The agent's behavior will change whether the errors are just random or indicate a major change of state. We need guided (model-based) learning to distinguish between errors that are just noise (e.g., they happen on no more than 15 percent of occasions), and errors that are important. The model-based learners have the advantage. They will downweight the errors when expecting noise and upweight the errors when expecting a change of state (Sarafyazd and Jazayeri, 2019).

If they have learned roughly when a change in state is likely to occur, then at that time they will start paying attention to errors. At this point, they believe that the errors are reliable and informative. They attend to the errors only when they expect a reversal to occur. Otherwise, they believe, correctly, that these errors are irrelevant and unreliable. They do not change their belief about the state of the world.

Our ability to distinguish between important and unimportant prediction errors plays a critical role in many social situations. For example, if I send an email and get no reply, is this because the recipient is lazy or difficult, or is the network unreliable? My interpretation of this evidence could have a critical effect on future interactions.

This problem has been explored using trust games (see chapter 9). In these games, if player A sends £1 to player B, then a benevolent banker will double it, and player B gets £2. And the same happens when player B sends money back to player A. Players will earn most if they cooperate and send all of their money to each other on every round. But each player has to trust the other to return money, and there is always a temptation to keep it. What should I do if I send a lot of money to my partner and he keeps it? If we both stop investing, both of us lose out. The problem, for the players, is how to achieve and maintain trust and cooperation.

Both theoretical analysis and empirical studies show that the best strategy for maintaining cooperation is tit-for-tat (Axelrod and Hamilton, 1981). In other words, you do what your partner did on the last round. If the partner cooperated (sending eight of his ten coins), then you cooperate (send back eight coins). If he failed to cooperate (sending only one coin), then you don't cooperate either (send only one coin).

But this strategy breaks down if there is noise in the system. This is analogous to the email problem. The money exchange sometimes goes wrong. My partner sends the money, but I don't receive it. I believe that he has ceased to be cooperative, and I retaliate in tit-for-tat fashion. This will cause my partner to retaliate in turn. Trust and cooperation are likely to collapse. Partners become trapped in cycles of noncooperative interaction. The problem occurs because the noise is treated as if it were signaling something important about the interaction. Paul van Lange and colleagues showed that cooperation indeed breaks down when there is noise in the exchange (van Lange, Ouwerkerk, and Tazelaar, 2002).

To solve this problem, players need to treat the uncooperative events as unimportant noise rather than important signals. One way to do this is via communication: the player is told that the money was sent but did not arrive (Tazelaar, van Lange, and Ouwerkerk, 2004; Rand, Fudenberg, and Dreber, 2015). Now the player knows that the event is noise. They ignore the evidence favoring their suspicion of lack of cooperation. Trust is restored. In this case, they have downweighted evidence from noise.

However, there is another way to deal with the problem. Rather than downweighting the noise, we can upweight our prior belief that our partner is trustworthy and cooperative. This belief also has a precision parameter. In Bayesian terms, belief has a strength that can be measured in terms of probability. This estimate of probability will also have a precision attached to it, and prior beliefs associated with high precision will be given more weight. If the prior belief has a high precision, then evidence against it will be ignored. In the trust game, if we have a strong belief that our partner is trustworthy, we will ignore the noise. In one study, van Lange et al. (2002) achieved this by making the partner consistently more generous and always returning more than was sent (tit for tat + 1). Such behavior actually creates the belief that this partner is trustworthy. As a result, cooperation is maintained when playing with such a partner in spite of noise.

As we saw in chapter 9, the same effect can be achieved by simply telling the player that the partner is very cooperative (Delgado, Frank, and Phelps, 2005). Players instructed in this way will maintain cooperation despite the bad behavior of their partner. Brain imaging reveals that the magnitude of the activity elicited by evidence (the uncooperative behavior) is reduced. The evidence has been downweighted.

Self-Help

As is evident from these examples, the balance between higher levels and lower levels of the hierarchy—the evidence and the prior belief—is constantly being adjusted to suit the tasks that we are performing. Both the reliability of the evidence from the data and strength of the prior expectation, have to be reevaluated all the time. However, there is another consequence of this interaction: higher levels can monitor and train the lower levels.

In the absence of the upper levels, the lower levels would be entirely subject to the vagaries of the external world. We would be prey to an overload of incoming information. We would not be able to ignore noise. To learn, we would be entirely dependent on training signals from the external world when often these signals come too late. The upper levels of the hierarchy allow us to escape form this tyranny. We can help ourselves navigate the world. How does this work?

Let us assume that, by default, agents try to use the lowest (Zombie) levels of the hierarchy, relying on habit and model-free systems, to control their interactions with the world. These levels run automatically. Hence, they are less cognitively demanding and work well with situations that have been experienced extensively in the past. Indeed, in many such situations, habits and model-free systems can achieve better performance that higher-level systems (Miyamoto, Wang, and Smith, 2020). Neither millipedes nor golfers should think too much about their actions (Beilock et al., 2002).

The upper levels of the hierarchy act as a so-called supervisory attentional system (Norman and Shallice, 1986). They effectively monitor the functioning of the lower-level systems and resolve conflicts between competing information. Whenever necessary, the higher-level systems will take over. For example, if prediction errors indicate that the lower-level system is not working well enough, a higher-level system is brought into play. Top-down biasing might also be used to bring a different information-processing strategy into play. For instance, if the problem arises from limited information, exploration might be required rather than the exploitation of existing strategies (Purcell and Kiani, 2016).

The upper levels of the hierarchy can also send training signals for the lower levels. They are not called "supervisory" for nothing! The model-based system can train the model-free system by replaying and simulating experience (Gershman, Markman, and Otto, 2014). And likewise, the meta-system, so important in humans, can train the model-based system by using confidence as an outcome signal (Gershman, 2019).

These kinds of influence can extend from the top to the bottom of the hierarchy. Human beings can be aware, at the meta-level, of the problems created by habits. For example, there is the driving to work problem. This route is followed so frequently that it becomes a habit. The problem arises when we need to drive somewhere else. Often, lapses of attention or preoccupations occur at choice points in the road, and we may find that we have driven to work by mistake (by habit). Being aware of this problem, and realizing that lapses of attention are outside our control, we can choose a route that avoids any part of our normal drive to work (Moran, Keramati, and Dolan, 2021). This route may be less than optimal, but it keeps us from being captured by a habit.

What's at the Top of Top-Down Control? Closing the Loop

We have made much of the idea that top-down signals play a critical role in the brain hierarchy that supports all our actions and thoughts. But where is the top? Where does the buck stop? Not in the brain: there is no location in the brain where there are only outputs and no inputs. There is, of course, a higher level that monitors and controls our behavior, but this is outside the brain (Roepstorff and Frith, 2004). This depends upon other people and, more generally, upon the culture in which we are embedded.

To interact with our culture, as Dan Sperber (1996) has suggested, we need to be able to make our mental representations public and also internalize public representations. This is what second-order modeling, the highest level of our brain's hierarchy, enables us to do. We can tell people about our thoughts and our reasons for action, and we can get advice from others. Being social means that we don't have to solve every problem by ourselves alone. We can draw on others' experiences. For example, we can be told how to avoid the driving to work problem.

And it is not just other people who provide this aspect of control. We are surrounded by material culture (Malafouris, 2013): gardens, roads, tools, books. We are surrounded by people behaving in characteristic ways: shaking hands, getting in line, eating with a knife and fork. Unlike the physical environment that forms the Umwelt of other animals, culture is our Umwelt. Our cultural environment provides the top level of monitoring and control, but it also provides the context for statistical learning and the development of habits acquired at the lowest level of the brain's hierarchy. In this way, the loop has been closed.

IV Culture—Sharing Experiences with Others and Modeling the Modelers

13 Signals from the Deep

This is the first of seven chapters dealing with some of the intricacies of human social interaction in the world of ideas: how our experiences are influenced by cultural beliefs and how we share subjective experiences with others.

Our present cultural beliefs make it hard to accept that much of what we do happens without any conscious thought. However, while we are unaware of the complex automatic processes that occur deep inside our brain, we immediately notice when one of these processes goes wrong.

We use the term *metacognition* to refer to the ability to reflect at a higher (conscious) level on the functioning of our lower-level (unconscious) cognitive processes. How does communication work between these levels? We suggest that a monitoring system enables information to rise from the hidden depths of the brain into conscious awareness. In response, we alter our behavior. For instance, if we feel that we are getting stuck on a problem, we can switch to a different strategy.

The messages that emerge from our brain's unconscious processes give rise to subjective feelings whose meanings we try to fathom. External factors influence our interpretation of these feelings, such as instructions or general cultural beliefs. Examples are the degree of regret when decisions go wrong, the degree of doubt or certainty when we make decisions, and the degree of fluency with which we perceive the world and select appropriate actions. While the interpretations of these feelings are subject to outside influences and may not always be correct, these interpretations can modify our behavior.

* * *

Who's in Control?

We like to think that we are in control of what we are doing, but this is not always the case. In everyday language, when we are fully in control, we say "I did it," and when we feel that we are not, we say "my brain did it" and claim that we ourselves were not responsible. When we feel in control, we can talk about what we are doing and why we are doing it. But where does this feeling come from? Somehow it emerges from the depths of the information-processing system embodied in our brain.

How do the lower and higher levels of this hierarchically organized system communicate with each other? This is still something of a mystery, but we have lots of evidence that they do. Here, we offer some speculative suggestions. We believe that any communication would be exceedingly difficult were it not for metacognition. Metacognition is not just the conscious capacity to think about thinking; it also has a place at the unconscious levels of the hierarchy, right down to motor and sensory systems. We can see metacognition as a remarkably clever system of monitoring that forms an extra layer on top of cognitive mechanisms. This means that these mechanisms, operating at different levels of the hierarchy, are not left to work blindly on their own; rather, they are coupled with *minders*. The point of the minders is that they can report function or malfunction to the next level up in the hierarchy. By connecting the levels, they serve as an important system of communication. But how do we envisage these minders?

In chapter 7, we described the amazing efference copy, which enables us to monitor our movements. This mechanism emerged almost as soon as organisms appeared that could move independently (Crapse and Sommer, 2008). We think of this as the blueprint for the minders (see figure 13.1).

For simplicity, we will take a shortcut and consider just two levels of the hierarchy: the lower depths and the top. The top is where our self feels itself in charge. It can become aware of what happens down below because it can interpret the messages that are sent by the minders. It can also send messages down to make adjustments, perhaps to make a bigger effort at the task at hand. If the messages from below indicate that everything is running smoothly, the self has the pleasant feeling of being in control.

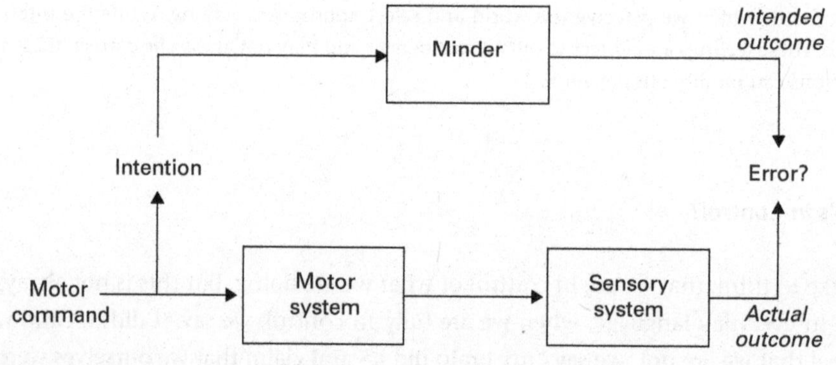

Figure 13.1
A minder for action: At a low level of the cognitive hierarchy, we initiate an action (e.g., push a button to turn on a light). A minder checks that our intended outcome was achieved. If it wasn't, a higher-level system must decide what to do about this. (Didn't press hard enough? Wrong button? Blown fuse?)

We can imagine these messages as something like lights and dials in a car that indicate how well the engine is running. The driver interprets these signals, feeling in control when the dials register that everything is running fine and stopping to take action when a warning light comes on. The monitoring systems in a car are extremely useful, as shown by the fact that they are getting ever more intricate. Nowadays, it is routine to expect that they automatically convey information for many components—whether the stoplights are working, the tank has enough fuel, the seatbelts are locked, the doors are closed, and so on. However, compared to the monitoring systems and signals between different parts of the mind, they are very limited. There is, after all, a lot more going on in the mental world than the car engineers have dreamed of.

Explicit and Implicit

In this book, we have used the terms "explicit" and "implicit" pretty much interchangeably with "conscious" and "unconscious." We realize, however, that they can mean different things to different people. Sometimes consciousness is used to refer to sentience, a state of having subjective experiences (i.e., perceiving an apple, feeling pain or hunger); and, as we mentioned in chapter 12, this is something we share with many other animals (Frith, 2021). But in this and the following chapters, we are concerned with a more sophisticated form of consciousness, typical of humans, and developing only slowly during childhood and adolescence. This consciousness is about *knowing* that we are having a subjective experience, not just having an experience. We can reflect on our experiences. This is a typical second-order or metacognitive process, and it is very much part of the Machiavelli thread.

We can imagine that a report is issued by a minder to the self, while the experience itself originated in the depths of the hierarchy. The self becomes aware of a subjective experience, which it can then report to other people. And this is probably the most amazing fallout from the whole process—the point where it matters that we are social creatures. Other people can learn about our subjective experiences and can eventually understand them as well as we do ourselves. The fact that private experiences can be shared with other people adds spice to our social lives. This deliberate sharing can give rise to surprising reinterpretations, not just during face-to-face interactions, but also when watching a film or reading a story. When misunderstandings occur, the consequences can sometimes be hilarious, but also serious. We will say more about this idea in the following chapters. Here, we can only give a hint. We will argue that the ability to report our experiences to others opens the door to the influence of culture. Culture can save us from being oddballs, and furthermore, it allows us to make sense of our experiences through stories, traditions, and numerous codes of practice.

We need to reiterate that implicit cognitive processes (our Zombie thread) are not less valuable for being unconscious. On the contrary, they produce much precise and well-adapted behavior (Jacob and Jeannerod, 2003; Goodale and Milner, 2004). These implicit processes are likely much the same in everybody—indeed, much the same for all creatures with brains like ours.[1] We can assume this because they exist only because they have earned their place over eons of evolution. They are doing really important work, looking after our vital functions.

"Implicit" and "explicit" are not merely buzzwords to distinguish introspective knowledge from the lack of it. There is increasing evidence that the processes they refer to not only are separate, but also involve independent cognitive and neural systems (e.g., for memory; see Kalra, Gabrieli, and Finn, 2019). Furthermore, while the ability to learn explicitly (e.g., via rules given by a teacher) improves greatly during development, there is little if any change in the ability to learn implicitly (Ghetti and Angelini, 2008). We could phrase this as the Zombie thread being steady and persistent, while the Machiavelli thread requires training and remains flexible. Still, while they are separate, implicit processes can inveigle their way into consciousness, like when we become aware of feeling anxious.

Listening to the Messages

Messages from the deep are a cause for wonder. They bubble up to the surface and can be tantalizingly vague while stirring up conscious interpretations. They often hit the jackpot, which is very pleasing when it happens, as in the tip-of-the-tongue phenomenon. Imagine that you are asked, "What is the capital of Slovenia?" You are sure that you know the answer. You are conscious of the goal—to retrieve a word—and you will be conscious of the outcome when it drops into your mind (Ljubljana, of course!). This can happen, perhaps only after some hours, but you are entirely unaware of the implicit process by which your brain retrieved this answer. And we all know the experience of knowing that we have the right answer, even though we can't tell how we arrived at it. These gut feelings have the advantage of being fast (Tversky and Kahneman, 1974; Evans and Stanovich, 2013). This makes them useful when we need to act quickly. But they can be wrong. And then we won't know why.

We can use our Bayesian framework to sketch out how implicit and explicit control loops interact with each other. We propose that if the implicit, lower-level action errors

1. That is, always allowing for individual variations. Such variations are inevitable consequences of the presumed genetic basis of the underlying mechanisms.

become too large, the minders will be sending an SOS to the explicit level. Higher-level processes can then act to modify the function of the implicit processes, for instance, by sending a command for them to slow down. This is illustrated in figure 13.2.

Stephen Fleming (2021), in his book *Know Thyself,* discusses the fluctuating feelings of confidence about the decisions that have to be made at every trial of a tedious signal detection task in the lab. Did I give the right answer or not? Even this trivial example shows the need for communication between Zombie and Machiavelli, or system 1 and system 2, using Kahneman's (2011) terms for implicit and explicit processes.

Confidence is a signal in need of constant reality checks. Fortunately, minders do this job by monitoring whether our confidence is justified, given our actual performance. To do this well, especially at the higher levels of the processing hierarchy, we need to gain insight into our own behavior, and to do this, we need to cultivate self-awareness. How might this work? Fortunately, Stephen Fleming and Nathaniel Daw (2017) offer a computational model that allows conscious and unconscious systems to work independently, as well as to communicate with each other (see also chapters 12 and 14).

Here is a vivid example of communication via messages from the lower depth of the processing system (Zombie level, or system 1) to the top level (Machiavelli level, or system 2).

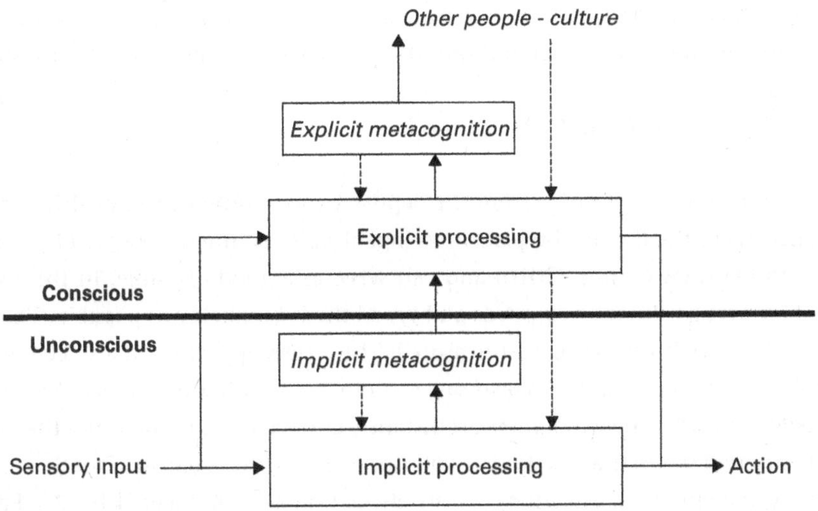

Figure 13.2
A hierarchy of implicit and explicit processes: implicit and explicit processing loops can function independently, but they can also interact. We can modify implicit processes explicitly if signals from the minder indicate that they are not functioning properly. Likewise, explicit processes can be modified by signals from our top-level minders (i.e., other people and our culture). Words in italics indicate the higher-level control processes (minders).

Imagine you are asked to say as many names of animals as possible in one minute. Try it! Typically, some names instantly trip off your tongue. You don't need to do any explicit cognitive work when performing this task since the words ("dog," "cat," "lion," and so on) will spontaneously enter your consciousness. System 1 is doing the job very well. We are not conscious of the process, although we are conscious of the goal and outcome.

Inevitably, after a while, the names will stop spontaneously emerging. System 1 seems stuck or exhausted. How are you going to get any further? You would do well to switch to your conscious system 2 and think about what you are doing. Should you give up, or is there something you can do? At the top, you know that you know more names. You also know that the flow has stopped, but, as in the tip-of-the-tongue phenomenon, you are getting distress signals from below. They might make you feel that it is worth persevering because you sense that you are very likely to succeed (Proust, 2014). Now your conscious self remembers a strategy that will likely produce more names: Use the alphabet! It can act as a prompt for more words. Appropriate signals go down from the top, and almost immediately, "aardvark" comes to mind.

Another strategy that your conscious self might come up with is to consider subcategories of animals (zoo animals, farm animals, birds, insects, and so on). Generate as many names as possible from one category and then, when the names stop coming, move to another category. When children are instructed in the use of subcategories, their performance improves, although not until the age of about twelve years (Hurks, 2012).

Messages Influencing What We Think and Feel

To find out how messages arrive from the depths, experimenters need stealth and cunning. The case of the deceived typists is an ingenious experiment designed by Gordon Logan and Matthew Crump (2010) and can serve as a good example. In this experiment, skilled typists listened to words and typed them as quickly as possible. The time that it took for each keystroke was measured. The letters appeared on a screen as they were typed. However, what the typists saw was not always what they typed. Sometimes they made errors that the experimenters instantly corrected, and sometimes the experimenters inserted errors that the typists had not made. (How fiendish is that!)

This experiment took advantage of an effect that was discovered by Pat Rabbitt (1966b). He showed that immediately after people make an error in an easy but boring task, they slow down ever so slightly on the following trial. He called the effect "post-error slowing" and related it to the "whoops" effect, which happens just as we realize that a response that is about to happen is not actually what we intended.

Indeed, as expected, the typists in Logan and Crump's study slowed down a fraction after they made an error, and this post-error slowing occurred even if the error had been corrected by the experimenters. However, they did not slow when the experimenters had inserted an error. We can conclude that the typists' own errors were faithfully detected by a minder and signaled to a higher level. This was followed by the instant top-down command to slow down before the next keystroke. Conscious awareness did not enter this process.

But what happened at the conscious level of the processing hierarchy? The experimenters frequently asked the typists whether they had made an error or not. Here, a very different picture emerged. Surprisingly, the typists accepted responsibility for the errors that had been inserted by the experimenters. They were led by what they saw, even though they did not slow down after these false errors. At the same time, they seemed unaware of their own real errors when they had been surreptitiously corrected. Yet here, they had shown post-error slowing.

This cunning experiment artificially produced a split between processing levels (here Zombie—there Machiavelli, each engaged by real and inserted errors, respectively), and thus allows us to look into the depths of the processing hierarchy. We can see a striking example of an implicit metacognitive process. The minder minding the typing movements detected an error, sent an SOS, and the minder at the next higher level initiated a slowdown. All this was dealt with automatically, and there was no need to transmit the information to the top to make it conscious.

In contrast, the minder minding what the typist perceived on the screen provides an equally striking example of an explicit metacognitive process. This minder did send information about words that looked wrong to the level of the conscious self. Recall that the errors had been deviously inserted by the experimenters. Nevertheless, the message received at the top level resulted in making the typists believe that they were responsible for the errors.

These results speak for the existence of both implicit and explicit metacognition. They also speak for a great deal of complexity as they go their separate ways. There is clearly more to discover in future experiments. Might they reveal a twilight zone, where bottom-up and top-down signals become intermingled and hard to distinguish?

It would be exciting to get a fuller picture of the communication between levels of the processing hierarchy and between conscious and unconscious processes. We need examples that tell us more about some of the minders that—behind our backs—carry information up and down. What kind of information is conveyed? What are the conscious feelings that are stirred up by the signals from the depths? We believe that there is already some experimental evidence to draw on, concerning, for example, feelings of regret, uncertainty, and fluency.

Be Warned: The Troubling Expectation of Regret

Many actions are hardly better than leaps in the dark, and we are only too familiar with the experience of regret when we realize that we made the wrong decision (see Frith and Metzinger, 2016).[2] This experience is explicit, as we are fully aware of it and can talk about it. But what value does it have for us? It always seems to be too late to do anything about it. But is it? Regret might first appear as a shadow in the twilight zone, and you may have to learn to recognize it. It is a shadow that can warn us *before* we commit to a decision. Once we have learned to heed this signal, we can anticipate the regret that we would feel if it turned out that we made the wrong decision.

Learning about this shadow can be quite rapid, as was demonstrated in an experiment with children aged six to seven years using a variant of the classic marshmallow test (McCormack et al., 2019), in which the children had to decide whether to wait for a short delay to win two sweets, or a longer one to win four. Those who chose the short delay were shown that they would have won more if they had waited longer and then were asked whether they regretted not waiting longer. The next day, the children were presented with the same choice again. Only those children who said they regretted choosing the short delay on day one changed their behavior on day two and tolerated the longer delay. Presumably, this was the effect of the anticipated regret that was invoked by the experimenter's question.

If you ever put in a bid at an auction, you will know that bidding is a rich source for feelings of anticipated regret. Imagine yourself bidding £100 for a rare book that you would dearly love to add to your collection. Your bid did not succeed, and someone else managed to get the book. You immediately wonder how much it went for. You will groan if you find that it went for £105. Next time, you will likely bid £110, or maybe even £120, for a similar book. In contrast, if the winning bid was £500, you will probably not bid at all for another one like it next time. The coveted book is clearly out of your price range.

There is another type of auction, in which you are not told the value of the winning bid. Here, there is less scope for anticipated regret, and you tend to bid less than you

2. Regret is not to be confused with disappointment. We feel disappointment when the outcome of a decision is not as good as we expected. We feel regret when we realize that, had we made a different choice, we would have achieved a better outcome.

would otherwise. This is not just speculation. Experiments have shown that there is a clear difference: People tend to make higher bids in auctions where they know that they will be told the winning bid. They can factor in the painful regret that they might feel if they lost out by a small margin (Filiz-Ozbay and Ozbay, 2007). No wonder that most auction houses advertise in advance that they will communicate the winning bids as soon as they are known.

A Toe in the Water: The Vexing Feeling of Uncertainty

We make decisions under uncertainty all the time. Was our last decision a good one? Will anybody tell us? It turns out that we can be guided to make our next decision even without feedback, via signals from the depths (Folke et al., 2016). For example, a feeling of high uncertainty creates low confidence. This feeling makes it very likely that you will change your mind when the same choice is presented again. People who are responsive to these signals are generally more accurate with regard to matching up their subjective uncertainty and their actual behavior. They are also better at knowing which task they should take on when given a choice. They can judge in advance whether they will be able to do the task well (Carlebach and Yeung, 2020).

What is the effect of feeling very uncertain? After making an error, we slow down a little, and are not even aware of it, as in the case of the typists. Slowing down gives us a chance to collect more evidence—that is, alter our speed/accuracy trade-off (van den Berg et al., 2016). In an experiment that involved making a decision between two essentially similar stimuli, the uncertainty was artificially manipulated by presenting variable evidence: that is, the stimuli were of different strengths and reliabilities. In this experiment, the participants tended to seek more information—they asked to see a stimulus again—even when their accuracy was high (Desender, Boldt, and Yeung, 2018).

This finding suggests that it is the subjective feeling of certainty, or confidence, more than the objective level of accuracy, that guides our behavior. Of course, in everyday tasks, the feeling of uncertainty tracks the accuracy of our performance. Hence, it is a good idea to seek more information when we are uncertain. When making difficult decisions, it is a good idea to confer with others when possible. In these cases, we are automatically drawn to taking advice from people who display more confidence. We say more on this topic in chapter 15.

One valuable effect of signals from the deep that feed our feelings of certainty (or rather uncertainty) is that they help us learn a new skill, even when we get no feedback about whether we were correct or not. For this reason, we can learn a difficult perceptual skill (in the lab, this often means detecting the orientation of stripes embedded in noise) through repeated presentations without ever being told whether we are right or wrong (e.g., McKee and Westhe, 1978; see also Zylberberg, Wolpert, and Shadlen, 2018). It is the feeling of certainty that shows us when we are on the right track, even if nobody is there to tell us.

This learning does not happen by magic. In fact, it is the feeling that bolsters our confidence, and that provides the necessary feedback. We note that this feedback is internal and based on translating a valuable message that monitors signals sent up from the depths of the information-processing system. The level of subjective certainty or uncertainty that is associated with each trial provides the signal that indicates whether a response was probably correct. Just as with external feedback, which may be derived from reward, regions deep in the brain (the nucleus accumbens and ventral tegmental area; see figure 12.2) encode prediction errors that are derived from this feeling of confidence (Guggenmos et al., 2016).

The relationship between the feeling of confidence and actual competence is fragile and can be biased toward either underconfidence or overconfidence. We will revisit this problem in the next chapter. For those who want to know how exactly metacognitive confidence is measured, we say the magic words "signal detection theory (STD)" (Swets, Tanner, and Birdsall, 1961). This method makes a distinction between sensitivity and bias. Bias here refers to being either overconfident or underconfident in our decisions. In contrast, sensitivity refers to how good we are at predicting whether our decisions will be good or bad. Do we know, or are we just guessing? Our sensitivity to the likely outcome of our decisions is an aspect of metacognition and is independent of our general state of confidence (Fleming, 2021). It is also independent of our ability to make decisions. In theory, you could be a good decision-maker but have no idea which decisions are going to be good or bad or why. Such a person would be difficult to work with, despite their ability, because we could not rely on their confidence reports.

In the Flow: The Pleasant Feeling of Fluency

A typical example is the comfortable feeling that some of us get when we hear people speaking in a dialect that is familiar to us. We may not necessarily be able to report the feeling, if it is hovering in the gray zone of consciousness, but it can be studied objectively. Perceptual fluency can be measured by the speed with which a sensory stimulus becomes a percept—in other words, how easy it is to recognize an object. Sheer familiarity is known to speed up recognition, and hence contribute to this pleasant effect.

Typically, an experiment might familiarize you with an object by showing it to you repeatedly. It is pretty certain that you will be faster to recognize an object the second time you see it. With repeated exposure, the increase in perceptual fluency can be substantial. But it is not only repetition that has this effect. There are other reasons for feelings of fluency. For example, different viewpoints can make objects either easy or hard to see (see figure 13.3). Objects seen in their so-called canonical view are perceived more fluently (Palmer, Rosch, and Chase, 1981).

However, the signals that we take to indicate perceptual fluency can be misleading. In a classic experiment (Kunst-Wilson and Zajonc, 1980), observers were first presented with ten abstract pictures. Each picture was shown extremely briefly and in a degraded form. The observers were then shown these pictures again, each alongside a new one that they had not seen before. Now the pictures were shown under good viewing conditions. Observers were then asked which picture they had seen before and which picture they preferred. They did not do very well. In terms of memory performance, observers were at chance, but, in terms of liking, there was a major difference. Around 60 percent of the items that they said they preferred, were pictures that they had seen previously. So the observers could distinguish between old and new items, at least to some extent.

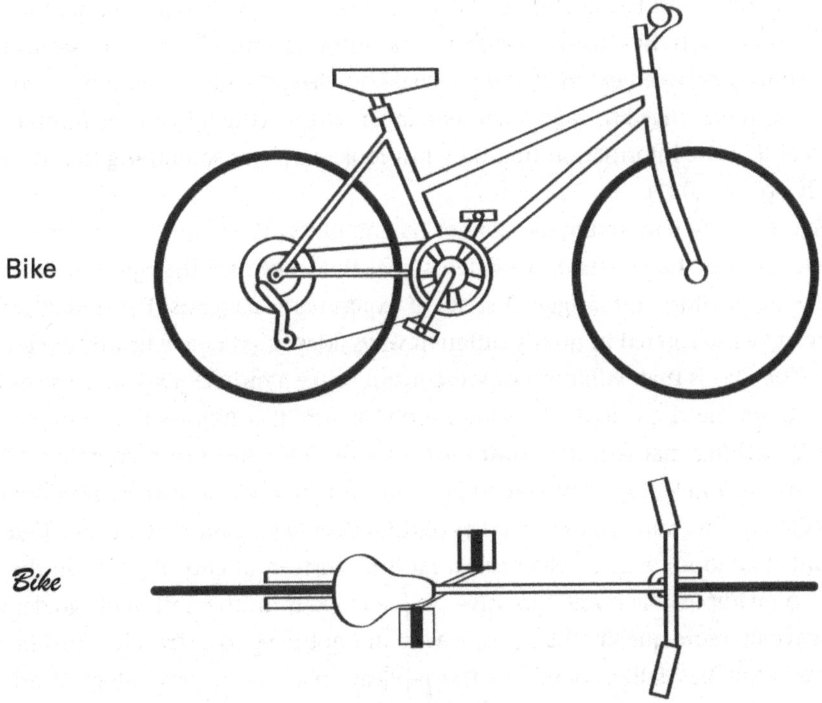

Figure 13.3
Some things are easier to perceive than others: the word and the picture in the top row are in a familiar form. They are easier to perceive than those in the bottom row.

But they interpreted the signals from the deep as reflecting liking rather than familiarity. Subsequent research has replicated this result and confirmed that such liking is indeed a consequence of perceptual fluency (Reber, Winkielman, and Schwarz, 1998).

There is also a form of fluency associated with action. This feeling reflects the ease with which we can select the appropriate action, such as choosing to press the right button with the right hand and the left button with the left hand. Here again, familiarity plays a role. Our feeling of fluency is high for an action that we have performed recently, and actions that we have practiced often feel more fluent and pleasant. When action selection fluency is high, we feel that we are in control of the situation.

Here again, the feeling can be misinterpreted. Action fluency can be increased by priming, such as watching a model performing the action. Afterward, we feel more in control of that action, even when our performance is no better (Chambon and Haggard, 2012). Conversely, it is also possible to reduce the feeling of fluency without a change in actual performance. This was done by zapping the premotor cortex with magnetic pulses by a method known as "transcranial magnetic stimulation (TMS)" (Fleming et al., 2015).

A deceptive boost to our feeling of competence and control is also present in a now-very-common activity—using Google to find information. This activity seems to blur the boundary between external and internal knowledge. Thus, people tend to overestimate how much they know and can remember without the help of the internet. They may feel that the information that they just looked up was something that they knew already (Ward, 2021).

We are clearly tempted by the lure of having pleasant feelings about ourselves. But we can also see the effects of a hidden signal that indicates the opposite to fluency, the feeling of effort and fatigue. A series of experiments suggests that signals from the deep can be interpreted in utterly different ways. Maybe you have heard of "ego depletion." The idea is that willpower or self-control is like a muscle. Excessive use of it tires you out. You need a rest after having exerted yourself! It follows that once you have done something that requires you to use a lot of self-control to keep going, you will feel fatigued. You will feel less able to put effort into a task that you are supposed to do immediately afterward. In early studies of this effect (e.g., Baumeister et al., 1998), participants had to force themselves to eat radishes instead of chocolate. Then they were given some impossible puzzles to solve. As predicted from the notion of ego depletion, they gave up more quickly than people who did not have to exert self-control first.

This result has fallen victim to the replication crisis in psychology (Carter and McCullough, 2014).[3] However, we can learn from such failures. Messages from the

3. When researchers have repeated certain well-known studies in psychology, such as ego depletion, they have failed to obtain the result originally reported.

deep are vague and their interpretation is not clear-cut. This could explain why different results were obtained. If participants believed that their vague inner states indicated fatigue, then they would give up on the task. If they did not, they would carry on. Are the feelings so vague that people can interpret them in opposite ways—as fatigue in one case and as an increase in energy in another?

This precise question was asked by Veronika Job, Carol Dweck, and Gregory Walton (2010). In this study, the participants were deliberately given opposite (fictitious) accounts of the effects of mental work. One group was told, "Working on a strenuous mental task can make you feel tired, such that you need a break before accomplishing a new task." Another group was told, "Working on a strenuous mental task can make you feel energized for further challenging activities." The first group showed ego depletion effects, but the second group actually improved their performance after working on the strenuous task. They interpreted the messages from the deep as a good feeling that kept them going.

It is sobering to realize that feelings emerging into consciousness can be as shadowy as ghosts and open to be modified by our prior beliefs, including beliefs that have been inserted from the outside. It is indeed an alarming thought that outside influences can shape the nature of our private feelings and subsequent actions. We cannot escape the conclusion that having private feelings and interpreting them are not the same thing. Metacognition is precisely about knowing that we have these feelings, and eventually about realizing that they are not entirely private. We might not even be able to interpret them correctly without help from our cultural Umwelt.

The ability to reflect on our subjective feelings and interpret them in the context of cultural influences provides a new slant on what makes us social. What makes us social is not just a matter of personal priors and predispositions. They may well suffice in the worlds of objects and agents, but we need more—much more—to take wing in the world of ideas.

We hope that after you read this chapter, we have shifted your interpretation of your own perceptual fluency for the better. Your familiarity with the term "metacognition" has increased, and with any luck, your belief in the energizing power of effort has increased as well. Hopefully, you will now be ready to read the next few chapters, in which we will argue that the outside influence on our private feelings is mostly for the good.

14 Consciousness and Control

Explicit metacognition lies at the top of the hierarchy in our cognitive system where clues about the functioning of the lower parts can emerge into consciousness. Metacognition may well be our human superpower. It creates an interface between the self and the surrounding culture and opens the door to the world of ideas. Explicit metacognition enables thinking about our own mental states, and explicit mentalizing enables thinking about the mental states of others. As a consequence, we can talk about and explain the motives of our own and others' actions and also predict possible outcomes over long time spans. Thus, it frees us from the constraints of the immediate environment. In the world of ideas, we can consciously compare subjective experiences, even though we don't have direct access to the experiences of other people. We can use verbal and nonverbal clues to find out if their cognitive system is working and can monitor their level of confidence and track their beliefs.

The meaning of these clues is not self-evident and has to be learned, including the recognition and use of deception. Our interpretations change as we discuss them with others and arrive at a consensus as to how we ought to behave and how the mind works. This consensus is an example of folk psychology and is one way in which culture affects the individual. The advantage of this consensus is that people from our ingroup will generally behave in ways that we expect them to behave, making interactions more fluent. The disadvantage is that the behavior of people from other groups, with different ideas about how the mind works, may appear irrational to us.

We know very little about the neural mechanisms underlying explicit metacognition, but the frontopolar cortex is likely to be important. This region lies at the top of the hierarchy of control and has been implicated in tasks requiring explicit mentalizing, as well as other forms of explicit metacognition.

* * *

Explicit Metacognition—the Human Superpower

In chapter 13, we suggested that there are minders that monitor ordinary cognitive processes. This monitoring happens automatically and is likely to be similar in all animals with brains. However, in humans, there is a level at the top of the information-processing hierarchy where we place explicit metacognition. Here, messages from the deep can emerge into consciousness and allow us to deliberately influence otherwise deeply hidden processes.

As we discussed in that chapter, some messages from the deep emerge in the form of vague feelings, such as the pleasant feeling of fluency. These feelings accompany our perception of the world around us and our choice of what to do next. They give us clues as to how well certain lower-level processes are currently functioning. However, they need to be interpreted, and this is where explicit metacognition plays a key role. It enables us to reflect on these feelings and discuss them with others, and this shapes their interpretation. Discussions can refine our interpretations, but they can also mislead. Explicit metacognition may be a human superpower, but it is not without problems.

We have made the point that we humans share the implicit and automatic social processes with other species. They are the bedrock of what makes us social. In contrast, explicit metacognition may well be uniquely human, not simply the icing on the cake. It allows us to reflect on our reasoning when we deliberate about what action to take and consider long-term consequences. Compared to implicit monitoring and error correction, this is a rather complex and drawn-out procedure, and by no means foolproof. Here, there is a bigger time window where we can consider strategies to improve our actions in the long term. This enables us to perform an amazing trick. We can distance ourselves from our immediate urges and pressures.

There is another gain, not dreamed of outside the superpower of explicit metacognition: we can use the same procedure that we use to fathom the signals from our own unconscious layers to fathom what might be happening in other minds. This is a tremendous gain when deliberating on the best course of action, since we can tap into the deliberations of other agents around us. This creates ideal conditions for pooling resources to develop better strategies, to make better decisions, and to achieve better models of the world. Here is a new interface between the individual and others, added to the forever-unconscious interface that suffices for copying and contagion, for alignment and affiliation. This new interface is where we can look for the origin of human culture and civilization.

What Is the Point of Consciousness?

Scientists have often wondered why we need explicit, conscious processes at all when we can achieve so much with implicit, nonconscious processes (e.g., Rosenthal, 2008). We believe that it is its social function that makes consciousness so valuable (Frith, 1995; Frith and Metzinger, 2016). Given our hierarchical framework, it is obvious that all our cognitive processes are subject to top-down control. But much of this happens behind our backs. Remember the typists mentioned in chapter 13, who slowed down after they made an error, unaware that they were doing this? Post-error slowing, a result of monitoring, happens automatically and enables us to do better next time. And this automatic kind of monitoring doesn't apply just to typing. Why do we need yet another layer of control that lets us become aware of some of these processes?

Evidently, there is something transformative about the layer of conscious metacognition at the top of the information-processing hierarchy. Why else would it have been favored by evolution? But what does it bring us that brains of other animals do not have? When we look at other animals, we may ask ourselves if they are conscious in the way that we are, but sadly, they are unable to tell us. We do not doubt that living creatures with a nervous system that allows them to sense the world and act on it, are sentient. They may have a model of the world, but can they reflect on it? Can they think that they are thinking or know that they are knowing? We doubt it. If we're right, then for better or worse, such creatures cannot enter the world of ideas.

We already mentioned some of the benefits to be had from the extra layer on top of the hierarchy. Explicit metacognition allows us humans to reflect on our models of the world, discuss them with others, and decide whether they are good or bad. But this layer is not just an add-on. It too relies on the descendants of efference copies, the monitors, and is deeply connected to all the other layers. Therefore, we propose that the emergence of explicit metacognition is an extension of the hierarchy that has vastly extended our social nature. For instance, through reflection on our actions, explicit metacognition has given rise to feelings of responsibility that are fundamental to the rule of law and the ability to govern large groups of individuals.

However, even though there are many benefits, we must not forget that computations at this level of the hierarchy are complex and costly to implement. Our brains much prefer to run things at automatic levels, using various simplifying heuristics. On the other hand, it is precisely through the effortful use of explicit metacognition that we have been able to develop many of these heuristics in the first place.

Top-Down Signals and How They Modify What We Are Doing

Top-down influences are amazing in that they work at all. If you have ever wondered whether there is such a thing as mind over matter, here would be the place to find it. Certainly, explicit deliberation can improve intuitive processes (Boissin et al., 2021). The self can issue commands like "Pull yourself together" or "Don't make the same mistake again," and often these commands are obeyed by deep layer processes.

Even more amazing is that these sorts of commands can come from other people, not just the self. We don't need to subject ourselves to hypnosis for outside instructions to work. Mere hints suffice if they refer to the behavioral norms of our community. These norms are fully explicit, and they are everywhere (e.g., no smoking; don't drive across a double yellow line; don't cut in line; give up your seat to someone who needs it more than you). We can marvel at the fact that they can alter the functioning of cognitive processes deep within us.

In the example of the verbal fluency task (say as many animals as you can in one minute) from the previous chapter, the use of the alphabet is a great strategy. It organizes the search space in which the system 1 naming process is operating. We are not conscious of how system 1 can now produce more animal names; we are conscious only of the top-down strategies that we are applying. At the same time, we are rather less concerned about where these strategies come from. Top-down strategies are rarely invented on the spot. It is far more likely that we have been taught them by others. For example: "I can tell you a trick: go through the alphabet and you will get more animal names that way." This one-off instruction is enough.

Now, if it is the case that we acquire most of our ideas and strategies from other people, we are forced to face an unwelcome truth: we (i.e., the self) are not in sole charge. The "top" in "top-down" is not in our own brains. Instead, the top is the culture created by the brains around us (Roepstorff and Frith, 2004). At the level of the conscious self, we are not quite as much in control as we like to think. However, given how deeply committed we are to the world of ideas, this should not be too surprising. It may even be something of a relief, as it lightens the weight of regret and responsibility. Marwa El Zein and Bahador Bahrami (2020) conducted experiments using a lottery where individuals could choose to act on their own or as members of a group. The results showed that being part of a collective provided a protective shield against the influence of negative feelings after a bad outcome.

The interface from individual brain to other brains is a two-way system. As discussed in the previous chapter, when signals from the lower depths reach consciousness, they do not usually remain within the head of a single individual. We love to talk about our feelings, including those of regret and uncertainty; we write and read about them in stories, poems, novels, and nonfiction. This exchange has far-reaching effects. It causes us to interpret the signals in similar ways to everyone else (Heyes et al., 2020). Therefore, we can align our minds just as we can align our physical bodies. Thus, if anything, we are even more embedded in the world of ideas than we are in the worlds of agents and objects.

How Do Top-Down and Bottom-Up Processes Work Together?

Reading and writing are prime examples of processes that have emerged through cultural rather than genetic evolution. Every individual has to be taught the shapes of the letters of the alphabet, as well as the sounds and meanings of written words. However, once this learning has taken place, these cultural priors become embedded in our brain's control system and enable us to read with ease. As it happens, the mechanisms involved in reading provide a nice illustration of our Bayesian model of information processing (see chapter 12).

An early cognitive theory of reading (Gough, 1972) assumed that reading is an entirely bottom-up process. Marks on the page are converted into letters, the letters into words, and the words into meanings. The problem with this approach is that reading, like all kinds of perception, must be able to deal with fuzzy and ambiguous stimuli. This problem is illustrated in figure 14.1. Here, we see two sentences that we read with ease but that contain some highly ambiguous graphic symbols representing letters that could be decoded in different ways.

Priors, which exist at higher levels of the hierarchy, constrain the computations carried out at the lower levels. Thus, our prior knowledge of the alphabet constrains how we interpret the marks on the page. At the next level up, our knowledge of language constrains how we interpret the letters. At an even higher level, our semantic knowledge constrains our understanding of the words. In the example shown in figure 14.1, the constraints at the level of meaning determine how we see the marks on the page. At a higher level still, our knowledge of particular individuals will constrain our understanding of the words that they use. This enables us to recognize irony, for example.

Talking about Confidence

When we observe others, we can be impressed by the speed with which they make decisions. We then judge them as confident, and by implication competent. We discussed this in the previous chapter. But here, we want to compare this unconscious evaluation with the conscious form. Patel, Fleming, and Kilner (2012) made such a comparison. They asked observers to estimate the confidence of actors who indicated their perceptual decisions by moving a marble to the left or the right. They also asked the actors to report their confidence in each decision verbally. In this study, there was a surprisingly good match. Observers were able to reliably estimate the confidence reported by the actor based on speed alone, with faster movements seen as more confident.

In another respect, however, there was a poor match. It turned out that there was an effect of bias on this judgment. This was discovered by making the observers perform

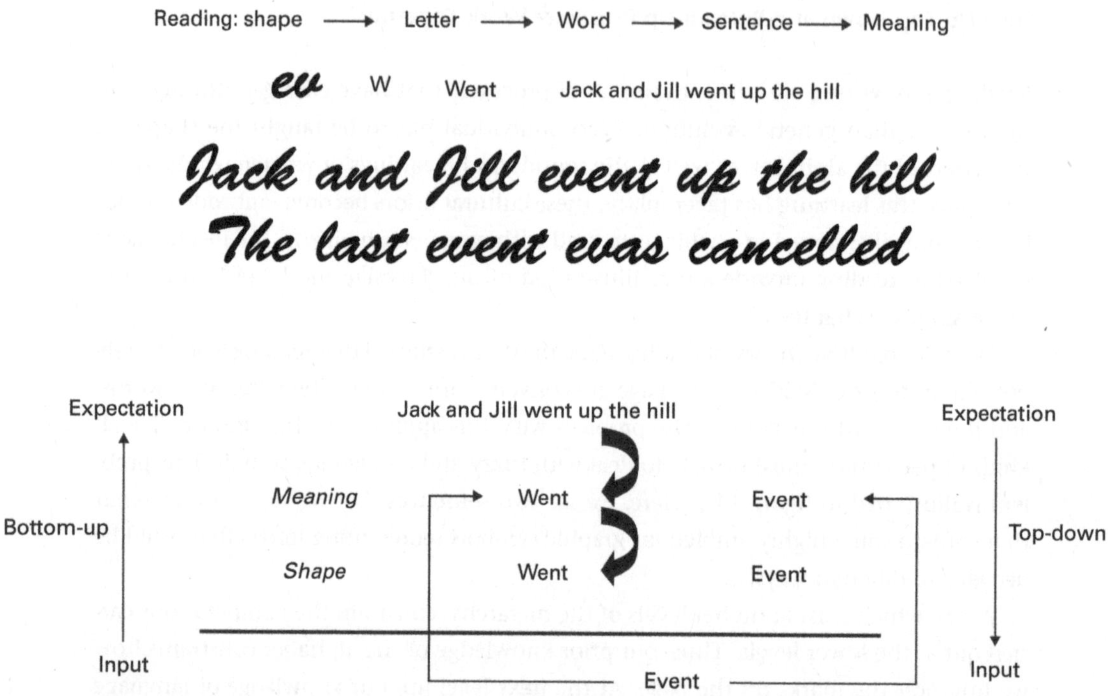

Figure 14.1
A hierarchy of constraints when reading: these two sentences, which look handwritten, can be read easily, but in fact they contain an ambiguous word, which can be decoded as "went" or as "event." If we used a strictly bottom-up process, we can't tell whether "ev" should be read as "ev" or "w." But our interpretation is constrained top-down by the sentence (*meaning*). In one case, the word must be "went." In the other case, it must be "event." In turn, this constrains the graphic symbol (*shape*) to be read as "ev" or "w."

the movements themselves. This showed that observers anchored their confidence ratings relative to their own speeds. Thus, an observer tended to rate an actor as more confident simply because the actor moved faster than the observer did. There is another experiment and another twist to the tale. When they were instructed to move more quickly after making their decisions, people report higher confidence—but now their decisions were often incorrect (Palser, Fotopoulou, and Kilner, 2018). Just making yourself behave faster has a direct effect on your feeling of confidence, but it is an illusion. It does nothing to improve your performance. These observations confirm that feelings of confidence can be decoupled from competence (i.e., actual performance).[1]

1. Outside the lab, we would do well to use the Machiavelli thread to factor in the fact that expressions of confidence can be misleading.

These experiments show that the experience of confidence (or its reverse, uncertainty) is highly malleable and easily manipulated. This is not surprising, as it waxes and wanes according to the task we do and the state we are in. As our skill increases, so should our confidence. We need to calibrate confidence against competence so that eventually, most of the time, high confidence will be a sign of high competence. This calibration is especially important when we decide from whom to seek advice or learn. If they are confident, then they are probably competent.

In one study (Campbell-Meiklejohn et al., 2017), people could learn about the state of the world through direct experience, but they could also learn from observing the decisions of others. Expressive faces indicated the confidence that these others had in their decision (see figure 14.2).

The participants took account of the decisions of others, but only when they were associated with a high degree of confidence. Clearly, we can be swayed by the expression of high confidence, taking it as a sign of a good decision, when we don't have any knowledge about whether the person in question is good at making the decision. But beware—this belief can be treacherous! As we shall see in chapter 15, we need to be able to see through the confident witness whose memory is inaccurate.

We discussed the feeling of fluency in the previous chapter. We are not always fully aware of it, but we can learn to use this feeling as a marker of a good decision. Younger children underestimate uncertainty and tend to be overconfident in their judgments (Beck et al., 2011). However, by age six, they are able to revise their judgments when necessary (Köksal, Sodian, and Legare, 2021). This is in line with the idea that being able to reflect helps them to learn to interpret the vague signals of uncertainty and fluency. This learning likely involves other people, listening and talking to them.

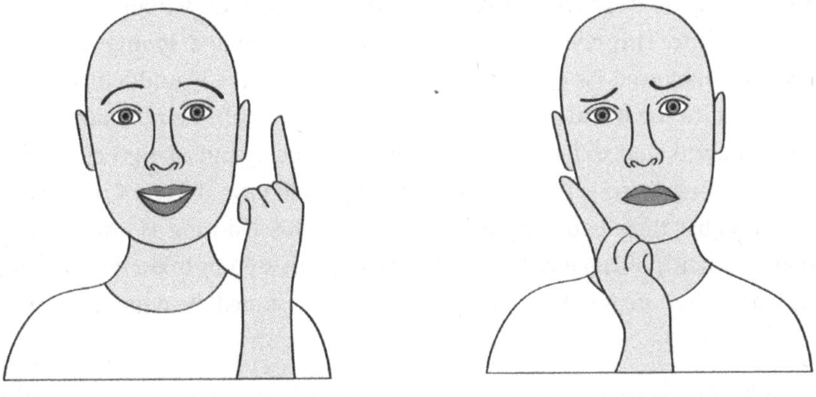

Figure 14.2
Expressions of confidence (left) and doubt (right).

However, even after years of learning, there is still no guarantee that the true implications of these feelings and their interpretation will be a perfect match.

Explicit Mentalizing as Part of Explicit Metacognition

How do beliefs about the state of the world emerge? As we have hinted already, beliefs can be created through instructions and cultural stories and norms. However, in the first instance, beliefs are created through exposure to evidence. The Maxi/Sally-Ann False Belief task is an example of this principle (see chapter 10). The key to the solution of this task, as precocious four-year-olds can tell us, is that Maxi was not there when the critical event happened (the change in location). He was not exposed to the evidence that his chocolate had been moved elsewhere, so he did not change his belief about the location of the chocolate. Thus, if we track the evidence that someone has been exposed to, we can estimate their most likely belief (Ereira, Dolan, and Kurth-Nelson, 2020).

Explicit mentalizing is a prime example of explicit metacognition (thinking about another agent's inner states), and it does not appear before the age of four years (Wellman, Cross, and Watson, 2001). At around the same time, reasoning with counterfactuals appears (Beck et al., 2006) and is thought to be involved in false belief reasoning (Rafetseder et al., 2021). Likewise, other abilities relating to explicit metacognition, such as telling lies, being able to monitor task performance and being able to report degree of confidence, start to emerge during this critical age (Hembacher and Ghetti, 2014). They are all abilities that are important components of metacognitive self-control, and they undergo a long process of development that continues through adolescence (Schneider and Löffler, 2016).

What about children below the age of four or so? To us adults, young children appear delightfully spontaneous and innocent, not blighted by knowledge, guilt, or responsibility. We cannot fail to be enchanted by them, but as they enter school age, this impression begins to fade. This could be because their innocence and spontaneity are being gradually overshadowed by conscious metacognition. There is undoubtedly a cost to entering fully into the world of ideas. It is plausible that with their budding consciousness, young infants start to figure out the worlds of objects and agents before the world of ideas. Thus, they learn early on that different people have different preferences and desires and adjust their behavior accordingly. But this learning is always about the immediate demands of the environment. A significant step in brain development has to occur before they are freed from the fetters of these immediate demands.[2] Then they

2. An example of being freed from the immediate demands of the situation at hand is the capacity to think of counterfactual situations (e.g., if Maxi had seen the transfer, then he would have known where the chocolate is).

begin to realize that different people can have different beliefs about the same situation and can use different names for the same object (Perner, Rendl, and Garnham, 2007).

These speculations are not contradicted by the existence of implicit mentalizing, which we assume continues to be used throughout life, especially when time is of the essence and when our processing capacity is threatened with overload. Implicit mentalizing is a gift from evolution (see chapter 10) and needs little if any learning. Explicit mentalizing is a gift from culture and needs a lot of learning.

We speculate that over the course of childhood, the information-processing hierarchy is being built up slowly but inexorably, with the upper levels of the hierarchy reached during the period of four to six years of age. This time appears to mark a critical step in cognitive development for children all over the world. It is not coincidence that this is also the time when children are traditionally deemed ready for school. The gradual development of explicit metacognition leads to a massive extension of communication, prosocial behavior, and moral rules. At the end of this long learning process, we achieve full admission to the world of ideas.

Still, explicit metacognition does not start from nothing. A longitudinal study of children's metacognitive insight, such as "knowing about knowing," was able to show that explicit metacognition at age six was correlated with implicit mentalizing, as tested two and half years *earlier*. This correlation held after accounting for individual differences in intelligence and executive functions, such as the inhibition of impulsive responses (Kloo et al., 2021). This bolsters the argument that explicit metacognition (knowing about the self) may well be rooted in early forms of mentalizing (knowing about others) (Carruthers, 2009).

Recursive Thinking

Recursion is frequently observed in nature and lends striking beauty to structures visible in crystals and plants (see figure 14.3). It is explored in mathematics and is a mainstay of computer programming. Therefore, it may not be surprising that it is a distinctive feature of explicit metacognition. But there is a health warning: thinking about thinking can induce vertigo. As soon as you start, you may find yourself sucked into a spiral. For example, you may reflect on the feeling of regret, and then you wonder if you should think about why you are thinking about your feeling of regret. This would be an example of second-order metacognition, and you could theoretically go on to third- and fourth-order metacognition.

Our cognitive capacity is unlikely to reach that far in the recursive spiral unless we put in enormous effort, but we do seem to regularly go up to second order, when we think about the minds of others (he thinks I think he wants to go to the football

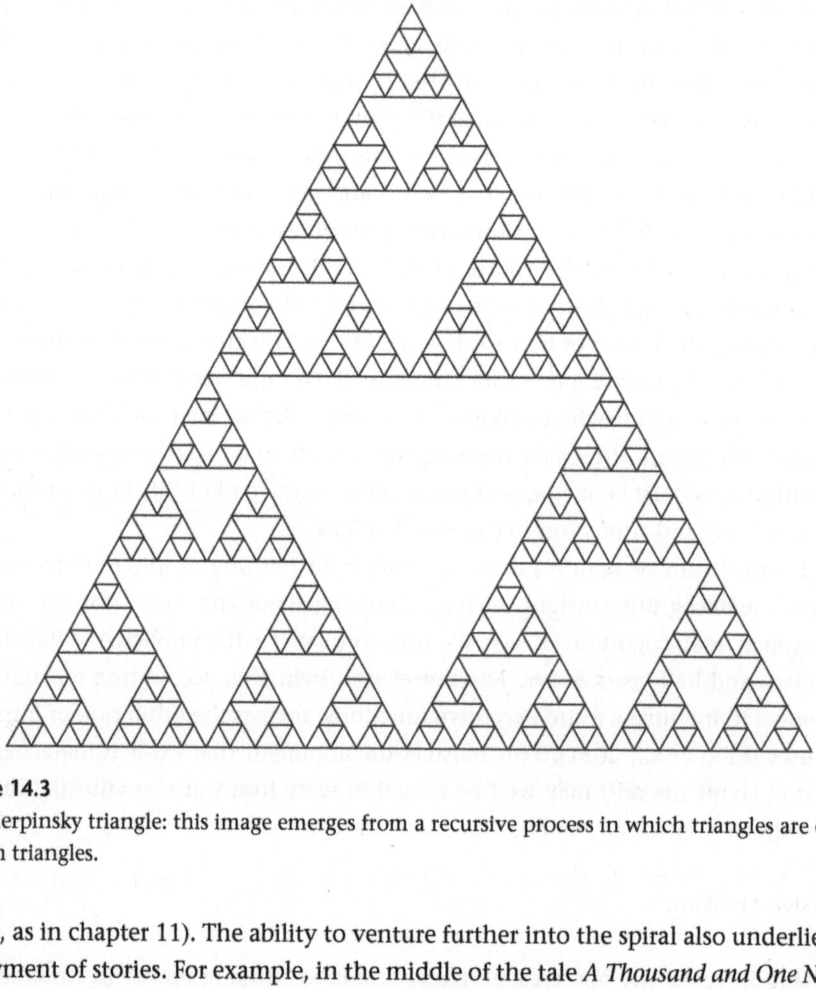

Figure 14.3
The Sierpinsky triangle: this image emerges from a recursive process in which triangles are drawn within triangles.

game, as in chapter 11). The ability to venture further into the spiral also underlies our enjoyment of stories. For example, in the middle of the tale *A Thousand and One Nights,* someone begins to tell the tale of *A Thousand and One Nights.*

It is widely thought that recursion has a critical role in human language (e.g., van der Hulst, 2010), enabling us to produce and understand long sentences such as "This is the cat that chased the rat that ate the cheese that lay in the house that Jack built." Recursion also is evident when we are trying to put into words how our own beliefs affect other people's beliefs, which then affect our own beliefs. We can easily get trapped in Russian nesting doll–type sentences, such as "I believe that she believes that he believes that I had told her already . . . ," and yet we can still track these embedded beliefs as we follow the plot of a story. This lends weight to the proposal that recursive language is a vital part of explicit mentalizing (de Villiers, 2007).

Folk Psychology

Folk psychology is all about explaining why somebody did something in the past and predicting what they will do in the future, and it generally involves explicit mentalizing. It overflows with mental-state language. Most six-year-olds can tell that Maxi *wants* to eat his chocolate and will be very *disappointed* when he doesn't find it in the place where he had put it. Our everyday gossip about why people behave as they do is based on our capacity for explicit metacognition. We assume that actions depend upon inner states such as perceptions and memories (Godfrey-Smith, 2005). The framework that we use feels like common sense and it enables us to continuously make inferences about other agents' competencies and motivations, their experiences, feelings and beliefs (Jara-Ettinger et al., 2016).

In contrast to implicit mentalizing, folk psychology allows us to go beyond simply keeping track of other people's mental states in the here and now. Instead of being only ever a small step ahead of what they are going to do next, we can factor in events that occur over longer time spans. This represents a major step toward freeing us from the demands of the immediate environment. We can treat the contents of mental states as premises for practical reasoning (Davidson, 1969). Like reading a map, we can locate different mental states in the past and the future. In this way, they allow us to make sense of our own and other people's behavior in a longer time frame.

Folk theories are theories that people hold about how the world works, no matter how unsatisfactory these theories are at explaining things. However, if they are shared between people in a group, they work quite well. Folk psychology theories differ markedly over time and place and entertain a range of explanations of why people feel and behave as they do. Witchcraft and magic were causal explanations at one time in history, while more recently, the scrutiny of intentions leads to preferred causal explanations. In the classic crime novel, different types of motives are often deliberately set against one another. For example, a hurtful blow was made with the intention to cause harm because the victim was a blackmailer. A Miss Marple–style detective then follows up a different theory. Perhaps the blow was accidental, and the real motive was to disinherit a competitor. We love classic detective stories precisely because behavior is often not what it seems at first glance.

Using our metacognitive stance, we can accept that a theory is wrong, and this makes no difference to the fact that it is a theory that many people hold. Being accurate is less important for these folk psychology theories than lightening the task of understanding each other. This is because folk psychology theories are, to some extent, normative.

People behave in certain ways, and we have learned this. But we have also learned that people ought to behave in these particular ways (McGeer, 2007). In consequence, our behavior is molded by social pressures, making it more mappable and predictable in the long run (Zawidzki, 2008). Therefore, cultural ideas are incorporated into our explicit mentalizing (Godfrey-Smith, 2012). We behave according to the theories that we have acquired about how people (should) behave.[3]

Behaving Rationally

One of our most basic expectations is that the people in our group will behave rationally. We disapprove of irrational behavior, in part because such behavior is more difficult to predict and control. The exception is with children, where we tolerate such behavior and are charmed and irritated by it in equal measure. This is another example of folk psychology: we consider that children do not have to behave rationally, but grown-ups *should* behave rationally. On the other hand, we are also not surprised if people from outgroups behave irrationally.

But how do we decide whether behavior is rational? In the case of goal-directed agents, this is often done in terms of economics, where a cost-benefit analysis is applied. An agent behaves rationally when it maximizes its expected utility, given its current knowledge (i.e., instrumental rationality). In other words, we choose the action that we expect will give the best outcome for the least cost. One example of such rationality is the expectation that goal-directed agents should always take the most efficient route to their goal, minimizing the cost of the action (Liu and Spelke, 2017; Csibra et al., 1999).

But this approach does not work with intentional agents. As we have seen in chapter 7, behavior that appears irrational in a goal-directed agent may indicate that we are dealing with an intentional agent. With such an agent, we have to take account of beliefs and desires, which may be different from our own. Faced with a choice between spinach and peas, Chris will choose peas. But he knows that there are some people, like Uta, who will choose spinach. In this case, he is willing (just about, anyway) to accept that this behavior is rational because it fits with his folk psychological theory that people choose what they like and that there is no accounting for taste. But defining rationality in terms of folk psychology can be very misleading.

Consider again the marshmallow test (Mischel, 2015). The child who eats one marshmallow rather than waiting ten to fifteen minutes to get two is assumed to not

3. Note the recursive nature of these thoughts.

have acted rationally (since two marshmallows are better than one) and shows lack of self-control. Furthermore, some studies have shown that such children tend to have worse outcomes in later life, such as poor educational attainments (Shoda, Mischel, and Peake, 1990). Clearly, this lack of self-control is detrimental, and programs have been developed to help children acquire more self-control (Murray, Theakston, and Wells, 2016). In our culture, exerting self-control is something we *ought* to do. And when we show lack of self-control, we are being irrational.

But is it *always* rational to wait for the two marshmallows? What if you live in an environment where marshmallows are rare and strangers cannot be trusted? In this case, it would be rational to eat the one marshmallow while you can. Celeste Kidd (2013) explored this idea by adding a condition in which the experimenter was shown to be unreliable before the child was presented with the marshmallow. After such an experience, most children did not wait for the second marshmallow.

A similar effect can be seen with the classic false belief task. In a study of mentalizing in Madagascar, Rita Astuti (2015) tested a four-year-old boy and his older brother. In the presence of both boys, a coin was hidden under one of two coconut shells. Then the older brother was sent away to fetch water. The coin was moved and the standard question asked, "Where will he look when he comes back?" But the most interesting observation concerned what the older boy actually did when he came back. "He went straight for the coconut where, in his absence, we had moved the coin to! As he grabbed the coin and ran off, he shouted: 'That's why you sent me to fetch the water!'" Folk psychology for this boy was not to trust the strange anthropologist. He anticipated an attempt to trick him. For him, it was rational to look in the "wrong" place.

Folk psychology, including how to reason from mental states, is indeed something we learn from "folk." And, obviously, different folk can have different ideas about the appropriate way to behave.

Explicit Metacognition and the Brain

We still know very little about the brain basis of explicit metacognition, and even less about its evolution. In relation to our ability to report confidence in our decisions, we believe that one particular region, the frontopolar cortex (BA 10, see figure 14.4), plays a special role in explicit metacognitive processes (Fleming et al., 2010). But what are the implications for brain imaging of our argument that explicit mentalizing is a prototypical example of explicit metacognition?

Explicit metacognition typically concerns the self, while explicit mentalizing typically concerns the Other. We have suggested that the medial prefrontal cortex (mPFC),

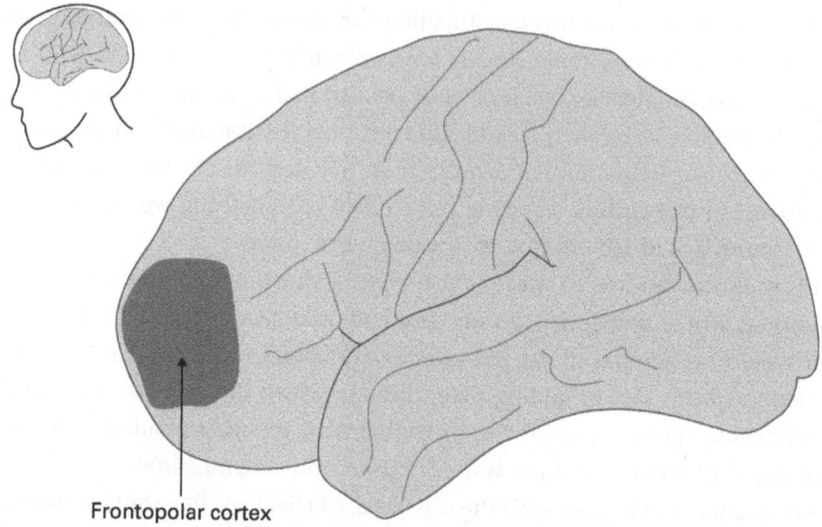

Figure 14.4
The frontopolar cortex: the left hemisphere of the brain, highlighting the frontopolar cortex. This region of the prefrontal cortex is the most recent to have evolved and is enlarged in humans compared to other great apes (Semendeferi et al., 2001).

the controller of the brain's mentalizing system, is concerned with the development of models of the minds of others so that we can predict what they are going to do (see chapter 10). However, there is also much evidence that mPFC is concerned with the self and is engaged when we think about our own mental states (e.g., Northoff and Bermpohl, 2004). Furthermore, there is evidence of a direct link between mentalizing and metacognition applied to the self. Self-reported sociocommunicative skills are positively correlated with perceptual metacognitive accuracy (van der Plas et al., 2021). At first sight, it is not clear why we should want to create a model of our own mind (Carruthers, 2011). Do we really need a model to predict what we ourselves are going to do? Surely we have direct access to our own plans and intentions. But do we? Maybe such direct access is only possible at a lower, subpersonal level.

The elegant computational account of the processes underlying explicit metacognition by Steven Fleming and Nathaniel Daw (2017) presents a framework in which the mechanism for monitoring the decisions of others would apply equally to monitoring our own decisions. In this framework, which is similar to our proposal of a hierarchy of control, explicit metacognition is a second-order process. First-order mechanisms are involved in decision-making, while separate, second-order mechanisms are involved in confidence estimation. In other words, an estimate of confidence does not directly emerge from the computations on which the decision is based (first order). Rather, the estimate of confidence represents a higher level of processing (second order), as it involves monitoring and making inferences about the decision-making process, and as it takes into account the attentive state of the decision-maker (Davidson, Macdonald, and Yeung, 2021).

Just as a decision involves making inferences about the state of the world, so estimating confidence involves making inferences about the state of the decision-maker (see chapter 12). For this computation, our access to our own decisions is as indirect as our access to the state of the outside world. As Fleming and Daw (2017) point out, there is a symmetry between evaluating one's own actions and those of another actor. Vuillaume et al. (2020, 1) reached a similar conclusion: "metacognition leverages the same inference mechanisms as involved in Theory of Mind." It will be most interesting to see if this kind of second-order computational modeling can be applied to mentalizing tasks, and if the same brain region, the frontopolar cortex, is implicated.

15 Making Decisions in Groups

Experiments have demonstrated that two people can make better decisions than either one can on their own. This effect depends partially on mechanisms for pooling information that are automatic and can be observed in many social animals. But in humans, explicit metacognition and the use of language allow more complex and flexible group decisions. For example, we have a sense of our confidence when we have chosen an option, and we can express this confidence to others, not only in the speed and vigor of our actions, as other animals do, but in the choice of words. By reporting confidence, we can improve group decisions by giving more weight to the person who is more confident.

Unfortunately, the confidence that we express may not necessarily reflect the probability that our decision is right. We need to take this dissociation into account in discussions and decisions by groups. There can also be a big difference between our private feeling of confidence and the signal that we send to other people, especially when we are competing with them. For example, we can misrepresent our confidence to gain an advantage.

Group decisions benefit when different people bring different knowledge and strategies to their decision-making. The more diverse the people in the group, the greater the potential advantage. However, there is a cost, as diversity makes communication less fluent.

* * *

How Do Groups Make Decisions?

Groups are amazing. They are more than the sum of their parts. At one extreme, a group is just a collection of individuals, like a shoal of fish, where group behavior emerges from rules of individual behavior. At the other extreme, a group may be a collection of individuals with shared goals and a shared view of the world. When learning about the outside world, groups can do better than individuals working on their own (Couzin, 2009). For instance, a shoal of fish can find a source of food more accurately than a single fish; a swarm of bees can agree on the best site for a new nest. But what mechanisms make this possible?

These phenomena depend on signal averaging and alignment. In each case, a weak signal must be discovered in a lot of noise. A shoal of fish can follow a light-level gradient that is too weak for a single fish to follow because the strength of the signal is increased by averaging the independent but weak signals from each fish together (Berdahl et al., 2013). This phenomenon also exists in human crowds and was demonstrated by Francis Galton, the nineteenth-century polymath and pioneer of population statistics.

In an audacious experiment, Galton took advantage of a competition with the aim to "guess the weight of the ox," conducted at a country fair where the closest guess would win a prize. Galton reckoned that each guess would contain some information about the true weight plus some error, by either overestimating or underestimating the true weight. However, the distribution of guesses would not be biased toward overestimates or underestimates. Thus, the average of the guesses might come closer to the real value than that of the closest individual guess. Galton was right (see Galton, 1907).

In both cases, fish and human, the advantage for the group arises from signal averaging. Each individual has some information about the world (the signal), but this signal is contaminated with error (noise). So long as the noise is different for each individual, they would do well to combine their sensations to enhance the signal (see figure 15.1).

But how is this averaging achieved? In the case of fish, this effect occurs because all the fish in the shoal tend to align their direction of movement. Each fish moves roughly in the direction of the light gradient, with random movements superimposed. When the fish align, the randomness cancels out, the signal-to-noise ratio is enhanced and the correct direction of movement is retained (Grunbaum, 1998).

In the case of humans, things are a bit more complicated. In Galton's "guess the weight of the ox" study, there was no physical alignment. The averaging was done by Galton himself afterward. What if there is no Galton to collect all the individual guesses and average them? Can averaging occur spontaneously? To answer this question, we can turn to decisions made by honeybees.

Signaling, Averaging, and Aligning for Action

Thomas Seeley, in his book *Honeybee Democracy* (2010), vividly relates what happens when a swarm of bees needs to make an urgent decision, such as finding a new nesting place. When it is time to swarm, a few hundred scouts go out to search for a new site. They collect information about a potential site. When the scouts return, they perform a waggle dance. The vigor of this dance indicates the quality of the site, with better sites reflected in longer dances. It is tempting to impute higher confidence to the more vigorous dancers.

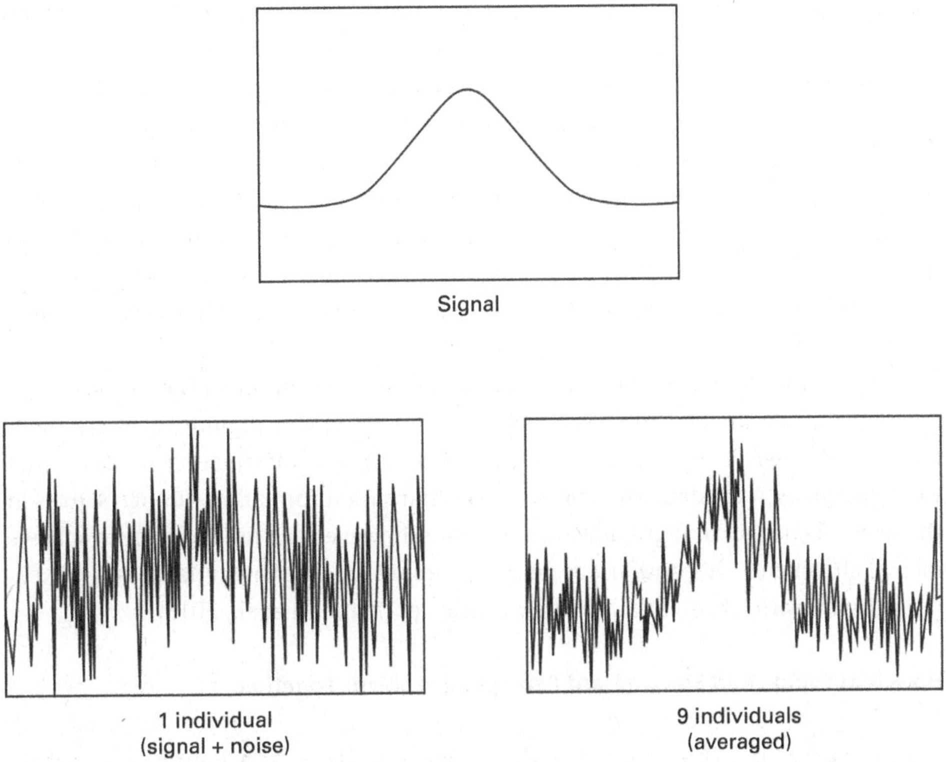

Figure 15.1
Signal averaging: Each individual has some information about the signal, but this is mixed with error (noise). By averaging across individuals, the noise can be canceled out, but only if the noise is different for each individual.

The longer dances give out a stronger signal and prompt more bees to go and examine the site being promoted. The new explorers then may confirm the quality of a particular site and return with their own dance. When a sufficient number of scouts are indicating the same site (quorum sensing; Pratt and Sumpter, 2006), they change their behavior and start to suppress the few scouts who still disagree. This decision-making mechanism has the advantage that many different sites can be assessed much more rapidly than would be possible for a single decision-maker, and furthermore, through the competition between the scouts, the best available site can be selected (Seeley and Buhrman, 2001).

One critical aspect of the bees' decision-making is that the vigor of the waggle dance communicates the value of the various options. Other animals also need to communicate the value of the various options when they make group decisions, and they will do so in their own way. One way is to represent the value of different options by the speed with which a particular option is chosen. An example is the behavior of homing pigeons, where a group will navigate more accurately than the best navigator working alone (Biro et al., 2006). When two pigeons fly together, at each choice point in the route, both typically follow the choice of the pigeon that responds more quickly. This is usually the pigeon with the greater experience, which effectively makes it the leader (Nagy et al., 2010). These pairwise interactions enable pigeons to learn which bird to follow. However, leaders rapidly lose their position when their knowledge becomes unreliable (Watts et al., 2016).

These examples show that group decisions can be made without a Galton, and without conscious awareness. The mechanisms are averaging and alignment, where alignment is achieved through communication. Bees transmit information to each other through signals in the dance of scouts. Pigeons transmit information through signals in the speed of changes in flight direction. Humans, however, can use language to achieve mental alignment. This notion was put to the test in an experiment designed by Bahador Bahrami with Chris Frith and other colleagues (Bahrami et al., 2010).

How Two People Can Solve a Hard Perceptual Problem Together

Picture a lab at the University of Aarhus. Two students sit at a right angle, each in front of their own computer screen. Each is shown an image of six stripy patches, and another very similar image shortly afterward. One patch in one of the images is an odd one out, as the stripes have slightly more contrast (see figure 15.2). The task is to decide together where the odd patch was. Was it in the first or the second image?

The task is dead easy if the odd patch is very different from the others. But here, the task was made difficult by making the difference only just detectable. This had the effect that in some trials, the partners gave different answers. However, they had to give a single joint answer. Before they decided which one to offer, they were able to discuss their decision as much as they liked. These discussions were recorded for later analysis.

There was a robust result: The decisions that were made jointly were on average better than the decisions made by each individual working alone. Moreover, these joint decisions were better than those made by the observer who was accurate more often.[1]

1. There is a proviso, however: both partners must have roughly equal ability to detect the stimuli. If they don't, then it would be better for the pair to always go with the answer supplied by the more competent of the two.

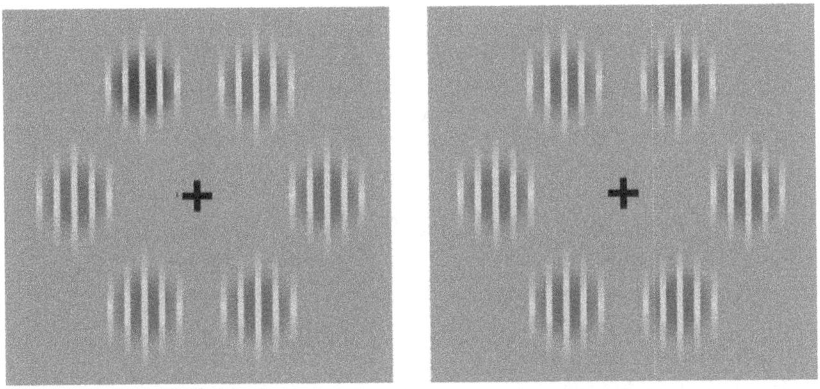

Figure 15.2
The signal detection task: these two images are presented one after the other. In one of the images, one of the six patches is slightly different. Is the odd one out in the first or the second image? (It is the first image, top left.)

Our claim is that this task is similar to the task of the shoal of fish. The group decision results in an enhanced signal to stand out better from the noise. The signal is a target, which is only barely visible. But what is the noise here? There is noise inherent in the visual system, and there is noise that arises from fluctuations in the attention of each of the observers. Their attention must remain very high because the task was difficult. Blink and it's gone.

We can assume that the observers had roughly equal ability to detect the stimulus, but that their attention fluctuated from trial to trial. This can explain why the pair can do better than a single individual. Their joint decision is always better if they go with the response made by the person who was more attentive at the time. The result is that over the course of the experiment, the noise is reduced and the signal is enhanced.

But how do the participants know which of them was the more attentive on a trial when they gave different responses? It was all in the discussions they had with each other. Kristian Tylén and Riccardo Fusaroli, experts in the computational analysis of language, saw the evidence very plainly when they analyzed these discussions (e.g., Fusaroli et al., 2012).

Reporting Confidence

What was the critical part of the discussions? It was about the confidence that each person attached to their answer. Mostly, the pair would go with the answer that was more confidently expressed (e.g., "I saw it clearly" versus "I just guessed"). While differences

CERTAIN

GROUP 43	GROUP 21
Saw it well	Sure
Saw it all right	Not 100% sure
Saw	More sure
Think I see it	Almost sure
Thought I saw	55% sure
Doubt that I saw anything	Sure—a little
Didn't see it	A little uncertain
Don't think I saw anything	Not quite sure
Couldn't see	Not sure
Didn't see shit	Unsure
Seriously, I didn't see anything	Far from sure
Don't see anything	Damn - I'm not sure
Didn't see any difference	Very unsure
I see nothing	Fucking unsure
Didn't see a thing	Too unsure
Only saw a blink	Totally unsure

UNCERTAIN

Figure 15.3
Verbal expressions of confidence: Each pair of participants developed their own verbal scale for describing how confident they were on each trial. The more quickly they developed such scales, the greater the advantage they gained from working together (Fusaroli et al., 2012).

in expressed confidence were plain to see, the different pairs used different words to describe their confidence[2] Figure 15.3 shows examples of two of the pairs of participants, with expressions graded from certain to uncertain.

The analysis of the discussions showed that over time, each pair developed a consistent way of talking about confidence. They did this readily, from scratch. But there are problems. What if they talk about confidence in very different ways and don't sort this out through their discussion? One person's "definitely, absolutely" may be another person's "possibly, maybe."

This problem is starkly revealed in a case study of forecasts made by intelligence officers of the Central Intelligence Agency (Kent, 1994, 75). Asked about relations between

2. Remarkably, this strategy was successful even when the participants were not getting feedback about whether they were right or wrong.

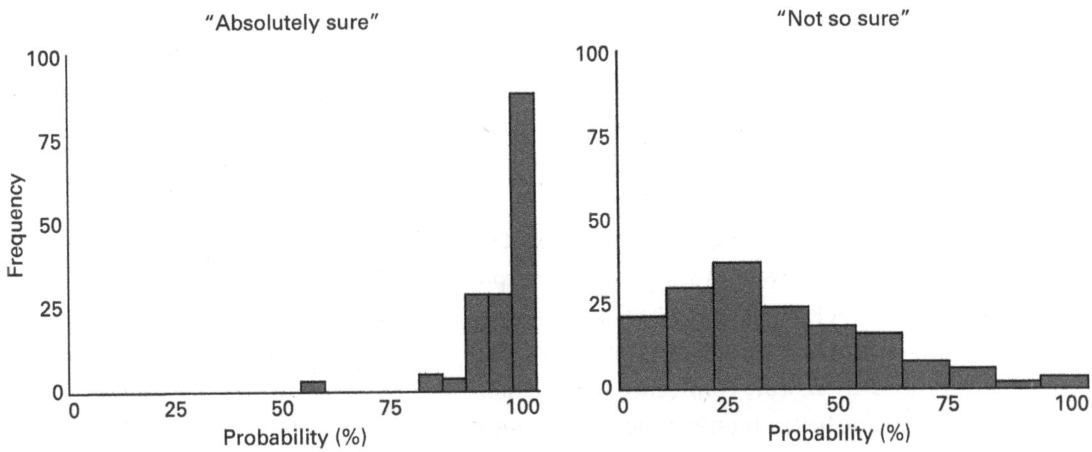

Figure 15.4
Variations in expressions of confidence: In a large study conducted on the internet, people were asked, if they were "sure" of their answer, and what the likelihood was that they were right. In the case of "absolutely sure," most people gave answers with between 85 percent and 100 percent certainty, but for "not so sure," the answers varied from 0 to 75 percent (Dan Bang, personal communication).

Yugoslavia and the Soviet Union, they had reported that "an attack on Yugoslavia . . . should be considered a serious possibility." But what did "serious possibility" actually imply? When asked to put a number on "serious possibility," it turned out that the intelligence officers, all expert forecasters, had very different probabilities in mind for the same term, ranging from 20 percent to 80 percent for the likelihood of an invasion.

The implications of this anecdote have been confirmed by Dan Bang. He showed that, in the absence of any discussion, people indeed attach a wide range of probabilities to the same verbal expression of confidence (figure 15.4).

How can we avoid the problem of people expressing their confidence in different ways? Perhaps we should force people to be explicitly aware of what they are doing and use numbers rather than words. In a study carried out in Asher Koriat's lab, observers were asked to rate their confidence in each answer on a numerical scale from 50 percent (guessing) to 100 percent and the response of the more confident person was chosen.[3] This explicit method also created an advantage for the pair (Koriat, 2012).

3. The confidence judgments were first standardized to remove the effect of individuals being consistently overconfident or underconfident.

Is an explicit strategy necessarily more effective for joint decision-making? Not so, according to a study carried out by Dan Bang and colleagues (2014). He found that the group advantage was greater when free discussion was allowed and there were no explicit instructions for talking about confidence. The correct answer was chosen by the pair more often when they "just talked," compared with when they were given explicit numerical scales to use.

We suspect that, by talking to each other about confidence, we can overcome the problem of the discrepant ways in which we may express confidence. Discussions are interesting as a form of joint action (see chapter 6), which means that we automatically tend to align our use of language. Indeed, observers readily converged in the choice of words that they used to indicate confidence (Bang et al., 2017). Further, over time, observers may well become better at estimating their degree of confidence. It seems plausible that simply by talking about confidence, they become more attuned to the subtle signals of confidence that emerge from the depths of the processing hierarchy (see chapter 14).

The Problem of Unequal Competence

There is no intrinsic value attached to confidence. Expressions of confidence are valuable only because they provide a shortcut to an estimate of someone's competence—how good the person is at collaborating and doing a particular job. This is a judgment that needs to be made in all cooperative endeavors. In any group, the competence of individual members varies depending on age, experience, and ability, and it is this very variation that makes groups thrive.

Each person will have their own special competences, and these can be taken advantage of, especially when the group is confronted with unforeseen circumstances. But to gain these advantages, we need some knowledge of the competences—both our own and that of others. An expression of confidence can be conveyed immediately, but it takes time and close observation to discover levels of competence.

The problems created by equating competence with confidence are evident in court cases where jurors have to assess whether to believe a witness. It is well established that eyewitness testimony is often unreliable and has led to miscarriages of justice (Albright, 2017), but this is still not fully appreciated by jurors. In one study (Fox and Walters, 1986), jurors gave greater weight to very confident witnesses even after they had been warned about the unreliability of this kind of evidence.

Confident witnesses are deemed more credible than unconfident ones. But once errors in testimony begin to emerge, a jury can start evaluating the competence of a witness and no longer needs to rely so much on confidence. Errors reduce the credibility of witnesses, but more in those who were confident in their erroneous testimony, and not so much in those who were less confident about it (Tenney et al., 2007). A person's reputation as a trustworthy witness will plummet if her or his confidence is found to be misplaced (Vullioud et al., 2017).

If collaborating partners are not of roughly equal competence, then group decisions are problematic. Here, the strategy of matching confidence will create a disadvantage. The person with the greater ability at the task *should* systematically report more confidence and be given more weight in group decision-making. But we can't rely on this. Indeed, it has been claimed that people who are more knowledgeable about a problem also know better how much they still don't know. Therefore their reports are often underconfident. On the other hand, people who have little insight in their lack of knowledge are often overconfident. This is known as the "Dunning-Kruger effect" (Kruger and Dunning, 1999).

Curiously, even if a difference in competence is obvious, the solution offered by the most competent person is not always picked. This might be because finding the best solution is not the only aim when two or more people are working together. Especially in very diverse groups, the predominant aim is often to achieve a smooth interaction and to give everybody a fair chance. But is this kind of goal a valid reason for making poor decisions in unequal groups?

This question was addressed by a somewhat devious version of the perceptual task described earlier. Here, one partner was deliberately handicapped by being presented with a degraded image, which resulted in more errors. The study was carried out in Denmark, China, and Iran, with the expectation that mismatched partners would act differently under different cultural norms (Mahmoodi et al., 2015). For example, fairness might be an overriding principle in Denmark but less so in China or Iran.[4] Surprisingly, no differences were found between participants from these countries.

The overall result was clear: too much weight was given to the less competent (i.e., handicapped) partner in all locations. This finding rules out the idea that cultural attitudes to trust or fairness are responsible for giving way to less able group members. Instead, it appears that the mechanism in play is impervious to such influences. Putting the responses of unequal partners on an equal footing may be a prior, inherited through evolution, that facilitates interactions with conspecific agents. We might call it an "equality prior." It would confer obvious benefits of smooth interactions when collaborations are the goal. However, its effect would be detrimental when attempts to build more accurate models of the world are the goal.

4. Data from the World Values Survey shows that 71 percent of people in Scandinavia endorse the statement "Most people can be trusted." But endorsement of this is much reduced in China (52 percent) and Iran (11 percent).

How Metacognition Helps Us to Make Better Group Decisions

The waggle dance of honeybee scouts indicates how good a new nesting site is. It is tempting to say that the dance represents what the scout "thinks" about the nesting site. But, of course, this is not the result of reflective introspection. But in what way are we different from bees when we make joint decisions? In chapter 14, we suggested that the signals that emerge from lower-level cognitive processes, such as speed and fluency are interpreted as indicators of confidence and are transmitted to others (e.g., Patel, Fleming, and Kilner, 2012). In this respect, we are no different from bees. But then things diverge. The communication system in bees is fixed and can only be applied to a small number of predetermined problems, such as finding nectar or selecting a new nest site. By contrast, humans can develop new communications systems on the fly (e.g., see Galantucci, 2005) and apply them to any kind of problem. Human group decision-making can take advantage of explicit metacognition.

When we communicate our confidence to others, usually in words, we may overrule the signals from our lower-level cognitive processes. The confidence that we report is not necessarily the confidence that we feel. Indeed, it is this possibility of decoupling that enables us to align with others (Bang et al., 2017). For example, if we are working with an underconfident person, we can change how we express our own confidence to match theirs. In this case, the confidence that we express will be lower than the confidence that we feel.

Playing the Confidence Game

If you search for confidence on the internet, you will find a multitude of sites giving advice on how to boost your confidence, such as "How many times have you told yourself that if you had more confidence, you'd be more successful?" Our basic belief seems to be that there are advantages to be gained by expressing a high level of confidence. This is borne out in practice. For example, self-confidence can increase influence over others, at least in the short term. Overconfidence helps people to sell their advice. Competition for influence drives overconfidence, independent of competence (Radzevick and Moore, 2011). But how long can the winning competitors sustain their position? This depends on the vigilance of their clients.

Another strategy for gaining influence is to exaggerate the precision of your advice. For example, a sports forecaster declared that a team had a "88.78 percent chance of winning its division" (Easterbrook, 2010). The precision that you attach to your advice is a marker of how confident you are. Confident people give more precise advice than less confident people, and this precision is perceived as a marker of competence. People prefer advisors who give more precise estimates (Jerez-Fernandez, Angulo, and Oppenheimer, 2013). We can play the same game here: By giving precise and confident advice, you can increase your social standing by 23.37 percent—and by reading this chapter, you can increase it by 37.23 percent!

However, the influence of competition on the expression of confidence can be more subtle. From the study of witness testimony (Tenney et al., 2007), we know that people really take notice when an overconfident person turns out to be wrong. It follows that if you want to ooze confidence, you better not make a mistake. On the other hand, when you feel ignored, you may choose to do so despite the risk of being wrong. Uri Hertz, with Chris Frith and other colleagues (2017), carried out an experiment to find out what happens when two advisors compete to give advice. They found that advisors were particularly likely to exaggerate their confidence when they were ignored, and unjustly so, since their recent predictions had been better than those of the other advisor.

We bother with our expressions of confidence because the feeling of confidence is derived from signals from the lower depths, but interpreted at the highest metacognitive level. This interpretation enters our awareness, and we can tell others about it. However, as we have mentioned already, the feeling of confidence that emerges into consciousness is not necessarily the feeling that we express. This is because, once the signal has become conscious, it can be manipulated in the hope of optimizing future outcomes. Unless you are completely transparent, others will not know whether you are sending a totally honest signal. Still, the witness who speaks with greater confidence is more likely to be believed.

Explicit signals are generally to the advantage of the group, as it allows everyone to align to achieve optimal decisions. It can also be to the advantage of the individual, especially when competing for influence. Our advice is to bear in mind the possible decoupling between the signals that you send or receive and the confidence that you or your partner feel.

Expressing Confidence in Science

All this applies to communications from scientists too. For example, when researchers write up their results in an article for a scientific journal, they use different language than when they write for a popular magazine. During the pandemic, scientists have discovered that politicians will follow their advice only if it is expressed with great confidence. When scientific results are publicized in the media, confidence is almost inevitably hyped. To counteract this, we ridicule statements such as "Drinking red wine makes you live longer."

Nuanced communication is a tricky issue. If results are contradictory, the reader wants to know whether group A's research is the one to go with rather than group B's. Can we assume that the expressions of confidence used by A and B are comparable? There is an unspoken code to avoid expressions of high confidence in scientific reports, even if the researchers privately feel very certain about their results.

An analysis of biomedical and popular science articles revealed that the percentage of words implying uncertainty (e.g., "could," "would," "should," "might," "can," or "may") is always much higher than for those implying certainty—by a difference of about 80 percent and 20 percent (Bongelli et al., 2019). This is only to be expected, since scientists are rarely able to draw decisive conclusions. Theories may be supported by empirical tests, but they are also undermined by findings that do not support them. In the end, a new theory replaces the old, and the cycle of testing starts again.

Naturally, we would say that in this book, we have taken care to screen out dubious hypotheses and potentially unreliable results. This is one of the reasons why we wrote the book in the first place. Still, we will always admit to only a medium level of confidence in our conclusions. You will find a lot of sentences containing "perhaps" or "possibly." Would you trust us if we made statements of high certainty? We doubt it.

The Wisdom—and the Folly—of Crowds

In our everyday lives, we constantly have to make decisions about what to do, and these decisions must be made in the face of uncertainty. Experience helps, but it will never be enough since the state of the world is always changing. By when we are with others, we can pool the experience and expertise of many and thereby make better decisions. This is the wisdom of the crowd, and it is in evidence throughout the animal kingdom (Couzin, 2009; Sumpter, 2010).

However, as we have all experienced, there are many cases where human crowds do not behave wisely. For example, if some members lack the expertise relevant to the decision, the automatic alignment of confidence reports lead to bad decisions (Bang et al., 2017). And these failures can arise even when problems are solvable and when all the members of the group have the same goal.

However, our ability to reflect on our subjective feelings and discuss them with others (in other words, our explicit metacognition) should, in principle, allow us to adopt tailor-made strategies to avoid failure and to make even better decisions than those that rely on unconscious processes alone. There are at least two ways in which explicit metacognition can help. First, the key automatic processes can be monitored and overridden when we know about the sort of problems they can create. Second, top-down strategies can be brought into play. Some of these derive from experience, prevailing norms, and instruction.

One example where reflection can help to avoid foolish decisions is to pay attention to the sources of information. Group advantages can be gained only when independent sources are combined. People in a group may base their decision on one and the same recent tweet. But they don't realize that everyone is using the same information. Rather, they imagine that there are different sources, and they are encouraged that all converge on the same answer.

Another danger to avoid is the information cascade, such as occurs in the buildup of a financial bubble. Individual traders are typically uncertain about the true value of the assets that they are trading. If they see other traders buying the asset, they assume that they are acting on good evidence. If the traders are particularly uncertain, then they may assume that the buyers have inside information that the asset is about to increase in value. Even though none of the traders actually has any inside information, they readily believe that the buyers do. They also start buying, and the price of the asset will go up well beyond its intrinsic value. In this way, small fluctuations in trading activity can end up having huge financial implications. The traders are inappropriately attributing knowledge to each other (De Martino et al., 2013).

It is not easy to be sure that sources are truly independent. Think of a committee in which many of the members have the same upbringing and background. They have all been to the same schools. They have read the same books and acquired the same prejudices. Therefore, they are not independent sources, and however expert they are, their decisions will not always be optimal. But this, of course, is an extreme case, which we hope rarely occurs in real life.

Or so we wish. Let us admit that this problem is not rare at all. However, it can be overcome if there are procedures in place for increasing the diversity of the group. We note here that we actively need to strive for this since there are powerful forces that act to reduce diversity. We all prefer to interact with people like us, our ingroup (see chapter 8). And here again, the internet is enhancing these effects. Google and Facebook deliberately target us with information that fits our preferences and those of people who are thought to be like us (Bozdag, 2013). We form information bubbles with like-minded people to confirm our own views.

But these forces have been with us long before the advent of the internet. Even if the members of our group are selected to be suitably diverse, these forces will act to reduce their independence. Inevitably, the members of a committee will start to see themselves as an ingroup. There will be an automatic drive to align with others and remain part of this group by demonstrating loyalty and similarity. This is likely to have a direct and damaging effect on the decision-making process.

What kind of damaging effects? Group discussions will tend to focus on information that is shared by all group members, often at the expense of information that is essential but only held by one group member or a minority (Kerr and Tindale, 2004). Shared information is more likely to be sampled because it is held by a greater number of people (Stasser and Titus, 1985). Group members will focus on shared information to increase their standing in the group. All this is due to the urge for alignment. We can signal our belonging to a group by emphasizing the knowledge that we have in common with others and ignoring the knowledge that is not shared. Likewise, others tend to judge us as more likeable and competent if they find out that we know the same things that they do (Wittenbaum, Hubbell, and Zuckerman, 1999).

The Benefits of Diversity

Diversity is a tool for groups that need to solve problems and create better models of the world (Sulik, Bahrami, and Deroy, 2021; Yang et al., 2022). It is worth pointing out that when we talk of diversity, we mean more than just diversity in the most commonly considered features, such as gender, age, and ethnicity. We think instead about diversity of perspectives, skills, experience, learning style, personality traits, and so on. At the same time, we would like to assume that everybody in the group has the same level of competence. As shown in the experiments discussed earlier, groups are not good at dealing with unequal competence, probably because we seem to have a default assumption of equity. This unfortunately weakens the group's ability to solve problems: too much weight is given to the less competent response.

A group of limited diversity is likely to see problems as simpler than they really are. Often, there are multiple solutions, and the first one you think of may not be the best. This is illustrated in figure 15.5, where the best solution means finding the highest point in a hilly landscape (Sanborn and Chater, 2016).

In figure 15.5, we can see that people with similar backgrounds believe that the peak they are standing in front of is the highest. In reality, though, there are higher peaks elsewhere. Having found such an apparently satisfying solution, people tend to devote their efforts to justifying it rather than looking for a better one (Mercier, 2016).

In a group with increased diversity, as shown in figure 15.6, there is a wider range of preferences and past experiences, and probably also a wider range of attitudes to problem solving. There is more likely to be a good mixture of exploiters, who take full advantage of the knowledge they have, and explorers, who continue to seek new

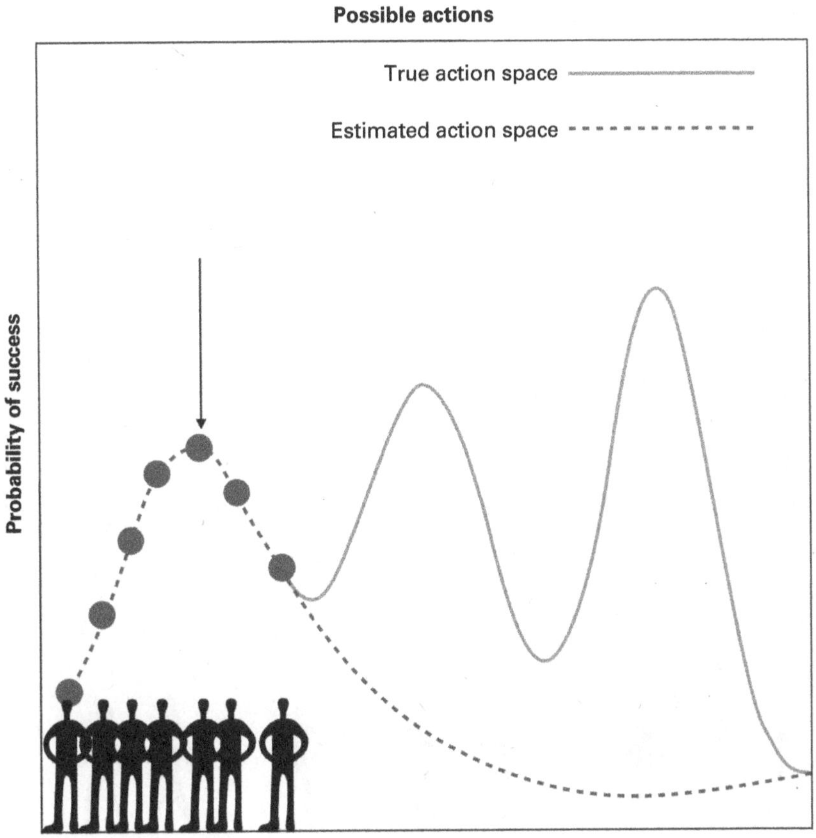

Figure 15.5
Searching for the best action—lack of diversity: these participants think that they've found the highest peak, but they have all searched in the same place (Bang and Frith, 2017).

knowledge (see chapter 12). With diversity, there is also less danger of solutions to problems being derailed by hidden biases because the individual biases are likely to be different and we are more sensitive to the biases of others than we are to our own (Pronin, Gilovich, and Ross, 2004). As a result of greater diversity, the range of possible actions will be more widely searched and the best solution is more likely to be found.

There are even more benefits. Diversity of equally competent group members will increase the likelihood of valuable discussion. This can increase the probability that information known only to a few will be considered, not just information known to all. Discussion gives room for opposing views and arguments.

There are many problems where the solution that first comes to mind is wrong (Tversky and Kahneman, 1974; Evans and Stanovich, 2013). One example is the bat

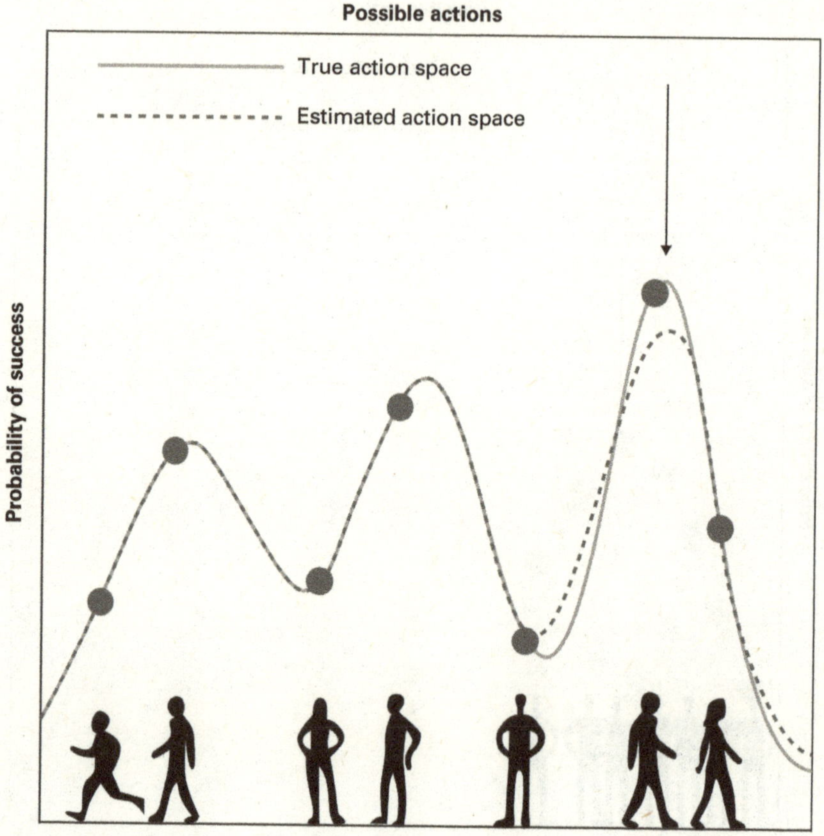

Figure 15.6
Searching for the best action—advantage of diversity: if these people can share their knowledge, they can find the highest peak (Bang and Frith, 2017).

and ball problem: "If a bat and a ball cost £1.10 and the bat costs £1 more than the ball, what does the ball cost?" Many people give the answer "10 p," but group discussion will usually establish that this answer is wrong, especially if somebody in the group disagrees with the answer that came to mind first (Trouche, Sander, and Mercier, 2014). A more complex problem that tends to lead to fast, intuitive, but incorrect answers, is the selection task introduced by Peter Wason (1968). This task demands logical reasoning, and only about 10 percent of individuals give the correct answer.[5] However, if the

5. All the cards have a letter on one side and a number on the other. The rule is: "If a card has a vowel on one side, then it must have an even number on the other side." Given the four cards **A C 3 4**, which two cards should be turned over to check the rule?

problem is presented to a group, then around 70 percent of groups come up with the correct answer (Moshman and Geil, 1998).

Where does this group advantage come from? The advantage occurs because people are naturally argumentative, and according to Hugo Mercier (2016), this is the basis of the human ability to reason. Here is an example: After one person has given the intuitive but wrong answer (the ball cost 10 p), another is likely to question it. This can lead to a rethink and a demonstration that the intuitive answer was wrong. "If the bat cost £1 more than the ball and the ball costs 10 p, then the bat must cost £1.10. So, together the bat and the ball will cost £1.20. This is wrong. The ball must cost 5 p."

Good arguments can achieve advantages for individuals too. For example, we tend to justify our own answers while questioning the answers of others. However, we can be tricked into questioning our own first answer if we are led to believe that it was given by someone else (Trouche et al., 2016). Argumentative discussion is a principle for lawyers who need to come to good decisions, and diversity ensures that such arguments will take place spontaneously. However, good arguments and good discussions take time, and some decisions can't wait.

The Problems of Diversity

With all the advantages of diversity, we must see the downside too. It can be summed up in one word: communication! We much prefer to communicate with people who are like us, precisely because it seems to need little effort. We have much shared knowledge and shared biases. We have developed a common language. We have the comfortable feeling of fluency when we hear the dialect of speakers from the place where we grew up.

With strangers and people who come from groups outside our own, we have to work harder to adjust to their dialect, style, and so on. It takes time to achieve the comfortable feeling of fluency and familiarity. There is no doubt that the lack of a shared language and culture can have a strong and often negative impact on decision-making at any level.

The reasons that make diversity good for group decision-making (namely, differences between members in background and attitude) are also the reasons that communication in a diverse group will be more difficult, at least initially. Communication is key to group interactions and the success of human groups. It is communication that we will discuss in chapter 16.

16 Communicating and Sharing Meaning

In this chapter, we discuss the kind of deliberate communication, verbal or nonverbal, in which both parties are aware that they are communicating. Such deliberate communication is uniquely human and is also known as ostensive communication. It is marked by the presence of ostensive signals that alert the receiver that the forthcoming message is relevant to them. These flexible signals range from a brief eyebrow flash to calling out somebody's name. To communicate successfully in this way, the individuals involved must take account of each other's intentions. Hence, there is an intimate connection to mentalizing. This connection makes it possible to convey deceptive information just as readily as truthful information, which manipulates the beliefs of others and is handy for achieving an advantage in competitive situations. Fortunately, in cooperative situations, truthful communication can dramatically improve learning and working together.

We know more about how meaning is transmitted between people than we know about meaning itself. Many but not all word meanings are inherently slippery and are interpreted depending on context. We propose that meaning is created through mutual adaptation and is a product of joint action, where signal-to-meaning mappings must work equally well for senders and receivers. A prerequisite is reciprocal alignment, which is manifest in copying and complementary action at many levels, including eye gaze, gesture, speech sound, grammar, and choice of words. The greater the alignment, the greater the success of the communication. We conclude that meaning does not reside in single brains. Rather, it is constantly re-created through interactions between brains.

* * *

"Hey, You"

We were introduced to the concept of ostension by Dan Sperber and Deirdre Wilson (1986) through their seminal book, *Relevance*. It has deeply influenced our own thinking, and we believe it to be fundamental for explaining human social communication. Before we unpack the general idea, we can summarize it as follows: behind every utterance is the intention to inform. This intention is often signaled by an ostensive gesture. For example, the shout "Hey, you" indicates that I have something important to tell you. Both the sender and receiver recognize that communication is taking place, and receivers expect the content of such utterances to be relevant to them A key aspect in this approach to communication is the importance of unconscious *inferences*. Just as with perception, the information presented in the utterance is never sufficient to determine meaning. We need to make inferences on the basis of prior expectations. Further, just as with perception, we are not aware of making these inferences. They are occurring in the depths, at the Zombie level. Most of the time, communication seems to us just as direct as perception (Recanati, 2002).

When you hear an utterance, you are likely to have responded to an ostensive signal. This signal tells you that a message is unmistakably meant for you, and often just you. Such a message is almost impossible to ignore. You may refuse to accept the communication or you may misinterpret it, but on the whole, the system works. Although ostensive gestures play a key role in initiating a communicative interaction, they are not necessary for sustaining an interaction once it has begun (Moore, Liebal, and Tomasello, 2013).

What are ostensive signals, and how are they sent and received? They are not fixed; they can use any of our senses, and they are always tailored to the situation. A speaker can make eye contact, call us by our name, or touch us. If they don't know our name, saying "Hello" or "Excuse me" might work. Hand gestures, particularly vertical movements, which are often exaggerated (as with lifting the arm high), can also indicate communicative intent. But usually, we prefer less flamboyant signals. In ordinary conversation, we tend to use nothing more than eye-gaze information as a sign that we wish to say something (Trujillo et al., 2018).

Thom Scott-Phillips and colleagues (2012) suggested that humans have developed a large number of signals specifically designed to indicate the intention to communicate in many situations. For example, if you wanted to request more wine at a party, you don't need to shout or carefully compose a request—you might just hold out your empty glass toward your host. Most likely, you would also show an eyebrow flash (see figure 16.1), a quick but effective gesture that neither you nor the host might even be aware of sending or receiving.

Human infants engage in communication with their carers soon after birth, using their face, body, and limbs. At six months, infants are already sensitive to ostensive gestures. They follow an adult's action with their eyes, but only when this action is preceded by a signal, such as direct gaze with eyebrow flash, direct touch, or infant-directed speech (Senju and Csibra, 2008). They seem to automatically understand that a useful communication will follow. Furthermore, at that age, infants are happy to watch two people apparently communicating with each other with electronic beeping noises and can use such beeps to learn to categorize objects (Ferguson and Waxman,

Figure 16.1
The eyebrow flash: the eyebrow flash is a brief facial gesture in which the eyebrows are raised and the eye widened. It is universally used as a sign indicating a desire for a social interaction with the person who is being looked at (Grammer et al., 1988).

2016). This suggests that the human brain has the necessary starter kit (or prior) for deliberate communication to get off the ground.

It is worth remembering that communication is not the same as language: communication can be nonverbal; language can be gibberish. There is evidence that this distinction has a basis in the brain. For example, Willems et al. (2011) studied neurological patients with pervasive language difficulties (severe global or agrammatic aphasia). Yet these patients could still successfully engage in communication when presented with a nonverbal task.

Ostensive signals are invaluable in language learning. Children are good at learning the names of things, even though this involves a mapping problem. Children solve this problem through recognizing the intentions of other people (Bloom, 2000, 2002). The child implicitly knows when an adult is naming the object for the child's benefit (e.g., the adult points at the object and says, "Hammer") and distinguishes this from the adult hitting his thumb with a hammer and saying "Blast!" If children were learning purely by association and eavesdropping, they would make many labeling errors. However, young children seem to have no trouble in sensing when a deliberate communication occurs. The exception is autistic children, who have problems with attributing intentions (Parish-Morris et al., 2007).

Depending on the occasion, ostensive signals can be part of the Zombie thread, such as the eyebrow flash, or part of the Machiavelli thread, such as "Greetings," shouted with arms extended. Indeed, it may be all there is to the message[1]. An early brain-imaging study using positron emission tomography (PET) was carried out in our lab (Kampe, Frith, and Frith, 2003). It compared the presence and absence of ostensive signals (without any message attached). Participants in the scanner saw a face either making direct eye contact or looking away, In another condition, the participants heard either their own name ("Hey, John") or another name. The effect of the two kinds of ostensive signal was the same. They triggered activity in the medial prefrontal cortex (mPFC), the site of the controller hub of the mentalizing system (see chapter 10). This study provided evidence that the mentalizing system is gearing up even before the start of an exchange. Mentalizing is bound to be involved since ostensive communication requires an understanding of the intention to inform rather than simply taking the utterance at face value.

---

1. This reminds us of Kurt Vonnegut's *The Sirens of Titan,* where aliens left a secret message when they visited Earth. The message was decoded after many years of arduous work by human scientists. What did it say? "Hello"! Perhaps it was a test message to establish a channel? If so, the aliens abandoned their wait for a reply.

The Secret Handshake

Irving Goffman (2008) speculated on how it would be possible for two strangers to meet in a busy hotel lobby in such a way that others would not even notice that they did not know each other before. The solution, formalized in game theory terms by Christoph Kuzmics (2018), involves the eyebrow flash during a direct stare. As we pointed out, the eyebrow flash indicates the sender's wish to initiate an interaction. But we need to consider the receiver's side as well.

The sender looks for a response indicating that the invitation has been received. In most cultures, the standard response to noticing that one is being stared at by a stranger is to look away. This indicates that if the gesture was an invitation to an interaction, then it has been rejected. In a busy lobby, people who are not expecting a meeting with a stranger will look away. Only the person expecting the meeting will return the stare. In this way, the two strangers will manage to discover one another and shake hands. This somewhat exotic example shows that ostensive gestures can make a reciprocal private communication possible with minimal planning, when this would not be possible with communication that is solely about information transfer.

A curious coincidence allows us to highlight the contrast between these two forms of communication. There is a method developed in computer science, which is also known as a "handshake." It involves a protocol set up to initiate an exchange of data between two computers. But this handshake is neither secret nor is it adapted to the situation. It consists of an inflexible routine. First, the computers need to be linked on the same network. Then, at the behest of a human agent, computer A sends a message to computer B requesting an interaction. Computer B returns a message indicating that the message has been received and understood. Computer A acknowledges that the message has been received and understood. At that point, data transfer can begin.

Some of us are old enough to remember the strange noises made by modems at the start of the connection with the internet. This was the sound of the handshake taking place. And we also remember the five-tone signal, the handshake used to initiate communication with the aliens in the movie *Close Encounters of the Third Kind* (see figure 16.2).

This example highlights the extraordinarily flexible nature of human communication. It also reminds us how much we depend on mutual trust when we communicate. We expect most communications to be true. We may not trust a mugger, but we do believe that he means it when he says, "Your money or your life." In contrast, when we examine data transfer between computers, we don't worry about trust, only about fidelity.

No doubt, we also use plain information when we can get it for free just by observing other agents. But information that another agent has preselected and sent to us

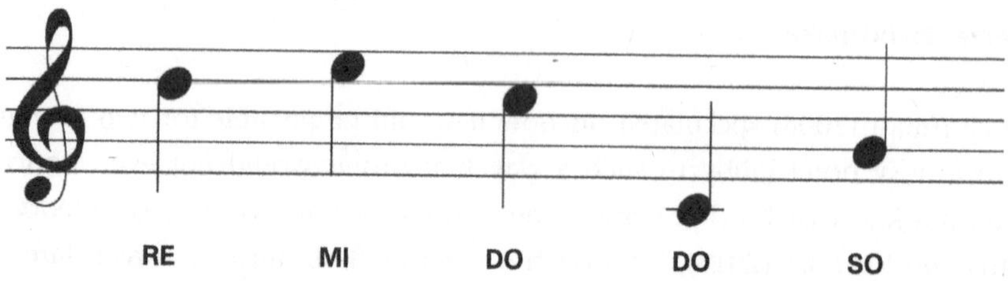

Figure 16.2
A handshake signal: the tune used by aliens to initiate communication with humans in Steven Spielberg's *Close Encounters of the Third Kind.*

with an ostensive signal is far more interesting. But there is a twist. The very fact that an agent wants us to have this particular information is likely to trigger our mentalizing ability in its Machiavellian color. We have to be on our guard. The question "Why did they say that?" seems to be hovering like a shadow behind each message.

Let us not forget that whenever we enter the arena of the mental world, there are not only ideas to explore, but there also lurks pretense and deception. And we are all playing this game. Who has not felt hurt by a sarcastic remark but hidden it by laughing? Who has not made a show of bravery when they felt anything but brave? As we argued in chapter 11, when it comes to gaining an advantage in competition, mentalizing reigns supreme. There, we highlighted lying and deception; here, we draw attention to their more subtle counterparts in rhetoric and persuasion.

Practitioners of the art of rhetoric take pains to distance themselves from deception. However, this does not protect them from being regarded with suspicion. In political debates and in the court room, we can observe blatant examples of insincerity and attempts to manipulate the audience's beliefs. Feigned emotion is a popular trick. And this works on social media. The presence of moral and emotional words significantly increases the extent to which a message will be shared (Brady et al., 2017).

The aim is to influence others in such a subtle way that they are not even aware of it. This is often achieved by advertisements, especially when they are wrapped in amusing and imaginative signals that grab our attention. What about the use of soft persuasion in books such as this, to make readers agree with the opinion of the authors? We do not exempt ourselves from such attempts, but we are aware of the duty to back our beliefs by evidence. We sincerely believe that our readers will benefit by gaining knowledge about human social nature. We also trust their use of epistemic vigilance to spot our biases and inevitable errors.

Communication Is Reciprocal

Another kind of ostensive gesture is pointing. Human infants start making well-coordinated pointing gestures at around eleven months (Butterworth and Morissette, 1996). Pointing gestures are an example of playful and pleasurable joint action, as they deliberately direct attention to points of shared interest (Liebal et al., 2009; Tomasello, 2010). While pointing (mainly using hands, but also head orientation and eye gaze) is extremely important in human social interaction and learning, it is rarely seen in other animals (Tomasello, Carpenter, and Liszkowski, 2007). Its use is by no means confined to young children and remains an essential tool for learning from others.

Communication enriches not only our ability to learn from each other, but also our ability to collaborate. In tennis, for example, winning doubles teams exchange around twice as many messages as losing teams (Lausic et al., 2009). When adults use communicative signals, such as smiles and eye contact, four-year-old children are more likely to cooperate (Wyman, Rakoczy, and Tomasello, 2012). Over and above these immediate gains, communication enhances the processes that drive the coherence of groups and their long-term success. Simply asking for help increases altruism and decreases selfishness (Andreoni and Rao, 2011). Communication, via face-to-face discussion, enhances group cooperation when confronting social dilemmas, such as the problems associated with free riding and access to public goods (Balliet, 2010).

Like any joint action, communication requires recognizing that there is a need for reciprocal adaptation and turn-taking. One person speaks while the other listens, and these roles are repeatedly reversed. Furthermore, conversation partners do not simply repeat each other; they exchange information. Even before they understand much of the content of a verbal exchange, infants, at around one year of age, can recognize such communicative turn-taking (Tauzin and Gergely, 2018, 2019).

We find it amusing if children aged around three claim that they cannot see others who cannot see them. If challenged, they may say that they cannot see someone whose eyes are closed, cannot hear someone whose ears are covered, and cannot speak to someone whose mouth is covered (Moll and Khalulyan, 2016). For young children, the basic nature of seeing, hearing, and speaking is for communication, and communication involves reciprocity. They do not consider communication to be occurring when one individual unidirectionally sends information to another while not being able to receive messages.

Communication is not just about achieving synchrony and mutual coherence. A genuine two-way interaction involves a form of reciprocity that consists of mutual adaptation. In the past, many studies of social cognition did not involve reciprocity.

For example, in studies of emotional mirroring (see chapter 4), people were confronted with a photo of the face of an actor who presented a certain expression. These studies showed that the expression on the face was often spontaneously imitated by the observer (Dimberg, Thunberg, and Elmehed, 2000). However, in real two-way interactions, emotional expressions are not always mirrored. Instead, we can see a form of reciprocal interaction.

For example, Carr, Winkielman, and Oveis (2014) showed that when a powerless person was confronted with a powerful person who presented an angry expression, they responded with a smile. They did not copy the angry expression. Perhaps they were trying to diffuse a difficult situation, hoping to mollify the angry individual. Expressing embarrassment is another technique for mollifying anger (Keltner and Buswell, 1997). Such interactions are examples of what we have called "closing the loop" (Frith and Frith, 2010, 165). In this case, the reciprocal exchange of facial expressions can be seen as a very simple communication system.

Alignment and Mutual Adaptation

As discussed in earlier chapters, alignment and mutual adaptation occur at many levels of joint action, and this is also the case during conversations. Thus, when two people talk to each other, they tend to align their rate of speech, their choice of words, and their grammatical constructions—all without conscious awareness (Garrod and Pickering, 2009). There is even some evidence that we comprehend better when we imitate the accent of the people we are talking to, but only if they don't notice (Adank, Hagoort, and Bekkering, 2010).

We can often observe individual differences in people's readiness to align their goals and means to reach them. For example, party bores are less likely to take account of the subtle signals of boredom that their listeners may send out. Verbally able autistic individuals may have no problem exploring possible signals for communication, but they often struggle to converge on shared signals (Wadge et al., 2019). The greater the alignment, the greater the success of the communication: better mutual understanding heralds better outcomes.

An example is the experiment described in chapter 15, in which two people had to come up with a joint decision when trying to detect a target that was only faintly visible (Bahrami et al., 2010). The different pairs spontaneously developed a scheme for communicating these feelings. Over time, individuals in a pair came to use the same words. Furthermore, they achieved a common understanding of what these words meant in relation to how confident they felt about their judgment. The more quickly a pair developed this scheme, the greater their advantage from working together turned out to be (Fusaroli et al., 2012). This is not an exceptional feat, unique to this experiment. Meaning is created spontaneously in social interactions all the time.

Alignments need not be directly concerned with language. For example, Daniel Richardson and Rick Dale (2005) studied eye movements while a speaker and listener were looking at the same picture. The more the listener's eye movements resembled the speaker's eye movements (with a delay of about two seconds from what the sender was saying), the better the comprehension of the listener. Analogous results have been obtained when measuring neural activity. In these studies, speakers tell a story while being scanned. The story is then played at another time to listeners who are also being scanned. The results showed that in the auditory cortex and angular gyrus, the listener's brain activity mirrored that of the speaker, with a delay of about three seconds. The strength of mirroring in the listener's brain was positively correlated with the quality of story comprehension (Silbert et al., 2014).

Many studies of alignment are conducted when sender and receiver are not communicating directly. The choice of such a paradigm is dictated, in part, by the practical problems associated with obtaining measurements from two people simultaneously. This is especially problematic for brain scanning. Then there is the problem of developing methods for analyzing the data so as to take full advantage of a simultaneous recording. In our opinion, such methods have yet to emerge (e.g., see Konvalinka and Roepstorff, 2012), but progress is being made (Wheatley et al., 2019). For example, one study showed that mutual understanding of novel signals synchronizes cerebral dynamics across communicators' right temporal lobes (Stolk et al., 2014). This interpersonal cerebral coherence occurred only within pairs who had a shared communication history.

Inventing Systems of Communication

Humans are remarkably adept at developing new communication systems on the fly. This allows us to share and update our knowledge. We revel in the flexibility of human communication, which makes alignment easy and pleasurable. This flexibility is constrained by several factors, including the cognitive biases imposed by how our brains work (Christiansen and Chater, 2008) and the external affordances and pressures imposed by the Umwelt, where the system develops (Nölle et al., 2018).

A study from the Donders Institute (Newman-Norlund et al., 2009) used nonverbal communication in a computer game. One partner had to indicate to the other, by moving a computer mouse, where a target might be located. When they believed that their partner in the game was a child, they slowed their movements and took longer pauses during the communicative part of the game. Similar strategies are associated with infant-directed speech (Brodsky and Waterfall, 2007).

Successful new communication systems can also be rapidly developed when only nonverbal signals are available (e.g., Galantucci, 2005). At the age of five years, children already can invent novel ways of communicating using gestures (Bohn, Kachel, and Tomasello, 2019). In a lab-based study of communication, people could successfully use the same sign for "odd-one-out," even when it signified something good in one context and bad in another (Misyak, Noguchi, and Chater, 2016). This is reminiscent of the easy use of now ambiguous terms such as "wicked."

Emojis have rapidly developed into a shorthand language adapted to the need to compress typing actions on smartphones. Just like spoken language everywhere, they are well suited to be used by one ingroup to differentiate itself from another. A frequently used example of an invented system are hobo signs, as shown in figure 16.3. They were supposedly used to communicate useful information within a fraternity while staying obscure to outsiders.[2]

The Elusive Problem of Meaning

So far, we have been tiptoeing around the probably most difficult question about the topic of this chapter: How does the receiver know what the sender *means*? Philosophers and experts in semiotics have speculated about this, and here we can only skim the surface. However, we would like to convince you that the creation of meaning, just like its transfer, requires mutual adaptation between the sender and receiver.

It seems plausible that intense experience in using language leads to shared knowledge about the conventional meanings of the words we use. But how do we get the knowledge in the first place? The popular answer from language development is that we acquire it through unconscious learning, by mere exposure to the world of objects and agents (see chapter 12), also referred to as "statistical learning" (Lany and Saffran, 2010). For example, without any conscious effort, our brain notices how words are related to each other through phrases such as "is a part of," "is a kind of," "belongs to," "is used for."

It is generally agreed that several routes lead us to make inferences about possible meanings for new words (Clark and Wong, 2002). One of them is a social route. As pointed out by Janet Pierrehumbert (2006), speech patterns become consistent through social interaction. This is evident in our everyday experience. For instance, when there are changes in the language used by our ingroup, we unconsciously adapt to them,

2. Remember that true and untrue statements are hard to distinguish in the world of ideas. It could well be that this story about hobo signs has no basis in reality.

Figure 16.3
Hobos (itinerant workers in the United States) were reported to communicate with each other by using a system of cryptic "hobo signs," which would be written in prominent places to alert future visitors. (Redrawn based on a photograph by Ryan Somma CC BY 2.0.)

modifying the way that we say things and the words that we use. For instance, we now feel wary of using the pronoun "he" and use "he or she" or "they" instead.

But words rarely stand alone. They have a context. Their meanings are not fixed but depend on what the speaker intends them to mean.[3] This would be entirely counterproductive for mutual understanding, but it works because the speaker and listener can

3. Irony (or sarcasm) is an extreme example where words can take their opposite meaning. For example, "Peter's so well read; he's even heard of Shakespeare" means that Peter is practically illiterate.

rely on a shared context, which primes the intended meaning. For example, the verb "to dust" can mean to remove dust from a surface—obvious if the context is cleaning—or to add icing sugar to a cake—obvious if the context is baking. However, it would be wrong to conclude that meaning is shifty and entirely context dependent. For the two meanings of the verb "to dust," there is a common feature: dusting conjures up a fine powdery substance in both cases.

Context is not everything. It is likely that there is also a fixed core to meaning, at least for objects in the physical world that to us seem constant. A moon is a moon regardless of its changing shape and position in the sky. We can refer confidently to the color red, even though how it appears depends on the light. The meaning in these cases seems outside us and untouchable by social context, and it can be conveyed by pointing gestures or verbal instructions from others. This now includes looking up Wikipedia, a rich source when we are unsure about the meanings of words. It is possible that this facility is helping to make the minds of Wikipedia users more similar to each other.

We note that meaning does not exist only in one-off creations. Even the most creative uses must somehow fit with past experiences and interactions. Such creative uses will also influence interactions over several generations. For this to happen, the brains in question crucially must share *common ground* (Clark and Brennan, 1991; Lee, 2001). Common ground is crucial to all collective actions, as it guarantees mutual understanding. It stems from what we refer to as "culture." In this sense, meaning resides between us and all around us, like the air we breathe. And it also exists around us in physical form in libraries and other locations, as well as on the internet.

How to assess when common ground has been achieved? This can be done indirectly, from the presence of synchrony as derived from eye movements and brain activity in people communicating with each other (Richardson and Dale, 2005; Hasson and Frith, 2016; Silbert et al., 2014). Some evidence for a link between common ground and alignment was obtained in a study where pairs of participants discussed paintings (Richardson, Dale, and Kirkham, 2007). Prior to discussing a painting, each participant heard either the same or different background information via earphones. Eye movement coupling was increased when participants had heard the same information.

In the early days of brain imaging, we sometimes asked, "Where is *meaning* in the brain?" But the more we have studied social interactions, the more we have realized that this is the wrong question. Meaning depends on use (Wittgenstein, 1953). It is something that emerges from interactions between people. We cannot know what we are talking about or thinking about unless we have interacted with others (Davidson, 1991). Now we might say that meaning is not in an individual brain; it hovers in the space between people. In other words, meaning is created and resides between brains that communicate, and the observations of mutual alignment give some substance to this idea.

How Is It Ever Possible to Understand Each Other?

We can communicate successfully only when we have a common ground. But how do we judge that the communication has been successful? This is a tricky question, but it can be addressed by looking at alignment and mutual adaptation.

The problem that we face here is sometimes referred to as "indeterminacy of reference," or, more generally, "indeterminacy of translation." This problem is the reason that we get frustrated so often when we look up a word in a dictionary. We suddenly realize that there is no one-to-one mapping relationship between words and meanings, let alone between words in different languages. But the problem is evident in the attempt to learn the meaning of a word in the first place. Here is the famous example provided by the philosopher Willard van Orman Quine: Imagine that you are in some remote and wild land and have no notion of the language spoken there. You see a rabbit run past, followed by a man shouting "*Gavagai*!" What does *gavagai* mean? Is he referring to the rabbit or saying "Let's go hunting" (Quine, 2013)? How can we know what the word refers to?

This problem of indeterminacy (lack of one-to-one-mapping) is not restricted to communication. In motor control theory, it is associated with the inverse model (Wolpert, Ghahramani, and Jordan, 1995). How do I choose one movement from among the many possible ones that I could make to achieve my goal? This problem is even more acute when I want to make inferences about the goals of another person. Remember the example of a person picking up a glass either to drink from or to throw the contents at an annoying party guest (Jacob and Jeannerod, 2005; see figure 12.5)? There is no fixed one-to-one relationship between the movements that we see and the intentions behind them, just as there is no one-to-one relationship between the words that we hear and the meaning behind them. The relationship is one-to-many.

Even when we are talking to someone face-to-face, and in our mother tongue, how can we be sure that we have understood them? Nevertheless, for much of the time, we feel that we can communicate perfectly easily. We believe this is possible because we can mutually adapt to each other and, thereby, reduce misunderstandings. Karl Friston and Chris Frith (2015) have developed what is essentially a Bayesian approach to the problem that depends on this mutual adaptation (see also Pickering and Garrod, 2007).

To some extent, we are all trapped inside our skulls and cannot know about the outside world except via the signals that are coming in through our senses. And this problem applies as much to the physical world as it does to the mental world. We suggest that the problem can be solved by making some rather free assumptions. We assume that we create internal models of the things in the world that are causing our sensations. We are answering the question floating at the back of our mind: "Why is this happening?" All this is part of the Zombie thread of information processing.

Here is an example. A friend says that she is in pain. "Why is she saying this?" might be the question that hardly reaches our awareness, but it still conjures up a reason for what is causing her to tell us this. What springs to mind is that she is asking for our help. Should we get ready to find paracetamol? On the other hand, she may want to let us know that she objected to what we just said. In this case, we might want to apologize to her or appease her in some way. Usually, the current context helps resolve these two possibilities (the models), but if not, we can ask for clarification. The new information may fit our currently preferred model. On the other hand, it might elicit a prediction error, which means that we must come up with a new model. As we shall see in chapter 17, the processes are remarkably similar to what happens between teacher and learner.

But this can happen only if there is sufficient commonality and goodwill to understand each other. Communication is bound to be problematic if the people involved subscribe to entirely different belief systems. This is the reason why communicating with outgroup members is much more stressful than with ingroup members and may lead to increasing misunderstanding and polarization.

Using Misunderstandings to Improve Our Ability to Share Ideas

The basic requirement for the successful sharing of ideas between two brains (at the Zombie level) is that each brain must model the other. In the predictive coding framework from which this approach is derived, a brain is trying to model that state of the world that is causing sensory input. But in an interaction with another person, the

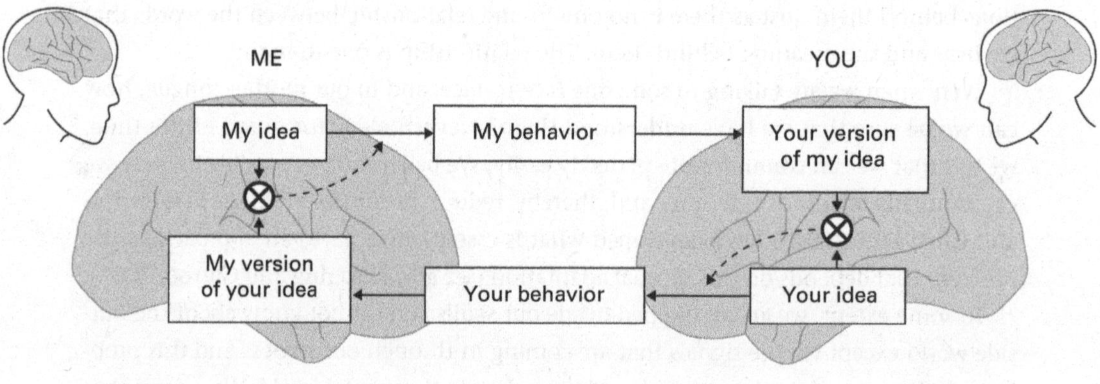

Figure 16.4
If my idea doesn't match my version of your idea (prediction error), then I must change my behavior. And you will be doing the same. Once the prediction errors are sufficiently small, we will have successfully shared our idea. (Figure 9 from Friston and Frith 2015).

sensory input is being caused by the other brain. And when there is mutual communicative interaction, the second brain is also trying to model the first brain.

Recursion looms! Must the first brain have a model of the second brain, which includes a model of the first—and so on ad infinitum (see chapter 14)? Fortunately, this infinite regress can be avoided if the two mind/brains are sufficiently similar and each brain models the sensations caused by itself and the other as being generated in the same way. In other words, if there is a shared idea or concept that both brains subscribe to, they can predict each other exactly, at least for short periods of time. It is this mutual prediction that indicates that successful sharing has occurred.

This solution is a necessary and emergent phenomenon when two or more (formally similar) active inference schemes are coupled to each other. Mathematically, the result of this coupling is called "generalized synchronization" (or "synchronization of chaos"; Rulkov et al., 1995). Such synchronization is an inevitable and emergent property of coupling two systems that are trying to predict each other (Friston and Frith, 2015).

In practice, what we are trying to do during our conversation is predict what our partner is going to say next. If what he or she says is unexpected (a prediction error), then we need to rethink our understanding and try again. As the conversation progresses our predictions become better and better, and we each go on in this way until the errors are so small that we are pretty confident that we understand each other. Even in the simplest interactions, these prediction errors are of vital importance because they signal misunderstanding. Here, both implicit and explicit signals are used. They can be vague and nonspecific, such as "er?" or "what?" There are similarly subtle signals for expressing the degree of understanding, such as "yes," "absolutely," and "OK." It is precisely the subtle signals of misunderstanding that play the major role in coordinating meaning during dialogue (Healey et al., 2018). Even if they are minimal, they avoid a slash-and-burn approach and help to keep conversations civilized.

The problem of infinite regress keeps lurking in the background when we try to explain reciprocal understanding in people who may not share a model of the world. But it is the same as the problem that emerges when we want to collaborate in simple competitive games like the battle of the sexes or stag hunt game (Skyrms, 2003; also see chapter 9). Before I decide to collaborate with you to hunt stags, I must believe that you will collaborate with me, that you believe that I will collaborate with you, and that you believe that I believe that you will collaborate with me (and so on).

The answer, of course, is to develop a shared model of the world. We should aim to finish our mutually adapting process with a single, shared representation of meaning toward which our conversation will converge (see also Lee, 2001). In the case of the stag hunt, our shared model would be that you and I (we) both are the sort of people who collaborate. This shared model is added to the shared knowledge and background beliefs on which our ability to communicate depends (Clark and Brennan, 1991).

But a conundrum remains. Through communication, we create shared models of the physical and the mental worlds. But we are able to communicate successfully only because we have these shared models.[4] Do we need to worry about which comes first? No. It is our drive to interact with others and discover shared meanings that squares this circle. As Robert Frost said, "You can never tell what you have said or done until you have seen it reflected in other people's minds" (Lathem, 1966, 71).

4. This problem is reminiscent of the hermeneutic circle, where we can't understand the whole without understanding the parts, and we can't understand the parts without understanding the whole.

17 The Power of Teaching

Learning from others by teaching is distinct from learning by observing others and is intimately connected with ostensive communication. Only human teachers deliberately communicate selected information, directly addressing their pupils and holding their attention. We suggest that mutual adaptation is key to human teaching, as it is for other kinds of joint action. The difference is that for teaching, we need mutual adaptation at a higher level than the level of action. Teacher and pupil both need to take into account the hidden states of knowledge in each other's minds. This requires modeling other minds, which is possible through mutual prediction.

Teaching has not always been appreciated as a form of social interaction, with all its complications. For example, education is used to influence and shape children's group identity, while the desire for affiliation with one's peer group exerts a force of its own.

Giving and receiving instructions are elements of teaching that transcend a school context. When we follow instructions, we experience a remarkable upgrade in the ability to learn from others. Simply being told what to do replaces the need to observe others and enables us to perform entirely novel tasks. A possible mechanism for this kind of learning involves downweighting the prediction errors associated with trial-and-error learning and upweighting prior beliefs about the situation. In this way, instructions create the prior beliefs about how the world works in our culture.

* * *

Different Ways of Learning from Others

Books and nature programs love to tell stories about parent birds teaching their fledglings to fly and mother cats teaching their kittens to chase mice. Alas, these stories are blatant examples of anthropomorphizing. Birds fly and cats rush after small objects, all without being taught. When a kitten chases a ball of paper, this is because the kitten's brain has built-in motor responses for hunting triggered by certain stimuli. Small, moving objects are such triggers. The kittens are learning, but the mother cat does not have to do a thing.

Most young animals, including humans, learn from observing their elders. This allows them to overcome their limited experience as individuals. Their nervous system is primed to copy what other agents are doing, while the agents display little if any interest in them. They tend to be oblivious to the fact that they serve as models and providers of information. We would not call these agents "teachers"; they are simply behaving as usual.

Teaching, at a minimum, requires agents to directly address their pupils and actively facilitate the transfer of information (Rendell et al., 2011). Such a basic form of teaching, although rare in animals (Kline, 2015), has been observed in ants, bees, and some birds (Hoppitt et al., 2008). Meerkats (*Suricata suricatta*) provide a particularly interesting example. Here, a relatively sophisticated form of teaching has been observed, called "opportunity teaching" since the teacher is giving the pupil the opportunity to interact with dangerous prey (Thornton and McAuliffe, 2006).

These amazing animals implicitly take account of the state of the learner when feeding their pups. They give dead scorpions to the youngest pups and offer live scorpions with their stings removed to the older ones. At a later stage, the pups are confronted with intact live scorpions, and the parent remains nearby to monitor the pups' handling of the prey. The adult meerkats do not have to think about this since the young pups emit begging calls that trigger the right response in the parents. These begging calls change as the pups grow bigger, and the parents alter their teaching accordingly.

As far as we know, the meerkats do not have any explicit recognition that they are teaching their young. It is precisely this explicit recognition that can elevate teaching to a different level in humans. Humans are unique in practicing the kind of teaching that involves the top level of the information-processing hierarchy, involving explicit mentalizing and mutual adaptation of knowledge.

We do not wish to claim that teaching is always superior to other forms of learning. Learning with and without a teacher happen side by side (Schulze and Hertwig, 2021). Different types of learning have their place, with unconscious and conscious processes often intermingled. For instance, we can learn spoken language very efficiently by mere exposure, Zombie fashion. However, we learn written language only by being taught explicitly, Machiavelli fashion.

An empirical study of adults learning to decode a new writing system either with or without explicit instruction found that explicit instruction was vastly superior to repeated exposure (Rastle, Lally, and Taylor, 2021). In another study, people were given a choice between learning by copying or by following verbal instructions. They much preferred to follow verbal instructions, so long as they trusted the instructor (Hertz, Bell, and Raihani, 2021). In general, explicit instructions are helpful for learning rule-based tasks. But there are many complex tasks where it is difficult to state explicit rules, and hence it is difficult to give instructions. In these cases, people are better off learning by copying or mere exposure, especially when it is necessary to integrate information from many sources (Rosedahl, Serota, and Ashby, 2021).

Formal teaching often relies on verbal instructions and is very variable across cultures (Legare, 2017)—and so are its aims. Ideally, teacher and learner have agreed on a common goal to coordinate their actions. We assume that this was the case even in preliterate societies. It is hard to believe that they would have succeeded in grand projects, such as the building of massive earthworks and stone circles, without fundamental knowledge of geometry and mechanics, and without transmitting this knowledge across generations. We presume that they had a form of teaching that accumulated knowledge and incorporated inventions and innovations over time. The fact that this was possible without the benefit of writing is astonishing.

Whatever other kinds of teaching are practiced, it is explicit teaching that enables humans to share stories that aim to explain the world and our place in it. Such stories are continuously and flexibly constructed through our ability to create and share meaning. Indeed, sharing them is crucial for creating a cohesive society. This is why almost every government in every country puts sufficient value on education to allocate it a considerable proportion of taxpayers' money. As we shall see, mass education turns teaching into a powerful engine for imparting information about all the things that matter in a particular culture.

In discussions of education, it is easy to remain fixated on the learner. We suspect that the extent to which teaching is based on mechanisms of *social* cognition has not always been acknowledged. In this chapter, we would like to overturn the view that teaching is a one-sided affair, with the teacher providing information and the pupil receiving it. Instead, we want to make the case that, for teaching to be more successful, information providers and receivers need to be working together through mutual adaptation.

A Novel Form of Social Learning

Before discussing advanced forms of formal teaching, we need to consider a form of implicit teaching identified by Gergely Csibra and György Gergely (2009). This is "natural pedagogy," defined as a spontaneous activity that is typical of human parents. Parents or carers of infants use ostensive gestures to show when they are teaching, and infants respond to these gestures. For example, infants will follow the direction of the gaze of an adult specifically if this is preceded by gestures such as eye contact (including an eyebrow flash) or infant-directed speech (Senju and Csibra, 2008).

But this is not all. These interactions are effective means of learning. Infants spontaneously assume that messages conveyed in this way are worth remembering. They interpret the information to be about durable meanings that can be generalized (Yoon, Johnson, and Csibra, 2008). In this way, they learn effortlessly, that apples are fruit and good to eat. In contrast, when infants observe an adult either eating or rejecting an apple, without there being an ostensive gesture, they learn that this adult likes or dislikes apples (Egyed, Kiraly, and Gergely, 2013; Henderson and Woodward, 2012). All this makes natural pedagogy a new type of social learning that may well represent an evolutionary adaptation in humans (Csibra and Gergely, 2011).

In addition to using ostensive gestures to indicate that useful information is forthcoming, adults spontaneously modify their teaching behavior to match the needs of the learner. A good example of the teacher's implicit sensitivity to the learner is seen in an experiment, which compared the language used by a mother to her infant and to her cat (Burnham, Kitamura, and Vollmer-Conna, 2002). When talking to her cat, the mother simply raised the pitch of her voice. She also did this when talking to her infant, but here she did something more. She exaggerated the vowel space as set by the mother tongue (see figure 17.1). This leads infants to attend to the most important variations in the sounds of their native language.

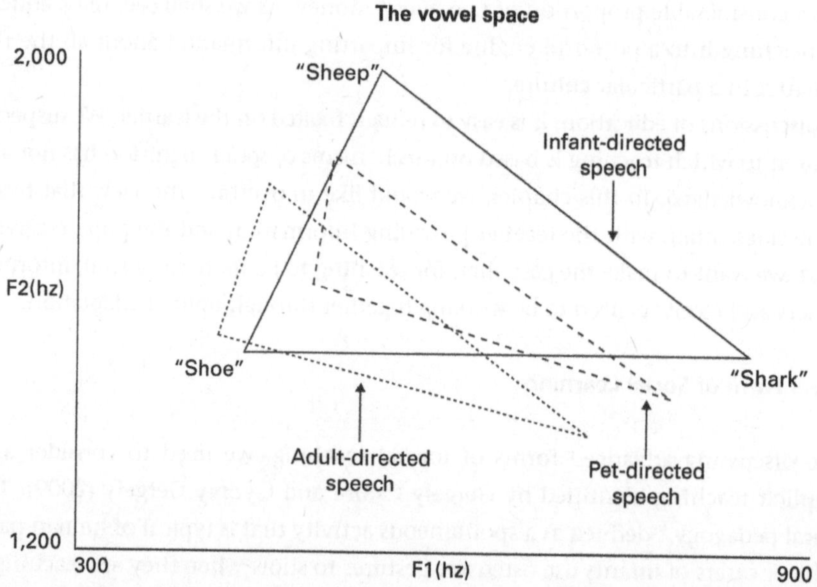

Figure 17.1
Infant-directed speech. The sound of a vowel is determined by two frequencies (F1 and F2), which place each vowel in a particular position in a "vowel space." When speaking to her cat, the mother simply increases the frequency. When speaking to her child, she enlarges the space between the vowels. Redrawn with permission from Burnham et al. (2002). Copyright 2002, AAAS.

Language directed toward children, known as "motherese," is not just uniformly modified, it is fine-tuned to the development of the individual child. For example, in a "guess the animal" game, parents give more informative clues about animals that they think a particular child does not know (Leung, Tunkel, and Yurovsky, 2021). Adaptation is also critical for learning a second language. In a much-cited experiment, Kuhl, Tsao, and Liu, (2003) taught infants Mandarin either by video or in person. The video was prerecorded, so no adaptation was possible. In this case, the infants showed no learning of the novel speech sounds. However, it would be wrong to conclude that teaching via video link does not work at all. A later study using live video chat showed that teaching in that way succeeded if there was the opportunity for mutual adaptation (Roseberry, Hirsh-Pasek, and Golinkoff, 2014).

Mutual Adaptation in Teaching Is Hard

Beyond natural pedagogy, adults can use more sophisticated forms of teaching and begin to use these when children acquire language. Thus, they can turn the process into one of joint action. In chapters 5 and 6, we presented evidence from a range of experiments suggesting that a key feature of working together is mutual adaptation in both partners. But for teaching, this presents a problem: Teachers and pupils inevitably differ in competence. The asymmetry is seen in extreme form in very young children. In the examples given previously, the adaptation required is the burden of the adult. In many ways, children have little choice but look up to adults as authorities who know much more than they do. For example, three-year-olds will believe what a teacher says, even when it is inconsistent with their own direct experience (Jaswal et al., 2010).

How do we overcome this problem of asymmetry? The teacher must adapt downward, the pupil upward. The teacher always has to be a step ahead, but not too far ahead. The influential Soviet psychologist Lev Vygotsky was known for drawing attention to the role of social interaction in teaching (Vygotsky, 1978). His groundbreaking idea was that there is a zone of proximal development, the difference between the level that the learner has achieved, where he/she can work independently, and the next level that he or she could reach under guidance. The pupil will learn best if both levels are identified correctly. In principle, this could be achieved through peer-to-peer teaching with teachers being only slightly more advanced than learners.

We can throw some light on the mutual adaptation between teacher and learner with the help of the deceptively simple tapping experiment discussed in chapter 6 (Konvalinka et al., 2010). Two players had to tap on a keyboard in synch, while also keeping to a particular tempo indicated by a metronome. They were wearing earphones linked to the keyboards in such a way that they could not hear themselves, but they could hear each other. Thus, A could hear B and B could hear A. Of course, the taps did not always occur at exactly the same time, and there was a continuous attempt by the players to adjust their speed to achieve synchrony. When A's tap was later than B's, then A tried to speed up, but, at the same time, B was slowing down, as his tap was earlier than A's.

This tricky adjustment was achieved quite unconsciously, yet continuously. It aimed at what we have called "closing the loop." In many examples of social interaction, one agent simply responds to another, but this is different with mutual adaption. There are always tiny gaps between responses, and these elicit a reciprocal response, so the loop can be closed (Frith and Frith, 2010). Usually, a leader-follower relationship develops. In the case of the simple tapping task, this happens when A concentrates on keeping in tempo and B is mainly concerned with maintaining synchrony (Konvalinka et al., 2014). The very rapid and continuous nature of the adaptation is possible only because it happens through automatic information processing that takes place outside conscious awareness.

The leader-follower relationship in the tapping task seems to us highly relevant to teacher-pupil interactions. If the teacher aims to transfer new knowledge to the pupil, he or she must adapt to the pupil's existing knowledge. The pupil, on the other hand, must be able to integrate the new information into his or her repertoire before moving to the next stage. But how does the transfer of information between teacher and pupil actually work?

Uncovering Hidden States

The kind of communication required for mutual alignment in deliberate teaching is neither simple nor direct. Mentalizing is needed to discover hidden mental states. In the case of teaching, these are the hidden states of knowledge in both teacher and pupil, and the ideal aim is that this hidden state in the mind/brain of the teacher is re-created in the mind/brain of the pupil. However, neither teacher nor student can directly perceive the hidden state of the other. This is another example of the tricky problems that must be solved for successful communication to occur.

Given that we believe that the brain is essentially a prediction engine, one solution is to let this engine estimate what the hidden state might be. One possible computational solution to this problem was discussed in chapter 12, where a study suggested that it is possible to read intentions from someone's movement kinematics (Kilner, Friston, and Frith, 2007). The engine makes an informed guess about the current state and then predicts what the person will do, given this hidden state. This process (see chapter 13) allows a continuous adjustment of guesses until the predictions are sufficiently accurate. At this point, the interpretation is likely to be correct.

Figure 17.2 (taken from Frith, 2007) suggests how this Bayesian approach to teaching might work. Note that the descriptions of what teacher and pupil are doing are

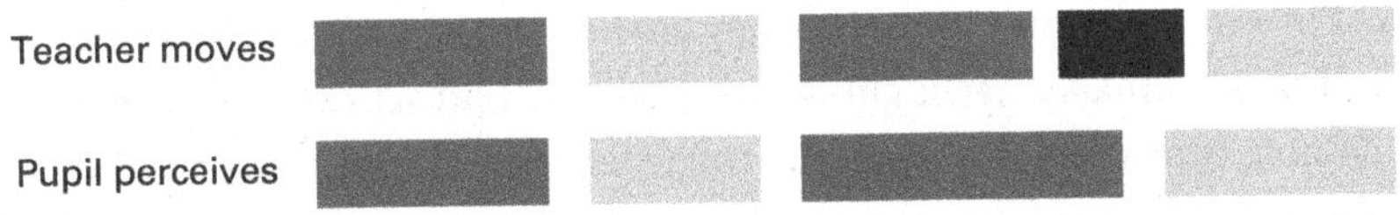

Stage1

Stage 1: *Observation.* The teacher performs a skilled movement comprising a sequence of five hidden control states (indicated by the different shades of gray). The student watches and tries to read the sequence of control states from the movement. This is his model of the hidden states of the teacher. His model is not completely accurate—he misses number 4.

Stage2

Stage 2: *Imitation.* The student imitates the movement using only four control states. The teacher watches and reads the control states from the movement. She sees only four control states. She knows that she used five control states. She identifies the difference between what she thinks is the student's representation of the skill and her own representation of the skill (indicated by the square box). This prediction error indicates that the student has not fully grasped what is required.

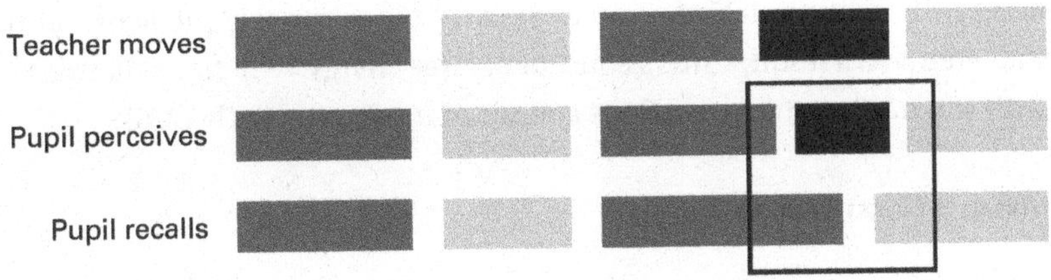

Stage3

Stage 3: *Teaching.* The teacher moves again and exaggerates the missing control state. The student now correctly reads the five control states. He recalls that he used only four control states the last time. He identifies the differences between what he thinks is the teacher's intention and his own intention. This prediction error indicates that he has not got the movement quite right. When he moves next time, he will correct the error, and the teacher will know that the skill has been successfully transferred.

Figure 17.2
Redrawn from figure 7.4 from Frith (2007). Copyright 2007, Wiley.

expressed in a way that presumes that they are fully aware of what they are doing. This, however, is most unlikely. Their guesses, predictions, and actions are mostly happening in the Zombie thread. In this example, the teacher and pupil have succeeded in sharing hidden states of a motor control system, which is most certainly impenetrable to conscious awareness.

The same principles can be applied to the sharing of ideas and concepts where the medium of exchange is words rather than movements. This is the system for sharing meaning that we described in chapter 16 (see figure 16.4). This system could be the mechanism that enables the form of communicative teaching that seems to be unique to humans. Consider the situation in which the teacher (let's call him Karl) is helping the student (let's call him Chris) to understand a new concept ("free energy"). Chris, the naive student wishing to be enlightened, will assign great importance to prediction errors to update his naive prior beliefs about the nature of this new concept, while the knowledgeable teacher will try to change the belief of his listener.

For the knowledgeable teacher, prediction errors indicate that his listener still has not understood. What Chris has just said does not fit Karl's understanding of the concept. So Karl will try to change Chris's understanding of "free energy" and may even modify his own understanding to improve communication. However, this is not enough. What Karl has just said does not fit Chris's current understanding of the concept. So Chris must update his representation of the concept. The ideal result of the interaction will be a generalized synchronization between the speakers in which prediction errors have been minimized. The emergence of such synchronization indicates that the concept has been successfully communicated—and each party can accurately predict what the other will say about it.[1] As a result, Chris's concept of "free energy" will have changed a lot, but even Karl's will have changed a little as a result of interacting with Chris.

What Makes a Good Teacher?

Natural pedagogy is accessible to every one of us, as it happens spontaneously, in Zombie fashion. Deliberate and explicit teaching, on the other hand, is a rather special skill and results in a significant upgrade of social learning. However, it comes at a cost and involves the Machiavelli thread. It takes conscious effort and is far slower and sometimes painful to achieve. We believe that improvements could be made if there were a more advanced science of teaching. As a start, it would be helpful to know what makes a good teacher.

1. Sadly, in this case, successful communication never quite happened.

It seems that from very early on, children recognize that some people are better teachers than others. Poulin-Dubois and Brosseau-Liard (1996) review evidence showing that preverbal infants are remarkably precocious in deciding from whom to learn. They prefer to learn from a competent teacher (here this was someone who had labeled objects accurately). At fourteen months, children copy the actions of a competent model more often than those of an incompetent model, where competence has previously been demonstrated in the mastery of a skill, such as putting on a pair of shoes (Kutas and Federmeier, 2011).

By four years, children can predict whether an informant will be accurate in the future. They seek and endorse information from accurate rather than inaccurate informants. One study was about accuracy in naming. For example, an inaccurate informant would call a toy car "a duck" (Koenig and Harris, 2005). Competence and accuracy are obvious criteria for being a good teacher, but there must be many others. From our framework, we would predict that teachers need to be sensitive to the knowledge state of the pupil and able to adapt their teaching to this state. Are some people better at being adaptable in this way?

Teaching adolescents has its own challenges. Sarah-Jayne Blakemore (2018) reviewed evidence that during this period, the need to affiliate with peers is especially strong. This can often result in a loss of common ground with parents and teachers. Conflict is likely whenever an older generation's demonstrations are different from the actions that are normative for the pupils' peer group. Various factors may affect the outcome of such a conflict, including the preference to learn from ingroup members (chapter 8). In one experiment, four- to five-year-old children preferred to imitate the consensus action of their group rather than the action performed by a reliable teacher. However, six-year-olds preferred to follow the reliable teacher (Bernard, Proust, and Clément, 2015).

In the age of mass education, the aim is to teach all children equally and without bias except for the case of learning disabilities, where special education is the answer. It is understood that some teachers need to be trained to respond to special needs, such as those caused by neurodevelopmental disorders. There is a constant demand to improve teacher training and to conduct basic research that may eventually lead to better methods of teaching. With the development of online teaching, it might be possible to create individually tailored matches between teachers and learners in the future. However, if artificial agents are to become good teachers, then they must be able to read the current state of knowledge of the learner and adapt appropriately and rapidly to the learner's responses.

Telling People What to Do

The problem of how we know whether important information has been transferred successfully does not just apply to teaching in schools. It applies more generally to any situation in which one person tells another what to do. By studying the effect of verbal instructions in the lab and in the scanner, we can make some progress in understanding the processes involved in explicit teaching. We would do well to understand them since explicit teaching greatly facilitates the acquisition of knowledge about all that is important in our culture.

Instructions can be indirect and become more powerful for that. Much of our knowledge of the physical and the social worlds comes to us through what other people tell us. We know what to do after we are told that black widow spiders are very dangerous, but false widow spiders are reasonably safe. And in the social world, we know what to do when we might find ourselves using escalators in Japan: in Tokyo, you walk on the right side of the escalator and ride on the left side, while in Osaka, it is the reverse.

However, telling people what to do is a problem that produces intense worry to us as experimental psychologists. This is because for every experiment, we have to carefully consider how to tell our participants what we expect them to do (Kihlstrom, 2002). The words must be the same for every participant, and they need to be utterly clear. This is one reason why we often restrict our subject pool to native speakers of English and people who come from the same cultural background.

But even obsessive care about the exact wording of the instructions is not sufficient. Is the task that the participants are doing the same as the one that we intended them to do? As an example, we can return to the joint task developed by Bahador Bahrami (2010) and discussed in chapter 15. The task was to detect a signal, and the experimenters tacitly assumed that the participants would maximize the accuracy with which they detected the signal. But this was not always the case. Sometimes they appeared to maximize the pleasure of the interaction instead. This meant that they accepted the decision of the less competent partner as a concession to fairness, even when they were pretty sure that the answer was wrong.

A moment's thought can convince us that the human ability to learn by following instructions can have very undesirable effects. For instance, instructions enable efficient coercion and oppression. They can make people do things that they would not do of their own accord. No wonder we put high value on the integrity of teachers and use strong monitoring systems and regular inspections that scrutinize their ethics and conduct. At the same time, the content of instructions needs to be monitored.

An enlightening study from Mahzarin Banaji's lab investigated how seven- to eleven-year-old children learn stereotypic attitudes toward outgroups. They compared the effects of explicit verbal instruction and implicit association learning (Charlesworth, Kurdi, and Banaji, 2019). The experimenters artificially created two groups that were called "Longfaces" and "Squarefaces." The children learned about the groups either through explicit verbal instructions (e.g., "Longfaces are bad, Squarefaces are good") or through implicit repeated pairings of pictures of Longface avatars with nasty things, such as snakes, and Squareface avatars with nice things, such as puppies. The results of this comparison showed that with verbal instructions, children rapidly acquired stereotypic social attitudes. However, this was not the case when they had only been exposed to repeated pairings of pictures.

How Do Instructions Get into the Brain?

In the early days of brain imaging, we were much impressed by a lengthy, heroic study from Japan in which two macaque monkeys were scanned while performing the Wisconsin Card Sorting Task (WCST; Nakahara et al., 2002). The WCST is a widely used test of frontal lobe function (Milner, 1963). Here, people have to discover a rule for sorting cards into groups. For example, should they use color or shape? Patients with frontal lobe lesions have difficulty switching to a new rule. Brain imaging revealed that the monkeys activated similar frontal regions to humans when performing this task.

However, for us, there was a much more interesting aspect of the study. Humans can be instructed how to perform this task in a few seconds (i.e., the time it took to say "Sort the cards by color or shape and discover when you have to switch"). For monkeys to be able to perform the task in a magnetic resonance imaging (MRI) scanner required about one year of operant training (Yashushi Miyashita, personal communication). The end result was the same. Both monkeys and humans responded appropriately to the stimuli appearing on the screen, and both activated the frontal cortex. But the journey to this end point was very different. Only humans could rapidly come to know how to perform this complex task through explicit instruction (Roepstorff and Frith, 2004).

We marvel at the fact that communications from others, including instructions, are not only understood but also exert top-down effects on brain function (for a review, see Cole, Laurent, and Stocco, 2013). However, as yet, there are few studies that have tried to explore how this is possible. An early study from Dick Passingham's lab (Sakai and Passingham, 2003) tackled the question by asking what happens in the brain in the interval between receiving an instruction and doing the task. The task was either to remember a sequence of squares or a sequence of letters. The instructions for one task were "Remember the letters" (testing verbal working memory), and for the other task, they were "Remember the positions of the squares" (testing spatial working memory). Brain activity was examined in the gap, lasting several seconds, between getting the instructions and the presentation of the letters and squares.

Note that the participants were doing nothing in this short time window except get ready to perform the task. But during this interval, the brain was busy preparing efficiently for the task at hand.[2] Before performing the verbal task, there was increased activity in brain regions associated with language (see figure 1.2 in chapter 1). Before performing the spatial task, however, there was increased activity in regions concerned with the perception of space (i.e., the superior parietal cortex). Importantly, in both cases, activity was also seen in the frontopolar cortex (see figure 14.4 in chapter 14), the region at the top of a hierarchy of control (Boschin, Piekema, and Buckley, 2015). It seems, then, that the instructions were readying the task-relevant brain regions via signals from frontopolar cortex. This study provides a start for thinking about how the brain translates an instruction to do a task into actually doing it.

Instructions allow us to do a task that we have never done before, and this still seems rather miraculous. In contrast, trial-and-error learning, particularly association learning, has been studied for so long that it is no longer seen as miraculous. We know a great deal about the neural processes that underpin this type of learning (see chapter 12). Li, Delgado, and Phelps (2011) conducted a direct comparison between these two kinds of learning. They presented visual cues that indicated the probability that a reward would be obtained. In the standard condition, participants learned the reward probability that was associated with each cue simply by doing the task. In the instructed condition, they were told the probability of the reward for each trial.

How did the difference between these two conditions play out in the brain? For the standard task, in line with many previous studies, activity was seen in the nucleus accumbens, known to be associated with prediction errors (see figure 12.2 in chapter 12). As explained in chapter 12, such prediction errors are a key part of the learning process since they indicate whether we have correctly recognized the significance of the cues. This activity was not seen in the instructed condition. How can this be explained? When you get a verbal instruction in this sort of task, you don't need to learn the significance of the cues! As you already know, prediction errors sometimes

2. This is referred to as a "task set," in analogy to a "learning set." Both terms describe the voluntary initiation of readiness to perform a particular task before task performance begins.

give no useful information. You can factor in that the cue signaling an 80 percent probability of reward will occasionally be wrong. You can think of it as noise and don't pay it any attention. Hence, the instructions reduced the response to prediction errors, most likely via inhibitory signals from the dorsolateral prefrontal cortex (DLPFC).

Very similar results were found in a study of trial-and-error learning about painful stimuli rather than reward. Here, people learned which of two pictures was followed by a painful shock (Atlas et al., 2016). Rather than learning by bitter experience, instructions enabled people to update their expectations immediately ("From now on, the blue square will be followed by a shock"). This immediate updating was seen in the activity elicited by the cue in the striatum and orbito-frontal cortex (the value system; see figure 12.2 in chapter 12). Here again, the effect was probably achieved via signals from the DLPFC.

Of special interest was the observation that activity in the amygdala (which is associated with fear; see chapter 4) was not modified by instructions. Perhaps this lack of penetration avoids the danger that might arise if responses to life-threatening events were completely switched off by instructions. Basic threat responses are too important to be meddled with!

Learning which picture will be followed by a painful shock is an example of Pavlovian conditioning. For this kind of learning, there is an interesting intermediate stage between trial-and-error learning and verbally instructed learning—learning by observing someone else learning by trial and error (this is social or vicarious fear learning, see chapter 2). Lindström, Haaker, and Olsson (2018) contrasted brain activity associated with direct fear learning and social (vicarious) fear learning. They found that, for direct learning, the unconditioned stimulus (US) enters the system via the amygdala and is sent on to the anterior cingulate cortex (ACC; see figure 4.2 in chapter 4). For social learning, the US enters the system via the anterior insula cortex and is passed on to the amygdala and the ACC. This is consistent with the idea that the conditioned stimulus (CS) in social learning is the pained facial expression that occurs when the person being observed gets the shock. The anterior insula cortex is activated when we observe others receiving painful stimuli (Singer et al., 2004).

This study was followed by a computational analysis of the difference between social learning and instructed learning (Lindström et al., 2019). Social learning could be explained in terms of Pavlovian conditioning, as would be expected if the social stimulus (a face wincing in pain) acted as a US. Instructed learning could not be explained in these terms. Instead, instructed learning altered the prior values associated with the cues, independent of any direct experience.

We are only at the beginning of explaining how instructions can reconfigure the brain, but the studies conducted so far suggest that instructions have their effect via regions at the top of our brain's hierarchy of control associated with prior expectations. Our human ability to teach new concepts provides a powerful means for creating shared models of the world. Shared models are fundamental to cumulative culture from prehistoric times to the present day. We will discuss reasons for these strong statements in more detail in chapter 19.

18 Culture and the Brain

Novel cultural concepts arise and spread with the aid of explicit metacognition. They often lead to the improvement of practices and strategies for behaving, as well as for thinking, in culturally appropriate ways. We use the example of written language to illustrate that cultural evolution, like genetic evolution, is not a linear process but a branching one, with many factors playing a role, including randomness. Cultural evolution is constrained by the cognitive capacities of the brain, but also by the need for compromise when developing cultural practices that individuals can realistically acquire and use for the good of the group.

The creation of cumulative culture rests on a whole raft of cognitive processes underpinning social communication. Prominent among these processes is overimitation, which contributes to the establishment of cultural norms and conventions. Rules and conventions need policing since they are important for stemming anarchy, which would be difficult to prevent if we relied solely on automatically functioning cognitive processes. Cultural norms influence subjective experiences and thus smooth out the differences between members of a group. This results in the sharing of knowledge and practices and the creation of common ground, which is vital for communication. Networks involving the prefrontal cortex (PFC) play a critical role in making the link between individuals and their culture. This link stretches from the worlds of objects and agents to the world of ideas. It depends on assigning cultural values to objects and actions and taking account of the beliefs of others.

* * *

The Roots of Human Culture

Discussions about culture can be very confusing. We often think of culture as referring to a way of life that makes the difference between our group and others. Some groups may be said to have a "consumer culture," a "dance culture," a "collectivist culture," and so on. But these differences are the tips of a giant iceberg of a universal human culture. Donald Brown (1991) has suggested that there are hundreds of universals of human cultural behavior and thought. Among universal behaviors, he mentions cooperation, gossip, sanctions for crimes, baby talk, and trading. Among universal thoughts, he mentions interpreting behavior, distinguishing normal from abnormal, true from false, and automatic from deliberate actions (see Pinker, 2002 for the full list).

But what underlies the universals? We have presented evidence for the cognitive processes that enable our social interactions in all the chapters of this book and pointed out that we share most, but not all, of these processes with other animals. And what underlies the differences? In chapters 3 and 8, we provided some evidence for the need to distinguish ingroups and outgroups and for the role of overimitation in making ourselves distinct from others. Briefly, the actions that you learned via overimitation differentiate you from other groups. This is summed up in the statement: "We are the people who do it like *this*."

When we talk of human culture, we have in mind *cumulative* culture—that is, transmission of knowledge and customs across generations with constant modifications and improvements. We can credit the human superpower of explicit metacognition for these processes since it enables us to communicate and teach in such a way that we can build better models of the world. We will not delve into the mechanisms of cultural evolution since these have been examined extensively in recent years by prominent studies, including Hannah Lewis and Kevin Laland (2012), Alex Mesoudi (2016), Cristine Legare (2017), Celia Heyes (2018), Joe Henrich (2018), Dan Sperber (2020), and Nichola Raihani (2021).

Marvelous Inventions

We have reasons to celebrate cultural achievements in art, science, and technology because they have made our lives longer, richer, and more comfortable. We believe that they not only enrich but also give meaning to our lives. We can dive into the toolbox of culture to find novel strategies to create and explore the mental world. A well-known example is written language, which we will consider in some detail in the following section. The domain of mathematics is equally (if not more) astounding. Concepts such as negative numbers, zero, and the decimal point were unknown about 2,000 years ago, but now we use them all the time.

Closely related to these mathematical tricks are the rules of logic, first formulated by Aristotle and developed by Descartes. One such rule is the syllogism illustrated here by a playful example from Charles Dodgson's *Game of Logic*, published in 1886:[1]

1. Dodgson is probably better known under his pseudonym of Lewis Carroll.

(1) Babies are illogical.
(2) Nobody is despised who can manage a crocodile.
(3) Illogical persons are despised.

Given that these premises are true, we can deduce that babies cannot manage crocodiles.

Verbal reasoning of this kind is not just fun, but a tool that enables fragments of knowledge to be integrated. This can be used to reveal new and often surprising information. But because it is fun, reasoning is often turned into a social activity in which people communicate and argue together. Through such interaction, fragmentary knowledge held by different people can be integrated (Oaksford and Chater, 2019). Hugo Mercier and Dan Sperber (2011) have suggested that the evolution of reasoning strategies was driven not by a desire to arrive at the truth, but by our wish to persuade others that we are right. We are very good at justifying why we did what we did and making it sound reasonable and correct—even if it is not. Arguing with others pits their justifications against our own, which is one way to counteract confirmation bias. Better decisions can be made when people work and argue together (chapter 15).

The tools of the cultural toolbox produce novel insights and strategies. The abstract concepts and rules of mathematics and logic help us take flight in the world of science and point us toward the inventors who have generated technologies that affect us in every part of our modern lives.

Creating Shared Cultural Practices: The Story of Written Language

We follow the evolutionary psychologist Celia Heyes (2018), who contends that we are born with propensities for thought and behavior that differ only subtly from those observed in other primates. However, when these propensities are subjected to the complex social environment that we all experience, they enable us to develop uniquely human mental gadgets, which are the consequence of cultural rather than genetic evolution. Reading is a striking example. But like genetic evolution, there is no linear trend, but rather a branching journey with random turns, which can still confer an advantage in a particular ecological niche. These ideas are reminiscent of the ideas of the great Soviet neuropsychologist Alexander Romanovich Luria (2012), who believed that the influence of culture on cognition starts at birth. Motor and perceptual functions are shaped through interaction with cultural artifacts. And this continues to be the case in every generation.[2]

2. A charming example is that babies today spontaneously swipe at pictures in books, having transferred this gesture from their early exposure to mobile phones and tablets.

Figure 18.1
Our environment is dominated by writing.

Literacy is a prime example of cultural evolution. Our urban environment is dominated by text in ads, smartphones, shopping centers, and streets (see figure 18.1). We process written language automatically and can hardly avoid reading text, even when we try to focus our attention on something else.[3] But writing appeared barely 5,000 years ago, whereas *Homo sapiens* emerged circa 300,000 years ago. Furthermore, writing did not become a universal aspect of human life all over the world until a century ago. Now literacy is indispensable.

How has literacy managed to become one of the gadgets that, as Heyes suggests, has incorporated itself into the brain? The ability to read and write is acquired by each individual from scratch—and yet not entirely from scratch. It can rely on a preexisting network of brain regions, inherited through biological evolution, which is perfectly suited to learning to read and write.

This discrete network, with a hub in the fusiform gyrus at the base of the left temporal lobe (see figure 1.2 in chapter 1), is a region of the brain that evolved over millennia for the identification and differentiation of objects, particularly objects made up of fine

3. The famous Stroop interference effect (where word reading trumps color naming) is based on this fact.

details involving line junctions (Dehaene and Dehaene-Lambertz, 2016). It represents faces and body parts, and it can represent many abstract shapes, including those of letters and words. Everyone who has become a fluent reader has adapted this region into a visual word form area. Still, it is only part of a much wider neural network that supports language. This network enables us to read effortlessly by linking speech to graphic symbols. The subtle anatomical changes that occur through becoming literate are an example of the brain's plasticity within an individual's lifetime (Dehaene and Cohen, 2011). All this is explained in detail in Stan Dehaene's book *Reading in the Brain* (2009).

The evolution of writing systems suggests that the influence of culture on the brain is a two-way interaction. Just as the ability to read rests on brain systems that originally evolved for the recognition of objects, so letter forms (and other characters used in writing) evolved to have the kind of shapes that are easily handled by this system. Of course, these shapes also depend on the materials available and other random factors that transform the development of writing systems in unpredictable ways. Changes that appear because of random events can be remarkably sticky. Think of the entrenched status of the QUERTY keyboard. It is possible that just as handwriting is being edged out by typing, so typing may soon be edged out by speaking words that then appear on a screen via speech recognition software.

Because we ourselves have been involved in research on literacy acquisition and its problems, we can't resist mentioning some of our findings. We were interested in how writing systems are shaped by the language that is being represented. For instance, it is easier to represent the sounds in Italian since there are only some twenty-five of them, while in English, there are about forty different phonemes. As a result, the relations between the letters of the alphabet and the sounds of English words are much more ambiguous.[4] In Italian, there is a much closer relationship between the letter strings and the corresponding sounds (e.g., Milano) than there is in English (e.g., Leicester). Indeed, learning to read is faster and less problematic in Italian than in English[5] These differences lead to detectable differences in the brains of English and Italian readers (Paulesu et al., 2000).

Writing systems also allow us to explore more fundamental differences between historically and geographically separated cultures. The alphabet, which links graphic

4. For instance: There was an old lady from Slough/Who developed a terrible cough. /She sounded quite rough,/But battled on through./I think she is better now, though.

5. Further, English dyslexics have a much harder time learning to read than Italian dyslexics (Paulesu et al., 2001).

symbols to sounds, is only one of several writing systems. The Chinese writing system links graphic symbols to meanings rather than sounds. The beauty of this system is that the symbols can be understood by people who are speaking different languages (such as Mandarin and Japanese). Syllabic systems exist, which can be exquisitely matched to the syllable structure of languages, such as South Indian languages. The Japanese writing system uses ideographic, syllabic, and alphabetic systems in parallel, a tall order for the learner. Interestingly, the speed of learning a writing system has hardly ever been the decisive factor in its adoption. Instead, political pressure and random events have led to the branching that we currently see with the different systems that are in use now.

Tensions Driving Cultural Evolution

There are always tensions between different demands that we cannot readily resolve, if ever. These tensions and the need to achieve compromise between conflicting demands may well be a driver of cultural evolution. Perhaps the most important is the tension between what is good for the individual and what is good for the group. In the case of literacy, for example, it is the individual who has to learn the mapping principles of spoken and written language to create messages. But it is the group that has to understand the messages. A good writing system has to compromise between being easy to write (for the individual) and being easy to read (by the group) (Frith and Frith, 1980). This compromise lies behind the complex orthography of English, where we can read "with our eyes", quickly apprehending visual features of a word, but write "with our ears", demonstrated by phonetic misspellings (Frith, 1979). A similar compromise can be seen in language more generally. David Saldana and colleagues (2019) suggested that linguistic structures evolved through two competing pressures: for ease of acquisition (learnability, for the individual) and for effective communication (expressivity, for the group).

Compromise is also needed to take account of the cognitive constraints imposed by the brain. Language is one of those faculties that challenge our virtuosity in decoding sound and articulation, but also our ability to produce grammatically correct and semantically appropriate words. Michael Hahn, Dan Jurafsky, and Richard Futrell (2020) suggested that the properties of language can be explained by the need to achieve efficient communication between humans in the face of these constraints. The syntax of languages must find a balance between these two pressures: to be simple enough to allow the speaker to easily produce sentences, but complex enough to be able to express abstract thoughts.

Cultural evolution is happening even within individual lifetimes. Our ravenous demand for communication is never satisfied, and this demand might possibly have driven modern technology to overcome our cognitive limitations. Thus, the internet and mobile phones have enabled communication at a click on a global scale. Previous cultural inventions to satisfy our demand for more efficient communication included many ingenious tools, such as the telegraph, Morse code, and the centuries-old and still (just) enduring postal system.

Costs and benefits of social communication are relentlessly calculated at all levels of the brain's processing hierarchy. Cultural practices can weigh in on these calculations and make individuals do what they might not choose to do on their own. An example is mass schooling, which demands a huge investment of time from individuals and cognitive effort to delay gratification. However, it lays the foundation for skills that pay off handsomely in the future. Moreover, it serves as a major vehicle for cumulative culture over generations. Education plays a role in the evolution of culture not only because it serves to transmit knowledge, but for another even more fundamental reason: it creates common ground and thus starts a virtuous circle. Ideas can be more easily transmitted if we already are part of a group with much shared understanding.

Smoothing out Differences between Our Individual Worlds

We might pride ourselves in being individualists with our unique taste in clothes, music, and art objects, but we acquire most of what we like and dislike by being immersed in our culture, from distinct foods to a distinct lifestyle. It is no exaggeration to say that our subjective experience is molded by the people around us, and they in turn are molded by cultural traditions that they were brought up in. We have already pointed to over-imitation as the most likely mechanism that creates differences between groups that all have their own recognizable subcultures.

We are fastidious about the type of music or art we prefer without necessarily being able to explain why. We can make up elaborate reasons for our preferences. Likewise, participants in erudite wine tastings give detailed descriptions of their subjective sensory experiences. However, there is another trend that puts limits on our subjectivity. In truth, it is not so much our need to feel distinct as individuals as our need to conform with our ingroup that creates and maintains these preferences and descriptions. This has the interesting effect to disguise individual differences in perceptual experience. But it also has the effect of reducing these differences. Thus, the mental world of people within the same group appears to them to be remarkably similar.

All this helps in creating confidence in the world out there, and in creating the common ground that enhances communication within our ingroup. Sometimes we can acknowledge or exaggerate differences between cultures, but we also like to think that our tastes and experiences go beyond petty tribal, national, and geographical boundaries to reveal our identification with all of humanity. And indeed, there are universal cultural effects. For example, while musical styles differ greatly between cultures, there is nevertheless a universal brain response to musical melody and rhythm (Boebinger et al., 2020).

Still, the mental world of individuals is never shared completely. When someone is born red-green color blind, their brain's starter kit for color is irrevocably different. Nevertheless, their color space will be molded by their culture to be as close as possible to everyone else's. The result is that some people are not aware that they are color blind until adulthood (see, e.g., Bradley, 1970). Typically, such self-awareness occurs as a result of communication failure. For instance, a friend of ours did not know that he was color blind until, as a medical student, he was told that he was filling in the wrong forms. This was because he could not distinguish between green and pink paper.

The situation is even more striking for the sense of smell. Underlying color vision, there are only three receptors.[6] Color blindness occurs when one receptor is missing. For smell, there are several hundred receptors, and it has been estimated that at least one of these will be missing in any person (Croy et al., 2016). So we are all smell-blind, but each in our own special way.

A similar story can be told about synesthesia. This is an unusual experience in which senses are combined. The commonest form is color-grapheme synesthesia, in which different words are experienced as having different colors (Grossenbacher and Lovelace, 2001). And there are many other forms, such as color with musical notes, taste with shape, and feeling a touch when seeing someone else being touched (mirror-touch synesthesia; see chapter 3). Whenever there is a lecture in which some form of synesthesia is described, someone is bound to come up afterward and say, "I thought everyone was like that."

Of course, there are individual differences in subjective experience. These are occasionally revealed, and they do not seem to have anything to do with anomalous brain systems. An example is the internet phenomenon of "the dress." In 2015, the photo of a dress generated a massive argument (about 10 million tweets' worth) about its color. Roughly two-thirds of viewers saw it as blue and black, while one-third saw it as white and gold. Although still not fully resolved, the most likely explanation for this

---

6. Except in the case of some women, who have four. This was discovered only relatively recently (Jordan et al., 2010). No one was aware of any differences in the experience of these people.

is that people have different expectations (priors) about how the dress is illuminated (Lafer-Sousa, Hermann, and Conway, 2015). A paper on the subject (free to access at https://www.ncbi.nlm.nih.gov/pmc/articles/PMC4921196) shows the critical photo. Its authors conclude that to perceive the "real" color of the dress, we must take account of the color of the light falling on it. If we think it is illuminated by yellowish artificial light, it is perceived as blue and black, while if we think it is illuminated by bluish natural light, we perceive it as white and gold. But it remains unclear where these differences in expectation come from.

Still, we hold to our claim that differences in our subjective experience are leveled by culture. Thus, we tacitly believe that we all perceive the world in the same way, and we are confirmed in believing that there is a world out there. This is a fundamental feature of the shared knowledge (*common ground*) that underlies culture. And it does not require the infinite regress (I think that you think that I perceive the world like this, and so on), associated with *common knowledge*. Instead, we each have a model of the world, which has been shaped by our culture. We take it for granted that a rough version of this model is true. It may not be literally shared by everyone, but we believe that a version of this model is shared at least by everyone in our own ingroup. Hence, the surprise when this very occasionally is revealed to be an illusion.

We don't know much about the cognitive mechanism involved, but we note that there are no changes at the bottom of the neural hierarchy. Our sense receptors have not been altered by these cultural effects. These effects depend on the decoupling of perception from sensation. There is more to be revealed by future studies. The explanation is not just at the level of current sensation and perception. There are also our prior expectations to consider. They determine our subjective experience (our perception). It is these priors that are modified by culture, and this happens without them ever emerging into consciousness.

Creating Shared Memories and Knowledge

It is not just sensory experience that we share in this way. We also have a shared model of the past (Seemann, 2019). The world that we remember is the world that our group remembers (Hirst, Yamashiro, and Coman, 2018). This effect extends from families ("She turned up unannounced, with three suitcases"[7]) to nations ("The Battle of Waterloo was won by Wellington").[8]

7. . . . with different accounts of whether the visit was announced and/or the number of suitcases.
8. . . . with a little help from his allies, who resent not being credited.

These are long-lasting effects of social contagion on memory. The process of creating a shared reality can be seen at work in interactions between speakers and listeners (Hirst and Echterhoff, 2012). In one study (Echterhoff, Higgins, and Levine, 2009), student participants were asked to present a description of an employee either to another student (ingroup) or to a company board member (outgroup). In both cases, the students fine-tuned the way that they presented the message to suit their audience's presumed attitude. Presenting messages in this way typically restructures the memory of the messenger, making it more like that of the group receiving the message. Here, however, this happened only when the students were talking to their own group (i.e., other students).

It has long been known that individual memories fit collective memory. This notion was put to the test using brain imaging in a study relating to memories of World War II. Pierre Gagnepain et al. (2020) analyzed thirty years of media coverage of the war on French national television. They scanned participants as they recalled displays shown in the memorial war museum of Caen. They found that the pattern of activity in the medial prefrontal cortex (mPFC) was better predicted by collective memory than by individual semantic memory. They concluded that collective memory, which exists outside and beyond individuals, organizes individual memories toward a common mental model.

Groups share knowledge as well as memories. There is more knowledge available to the group than to any one individual. And working together, we can do much better than working alone. We also happily outsource our knowledge to others. We may not know how to repair our car, but we do "know a man who does". However, although we know that we wouldn't be able to repair a car engine, we fondly believe that we have a rough idea of how a car engine works. This is an example of the We-mode (see chapter 5). By being part of a group, we enlarge our knowledge space and treat the knowledge of others as if it were our own (Wegner, 1987). Of course, the belief that we know so much is often an illusion. For instance, many people believe that they know how everyday appliances work, such as zip fasteners and flush-toilets. But, as Leonid Rozenblit and Frank Keil (2002) have demonstrated, when we are specifically asked to explain how these things work, we usually fail. This has been called the "knowledge illusion" (Sloman and Fernbach, 2018). Our shared model leads us to believe that we ourselves know things if we are confident that others in our group know them.

Spreading Cultural Concepts and Practices

Cultural evolution, like genetic evolution, depends on variation and selection. Variation is provided by the roll of the dice that results in individual differences within groups, as well as between groups. It is widely accepted that individual selection, and possibly group selection, are responsible for changes over generations.

Many unpredictable factors are involved in spreading cultural models of the world. Only some ideas, be they mathematical concepts, behavioral norms, artistic creations, or writing systems, become popular and widely distributed. Large-scale historical influences, such as trade, conflict, and conquests, are bound to play a role, but so do special individuals, whose personal taste may create an enduring style.

But how, in principle, can an idea arise and then spread and dictate the behavior of individuals and groups? This question is still wide open to investigation. Leaving aside coercion, we suggest that this depends, fundamentally, on a two-way process. On the one hand, the individual's mental representation must become public. On the other hand, a public representation must become internalized by the individual (Sperber, 1996).

It seems likely that explicit metacognition, which we like to refer to as our human superpower, plays a critical role in this process (Shea et al., 2014). As we saw in chapters 13 and 14, there are channels through which our unconscious cognitive processes can surface, while at the same time, conscious processes can affect what goes on at the unconscious levels of the system. Those internal signals that become conscious can then be communicated to others with the implication that our thoughts can affect other people's cognitive processes. Likewise, what others say to us can be converted into a form that can change the way that our own cognitive processes work.

As an example, we have discussed teaching and learning from others via verbal instruction (chapter 17). We highlighted the unique human ability to become aware of and to communicate information about our cognitive processes to others (chapter 14). Is it this ability that has allowed cultures to take off? Quite possibly. The appearance of this ability, an explicit form of metacognition, doesn't just help the individual owner of the skill. It also helps the other members of their group with whom the individual discusses problems and coordinates actions. By sharing and discussing our cognitive processes, we can develop novel forms of adaptive group behavior that can spread and make culture cumulative.

In the most basic case, people select what metacognitive information to broadcast to each other to optimize their division of labor in the shared space of the task that they are currently carrying out. But such communications can have a more lasting effect, as in learning new strategies for performing novel tasks. An example we have used before in this book is what you should do when you run out of animal names when the task is to name as many animals as possible. Somebody suggests, and thereby implants in your mind, that a good strategy is to think of subcategories of animals. Now you can use this strategy in other tasks too, such as when you have to come up with as many names as possible of scientists whom you might nominate for a prize. You would do well to think of different disciplines, genders, geographical locations, and so on.

Much of our education is concerned with learning about the various cognitive strategies that our ancestors have discovered over a long time. Learning about behavioral and cognitive strategies and instructing others about them, opens new doors within the world of ideas. All this seems to be a feature of human culture, and it depends on our ability to communicate and to mentalize. This is tantamount to saying that human culture is enabled and continuously fostered by explicit metacognition. Mass education ensures that useful behavioral and cognitive strategies are passed to the next generation and thus maintained. But how do we account for the emergence of new and better strategies?

New strategies can emerge by accident. For example, errors of copying in the making of artifacts result in variations, which can be amplified or reinterpreted (Eerkens and Lipo, 2005). The larger the population, the more likely that such events will occur (Fogarty and Creanza, 2017). However, the increasing speed of the evolution of new strategies, and indeed inventions, indicates that many new strategies have been developed by design rather than by accident. Curiosity, risky exploration, and the pleasure that we take in novelty are probably the main drivers.

Why do some models and cultural practices survive for longer than others? What makes some ideas better than others? Clearly, they should fit the way our brains work, and clearly, they should not be too cognitively taxing. But to survive, in the long term, they must confer some sort of advantage (Buskell, 2017). For novel ideas and strategies to become part of our culture, there needs to be some process through which they become shared by the group. Such processes depend on both the face-to-face interactions of individuals and the nature of the network of which the individuals are a part. Memories that are mentioned frequently in individual conversations become part of the collective memory of the community. This collective memory spreads more readily when there is a small degree of separation between all the members of the community (Coman et al., 2016).

For ideas to be transmitted accurately, we already need to be part of a group with much shared understanding. Thus, ideas will be mostly associated with groups rather than single individuals.[9] This also applies to the variation and innovation of ideas.

---

9. We don't wish to deny that there are exceptional individuals, artists, scientists, inventors, and others who have revolutionized how we interpret the world and enrich human life and experience. These individuals benefit from contributions of their group (e.g., in terms of resources, recognition, and reputation).

Variations in ideas and practices are likely to be transmitted when they are improvements, and they are more likely to come from people working together rather than as individuals (Godfrey-Smith, 2012). As we can attest from experimental studies as well as from our own personal experience, two (or more) heads are often better than one. Further, it has been suggested (Henrich, 2004) that cumulative culture is better promoted by large social networks. Large networks are likely to be associated with the increased diversity that can enhance innovation and problem solving (Bang and Frith, 2017).

Taking this idea further, we might speculate that there could be advantages for groups with norms that foster group diversity. For example, Joseph Henrich (2020) proposed that by banning cousin marriage, the Western Christian Church broke down kin-based institutions. This curbed the accumulation of inherited wealth in a few powerful families, while encouraging the eventual emergence of individualism and nonconformity. In contrast, societies with kin-based structures tend to promote cousin marriage and ingroup loyalty, often resulting in increased nepotism and conformity.

We conclude that it is the group rather than the individual that enables the survival of practices and ideas. For example, groups containing many altruistic people are likely to do better than groups containing few altruistic people, even though within the group, the altruists may do worse than others (Sober and Wilson, 1998). Group selection, at the level of genetic inheritance, remains controversial. However, this is not the case for cultural inheritance.

How Culture Gets into the Brain

How, precisely, is the influence of other minds (i.e., culture) implemented in the brain? We would love to know the answer! We know that the PFC plays a key role in top-down control and explicit metacognition. But we do not yet have sufficient information to even sketch out the mechanisms underlying the creation and updating of the high-level priors that reflect our culture. We know about only some of the components (see figure 18.2). Based on our speculations in earlier chapters, here are some of our best guesses.

At the lowest level of the hierarchy, behavior depends on model-free learning systems in the midbrain and striatum (see figure 12.2 in chapter 12), which seek maximum gains in the short term. This low-level system is modified by high-level model-based systems, associated with the PFC, which take into account the current state of the world (see, e.g., Buckholtz, 2015, and chapter 13). We presume that culture influences the brain via the high-level control systems in the PFC, with three critical stages. First, cultural norms for behavior must modify the values associated with various actions

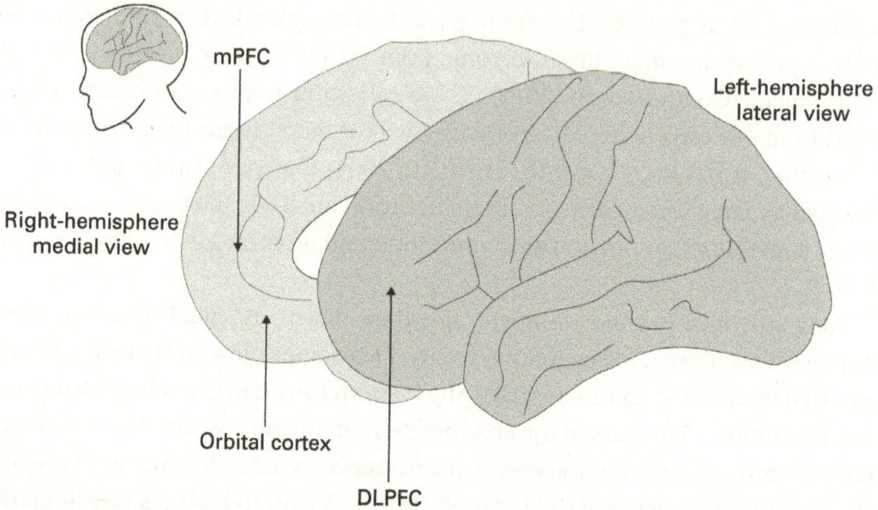

Figure 18.2
Regions where culture interacts with the brain: the orbital cortex, dorsolateral prefrontal cortex, and medial prefrontal cortex.

and outcomes (the orbital cortex). Second, the response space must be sculpted to give higher weights to those actions appropriate in the cultural context (the dorsolateral prefrontal cortex, or DLPFC). Third, when the behavior of others is relevant, some account must be taken of their status and intentions (the mentalizing system, including the mPFC).

We have pointed out the importance of norm-guided behavior, but how do we internalize the unwritten rules that govern our behavior? Interactions between the orbital cortex and striatum (nucleus accumbens) enable the learning of norms through direct experience and observations of the behavior of other people (Morelli, Sacchet, and Zaki, 2015). The orbital cortex is also sensitive to fairness in social exchanges (Gu et al., 2015) and is activated by unexpected deviations from current norms of behavior. The amplitude of the norm prediction errors observed in the orbital cortex (and the insula; see figure 4.3 in chapter 4) correlates with subjective feelings about the fairness of the exchanges (Xiang, Lohrenz, and Montague, 2013). In addition, the insula seems to play a crucial role in learning to adapt when reality deviates from norm expectations (Gu et al., 2015).

Our cultural norms determine which actions are appropriate or inappropriate in various contexts. This depends on a process that Chris Frith that characterized as "sculpting the response space." This creates a context that ensures that the most appropriate actions are the most likely to be performed (see section 24.8 in Frith, 2000). The

DLPFC plays a major role in this sculpting process. It is engaged during norm-guided behavior that ensures that the appropriate behavior occurs. DLPFC dysfunction, on the other hand, is associated with antisocial behavior (Buckholtz, 2015). We see here a neural mechanism that aligns behavior with social norms, especially when punishment is possible (Ruff, Ugazio, and Fehr, 2013).

Influences of culture implicate mentalizing ability. This system comes into play whenever we need to take account of the mental states of others. This occurs during vicarious learning from observing others (Morelli et al., 2015), as well as when we need to take account of how our behavior is viewed by others (i.e., our reputation; see Izuma, Saito, and Sadato, 2010).

There is a spatial gradient in activity within the PFC, such that lower parts (e.g., the orbital cortex) predominantly represent values for the self, while the dorsal parts (e.g., the mPFC, which is part of the mentalizing system) predominantly represent values for others. Prosocial individuals show stronger coupling between the mPFC and the striatum when they make choices for another person rather than for themselves (Sul et al., 2015). When people learn to avoid actions that harm others (rather than the self), there is a stronger connection between the orbital cortex (representing value) and the temporo-parietal junction (TPJ), the connector part of the mentalizing system. This effect is especially marked in people with high empathy (Lengersdorff et al., 2020).

These are strong hints as to the nature of the brain mechanisms that enable human cultural norms to modify thought and behavior. But these processes are independent of the mechanisms through which these cultural norms emerge in the first place. The way that our brain works places constraints on the kinds of norms that will emerge. However, within these constraints, anything might happen. There is, therefore, a need to monitor and police the system.

If we see someone making an unfair offer (a norm violation), then activity is automatically elicited in the amygdala, as well as in the striatum (Haruno, Kimura, and Frith, 2014; see also figure 1.2 of chapter 1). It seems that observing norm violations elicits an emotional response, especially in people with a prosocial orientation, who believe that resources should be shared equally. Emotionally graphic descriptions of harmful acts boost this amygdala activity and, at the same time, increase the severity of punishment. This effect is linked to strengthened connections between the amygdala and lateral prefrontal regions (Treadway et al., 2014). However, when the actor does not intend the harm, the mentalizing system suppresses amygdala activity, and the effects of graphic descriptions on punishment severity are abolished.

People punish outgroup members more than ingroup members for the same norm violation. When some ingroup members stray, they become the topic of gossip, which has a nose for any such violations. Indeed, gossip polices norm-based behavior (Beersma and Van Kleef, 2012). However, when ingroup members violate norms, we try to explain the violation via mental state reasoning. In contrast, we don't feel the need to understand outgroup behavior in such terms (Harris and Fiske, 2006). Hence, outgroup punishment is altogether more severe. It is associated with enhanced activity in the DLPFC. This activity, as well as the associated punishment, are reduced for ingroup members via signals from the mentalizing system (Baumgartner et al., 2011).

Given these limited observations, we can summarize what little we know about brain and culture by linking the three major regions of the PFC with the three aspects of the worlds around which we have structured this book. The orbital cortex is concerned with the subjective values of *objects*. We can learn values from direct experience, but a far greater effect comes from other agents via observation or instruction. Our subjective value system, instantiated in the orbital cortex, is sculpted by our culture to create value norms. The DLPFC is concerned with the actions of *agents,* and here again, our culture sculpts the action space instantiated in the DLPFC to ensure that we perform the most appropriate actions and that we recognize when our actions are inappropriate. Finally, the mPFC is concerned with the world of *ideas*, particularly the minds of others. Using the mentalizing system, of which the mPFC is a part, we can recognize that others may have different ideas about what are appropriate values and actions.

19 Getting Along Together

To be socially adapted, we need to balance our deep-seated need to cooperate with our urge to compete. When cooperating, we need to make constant compromises between concern for the self and concern for the group. Cultural norms help us in this endeavor, and we follow them because of our desire to affiliate. In Western societies, we are imbued with a strong sense of agency, and following from this a sense of responsibility for our action, sharpened by feelings of regret, guilt, and shame. Our belief in free will is as indispensable for our model of ourselves as people who can apply top-down control over our selfish urges.

What are the practical implications of the research that we survey, which might help us to get along better together? We are impressed by the way that behavioral rules and norms can smooth interactions, and we are impressed by the way that legal systems save our societies from chaos. By being part of an ingroup, we can resolve some of the clashes between selfishness and altruism. We can benefit from the presence of competing outgroups because in their absence, groups can be exploited by free riders. We can collaborate better, make better decisions, and create better models of the world, so long as our group is sufficiently diverse. As individuals, we should adopt diverse points of view and strategies for action.

Conflicting interests remain a problem that we seem to be stuck with. However, there are indications that the polarization of ideas can be reduced by forming small groups that are prepared to discuss differences and may eventually achieve consensus. Looking at the bigger picture, we cannot help but be amazed at the skill in which we all navigate the highly complex worlds of objects, agents, and ideas.

* * *

Can We Become Better People?

We are individuals as well as members of a group. But which comes first? Can we ever resist the temptation to maximize our own benefits at the expense of the group? This question becomes urgent when we recognize the huge disparity between rich and poor. It becomes even more urgent when we consider our impact on planet Earth, endangering the survival of other species and risking our own future. Some of these problems can be seen as the results of selfishness and runaway competition. And yet most people are easily persuaded to be cooperative and fair.

As the early chapters of this book make plain, people have a deep-seated need to be with others, to be liked by others, and to be like others. We seek out cooperative people and value our friends. We love our team and feel proud of its success. But there is a counterpart of these feelings: we hate the other team. Ingroups automatically create outgroups, and opposition between them means conflict. Can we ever resolve this dilemma? We doubt it.

If we can't avoid potentially fatal splits, can we achieve a balance between the value we put on individual freedom and the value we put on society? This might make us better at living together and managing our negative feelings toward outgroups. Here, we reflect on some of the evidence presented in this book and follow up on the most promising leads. We have highlighted the value of diversity, as well as the importance of metacognition in providing an interface between the self and society. This is where we can monitor and regulate our attempts to resolve conflicting aims. And, just possibly, this is where we may find the space in which our mental world and the resulting behavior can change for the better.

Freedom of the Individual

One conflict that we can address here is the rampant individualism in today's Western culture. The emphasis on freedom and individualism has tipped the balance in favor of the self rather than society. But this does not have to be the case. Indeed, the belief that we are free and in charge of our actions is a potent example of how culture interacts with individual minds. The question is whether we can we combine our selfish desire for freedom with our desire to be part of a well-functioning society.

But where does this important sense of freedom come from? As students of social cognition, we look at our sense of agency. By this, we mean our feeling of being in control of our actions. The freedom to do what we want is exhilarating and seems to be the basis of our individual identity. When we perform an action, the self, seated at the top level of the hierarchy, receives information about whether we control our movements and have caused the outcome. This has consequences for the experience of free will, and this experience can be studied in the lab.

Cognitive neuroscientist Patrick Haggard has studied the experience of free will for many years. He made the exciting discovery that the time that passes between initiating an action and perceiving its outcome is experienced as shorter than it really is (Haggard, Clark, and Kalogeras, 2002). Time appears to shrink when we perform a voluntary action, an effect that he termed "intentional binding." A typical experiment goes like this: The participant decides to press a button that causes a tone to occur a few seconds later. In another situation, the participant's finger is pressed down by an apparatus that exerts an external force. In both cases, the participant looks at a clock to tell the time at which the button was pressed and when the tone occurred. As shown in figure 19.1, in the voluntary condition, time shrinks compared to real time. In the involuntary condition, it expands.

Intentional binding is amazing not because it happens at all, but because it provides an objective measure of the subjective sense of agency: the stronger the experience of agency, the greater the binding.

The studies by Patrick Haggard and his colleagues show that our experience of action has two components: the expectation of the outcome associated with the action and the actual outcome. There is a greater sense of agency if the expected outcome occurs. Conversely, the sense of agency is reduced if the unexpected outcome occurs (Moore and Haggard, 2008).

Our sense of agency arises from signals from the deep concerning expectations and outcomes, creating an important form of explicit metacognition. Our experience of agency becomes very relevant when we attempt to *justify* our actions. For example, we can claim that we had no sense of agency: Our action was accidental rather than deliberate. As always in the world of ideas, this claim might be made in good faith or it might be a lie. Our sense of agency becomes part of the Machiavelli thread.

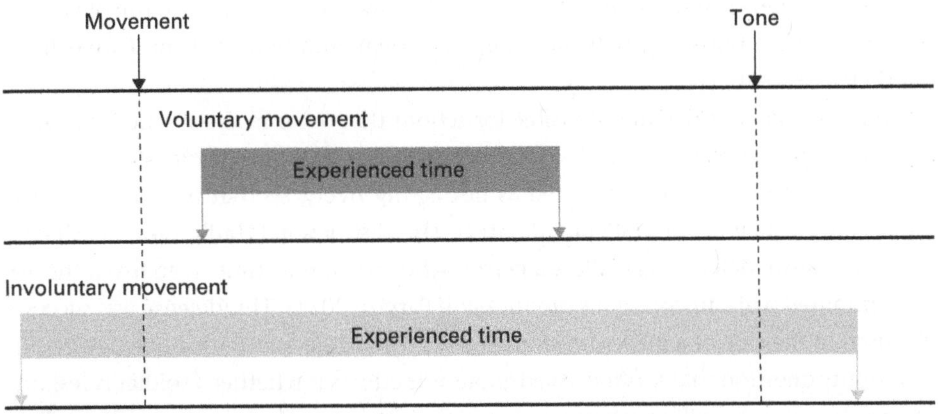

Figure 19.1
Intentional binding: When we press a button to cause a tone, the time between movement and outcome seems to shrink (more binding). When our finger is pushed onto the button by an external force and causes the tone, the time seems to expand (less binding) (Haggard et al., 2002).

Our individual sense of agency may be overblown, but it may be justified by a far-reaching and beneficial consequence for the group. It makes us feel responsible for the outcome of our actions. We feel less responsible when the outcome was accidental and not intended. In such a case, we are less blameworthy. Children rapidly learn about this distinction and try to gain advantage from it ("You kicked me!" "It was an accident!").

We discussed in chapter 14 an interface in the top layer of our mental hierarchy. This is where we find the portal to the mental world of others. It follows that our sense of agency is not isolated in an impenetrable fortress. It can be altered by being indirectly influenced and directly instructed by other people.

If we are made to believe that we are causing an outcome when in fact we aren't, then intentional binding increases (Dogge et al., 2012). This result demonstrates that our sense of agency is malleable and susceptible to outside influence. This seriously undermines the idea of the self being in sole charge. But such a result hardly impinges on our belief in being free agents. As we shall see, this malleability may in fact be a good thing.

The Voice of Conscience

Most of us believe that with freedom comes responsibility. Our concept of responsibility is intimately linked with our belief about free will (Nahmias et al., 2005). This is because our sense of agency automatically picks out the outcomes of actions for which we are responsible. Thus, it is effectively an internal marker of whether we are guilty and deserve punishment. In this sense, it serves as the voice of conscience. Given that the world of ideas allows minds to penetrate each other, we can also use our sense of agency to interpret the actions of others, put blame on them, and punish them if we believe that they deserve it.

People are not considered responsible for actions that occur in the absence of a sense of agency. When people's behavior is associated with an unconscious state, such as sleepwalking, they will be considered as not acting freely, so that they are not held responsible for their actions (Shepherd, 2012). This distinction is built into English law. The "automatism defense" can allow a person who commits a crime to go free if the act was committed while unconscious (Rolnick and Parvizi, 2011). This defense was successfully used in the case of a man who strangled his wife while asleep (de Bruxelles, 2009).

A thorny question that is often asked about war crimes is whether a soldier is responsible for harming a civilian when he claims that he was only "obeying orders." This question was tackled in the lab using the phenomenon of intentional binding. The answer was clear. There is a big difference between inflicting harmful acts when coerced to do so, compared to those carried out voluntarily. When participants obeyed an order to administer a painful electric shock to a victim by pushing a key (and earning money to do so), they showed less intentional binding than when they performed the same action (administering a shock and earning money) of their own accord (Caspar et al., 2016). Thus, when obeying orders, people subjectively experience their actions as more like passive movements, and unlike fully voluntary actions. The question remains whether they are less culpable, not only in their own eyes, but also in the eyes of others.

Responsibility can be overclaimed. When we observe a link between an action and an outcome, this may simply be a matter of correlation. The action may not have caused the outcome. As Daniel Wegner (2003) demonstrated in a series of elegant experiments, people often believe that they have caused something simply because an expected outcome occurred shortly after their action. For the same reason, we may blame others who have done nothing wrong but were in the wrong place at the wrong time.

How do we go beyond correlation to demonstrate causation? As we saw in chapter 12, one solution is to consider counterfactuals (Pearl and Mackenzie, 2018). To answer the question "Did my action cause the light to go on?," we must consider the counterfactual question "If I had not pressed the button, would the light still have come on?" The relevance of this kind of reasoning to our feelings of agency and responsibility was recognized by the ancient Greek philosopher Epicurus. For him, one key component of agency was the sense that "I could have done something else." This counterfactual element of our experience of agency most often emerges as a feeling of regret: "Things would now be better if I had done otherwise."

The idea that agency is the basis for moral responsibility also goes back at least as far as Epicurus. He suggested that we acquire the idea that we are causal agents through the observation that human beings, including ourselves, are praised and blamed for their action (Bobzien, 2006). Indeed, in most societies today, children are rewarded and rebuked for their actions from quite a young age. Thus, they grow up embedded in a cultural niche where it is assumed that they control their actions and have the ability to do otherwise. They learn that they themselves will be held responsible for their actions, and at the same time, they learn that others will be held responsible for theirs. This is consistent with our proposal that explicit metacognition, of which our sense of agency is part, is strongly influenced by the people around us.

Feelings of regret also address the conflict between the self and the group and often morph into the more obviously social emotions of shame and guilt. We feel shame when we have done something that will devalue us in the eyes of others, and we feel guilt when we recognize that we have done something wrong in a moral sense, whether or not others know about it. In both cases, our reputation will suffer, as discussed in chapter 9. The anticipated intensity of emotions like shame and guilt encodes the social cost of the various possible actions and tells us which actions will cause our reputation to go to pieces. Data from many societies from across the world show that the shame experienced after a misdeed closely tracks the magnitude of the loss of reputation reported by observers of those same acts. This effect is not observed with other negative emotions such as anxiety or sadness (Sznycer et al., 2018).

While it may be a cultural universal that the emotion of shame emerges to link bad choices with loss of reputation, cultures still differ in the degree of shame that is linked to any particular action. An extreme example of such differences would be the honor killing of a family member who has behaved in a shameful manner. In Iraq, those involved in an honor killing received six-month suspended sentences, while, in Sweden, the court ordered life imprisonment for those involved in the same killing (Wikan, 2004, 39ff).

The development of feelings of guilt and shame are a good example of how we become aligned with our group. And once these shared models have been installed, they will exert top-down influence on individual behavior (see chapter 14). One important effect of these emotions is that they are a form of self-punishment, and thereby they reduce the need for direct punishment. The voice of conscience causes us to suffer emotionally because we have violated group norms (Frith and Metzinger, 2016), and cultivating the voice of conscience is likely to increase our chances to become better at balancing selfish and altruistic aims.

The Rules of Behavior

In everyday life, we don't just go about doing what we like. We are powerfully constrained by behavioral norms, and however much we value our freedom, we can be grateful for this constraint. Norms are essential to lessen the friction of living together. A simple example is the rule to drive on the right side of the road (or on the left in some countries). Without such a rule, we would crash into each other.

Because norms are so important in regulating life in groups, much has been written about this topic. Indeed, there is even a review of reviews of the extensive literature on social norms (Legros and Cislaghi, 2019). Remarkably, as far as is known, no other animals invent or enforce social norms.[1]

The creation of laws and a system of justice stands out as one of humanity's most astounding cultural achievements. Without such a system, our societies would collapse

1. Habitual behavior, in contrast to normative behavior, can be observed in all animals.

into chaos. One of its main tasks is to hold lawbreakers to account, but punishment is costly and most people hate to administer it. Therefore it needs its own monitoring systems. Robert Axelrod (1986), when writing about the evolution of cooperation, suggested that if a punishment system can employ metanorms by calling for the punishment of those who fail to administer punishment, the lower-level behavioral norms will be stabilized.

The law aspires to be fair and impartial. Everyone, including the most powerful people in society, should be subject to and equal before the law, but typically, these people may try their best to place themselves above the law. Making the justice system work is an exceedingly complex task and always a work in progress.

In addition to explicitly stated rules, there are unwritten rules that describe how most people in our group behave most of the time. For example, we expect guests invited to a birthday to bring a gift and are perplexed if they come empty-handed. These rules are revealed in statements such as "We *always* do/eat/play, etc." They are also revealed in the surprise at social gaffes, such as appearing in shorts at a formal event. Such rules may be ridiculed as trivial, but in truth, without them our interactions would be far less smooth.

Even strong-willed three-year-olds can override their own preferences and follow group norms (Li, Britvan, and Tomasello, 2021). From childhood onward, not conforming to the norms of one's own group is considered a disgrace. For example, if a puppet joins the game and makes a move that violates the rules, children aged two to three will spontaneously protest and tell the puppet how to behave properly (Rakoczy and Schmidt, 2013). Studies from Susan Gelman's lab used made-up rules for invented characters living in invented worlds. They showed that from age four, children explicitly disapprove of nonconformity ("Hibbles are not supposed to play games with that kind of toy"). Indeed, children considered somebody who is not conforming to a norm to have committed a moral transgression (Roberts, Gelman, and Ho, 2017). Third-party punishment of selfish behavior that is against the norm is performed from middle childhood onward across different societies (Marshall and McAuliffe, 2022).

Our recommendation is simple. If we want to be better at getting along with each other, respecting norms and behaving accordingly are excellent schemes. Often these norms serve a useful practical purpose (cover your mouth when you sneeze),[2] but often they don't (shake hands as a greeting). Normative behavior can become habitual and

2. Our three-year-old granddaughter rebuked Uta when she covered her mouth with her hand and demonstrated the newly promoted, COVID-appropriate action of sneezing into her elbow.

part of the Zombie thread. But often it follows the Machiavelli thread, for example, by exaggerating or circumventing conventions. There will always be nonconformists and rebels. They can play an important role as founders of new groups with their own norms. Norms change, and new norms are quickly generated as needed. An example can be seen in the adoption of wearing masks and observing social distancing rules at the height of the COVID-19 pandemic.

Cultural norms specify our obligations to our ingroup, especially to our kin, and this plays an important role in the moral judgment of selfish behavior (McManus, Kleiman-Weiner, and Young, 2020). If a high value is put on kinship loyalty, then if you had to choose between helping a member of your family or an outsider, you would be judged as immoral and untrustworthy if you did not help your family first. In contrast, religions often strive to extend their reach well beyond people's original ingroups. They can forge new norms of behavior and can create new role models, such as the good Samaritan, an outgroup member who is praised for aiding a stranger.

Norms observed by groups are not always benign and can sometimes have terrible consequences. An example is the custom of foot binding in nineteenth-century China. Extreme examples can be found in some cults, which may demand self-mutilation or even human sacrifice in their rites. Jim Jones was the charismatic leader of a modern-day cult (the Peoples Temple, 1955–1978). The members of this group had many shared beliefs, including the expectation that an unknown enemy would descend on them and kill them mercilessly. In 1978, Jones persuaded his group of over 900 people to commit mass suicide (Sorrel, 1978). This horrific example shows that it is possible to overpower basic instincts, such as avoiding harm to the self. Members can become ever more tightly bound to their group, and then the balance entirely tips against their individual needs and their freedom of action.

Why is normative behavior so pervasive in humans? It seems plausible that one driver is our existential need for affiliation. The intense need for affiliation may be the single most important element that makes us social. It connects us with all other social animals. We need to behave, move, feel, and think like the others in our group. But for humans, this is not enough. We need to adopt behavior that identifies us and the group we belong to. We want affiliation only with our ingroup and differentiation from the outgroup. Thus, our spontaneous tendency to overimitate what other group members do may well be the engine that powers our adherence to even arbitrary rules of behavior.

How are norms for cooperative behavior promoted? Religions are powerful sources of behavioral norms. The mechanisms involved are complex, and we pick out just one strand, storytelling. Storytelling is certainly one mechanism to promote group identity and bind group members together. Stories are easily remembered and transmitted to others. They tend to contain a moral lesson that is readily comprehended, such as the tale of the boy who cried wolf. All members of the group know the same stories and the lessons they teach about how to behave and be a better person. For instance, hunter-gatherer societies tell stories with messages that enhance cooperation, which is important in a foraging economy, and it has been shown that the presence of good storytellers is associated with increased cooperation (Smith et al., 2017).

What is the effect of normative behavior on the cognitive factors underlying information processing? One such effect is that it modifies our priors, including our perceptual biases. When we stay within the rules of our group, we become chronically biased toward sensory information that is congruent with the norm, even if it is different from our individual experience. A learning experiment showed that when people were asked to name one of two colors in a speckled square, over time they changed the name they first used to be in line with the color that was allegedly used most frequently by other group members. This happened in the learning phase of the experiment and then persisted in the extinction phase (Germar and Mojzisch, 2019). But there is a much bigger story to tell about priors and the top-down force they exert on our behavior.

The Case for Free Will as a Cultural Prior

One striking example is the belief in free will. Is it a delusion, and if so, should we ditch it? We, the authors, like most people in our culture (Sarkissian et al., 2010), tacitly believe that we have free will. If asked to explain why, we reveal that we subscribe to the long-standing idea, associated with philosophers such as Plato and Immanuel Kant, that there is a form of top-down control that enables us to overcome our basic urge to be selfish. Having such control over our actions is fundamental to ideas about moral responsibility, many of which are enshrined in law. Free will is the key to this control (Nahmias et al., 2005). Hence, we propose that believing in free will is a cultural prior.

But how is this belief inserted in our mind? As we have mentioned repeatedly, the hierarchy of control does not stop inside the person (Roepstorff and Frith, 2004). There is a portal between the inner and the outer worlds. In most situations, the highest level of top-down control comes from outside the individual. For most purposes, it is not a single individual but the pressure of the surrounding culture that acts at this highest level of control, and we presume that it acts equally on all members of the group. This has the effect of making us more similar to each other. In this way, regular patterns of action are installed, which make it easier to coordinate with others (see McGeer, 2007). We all know how we *should* behave, and we often smugly claim that's how we behave all the time.

In chapter 18, we discussed how teaching can instill cultural models and cultural assumptions into individuals who then incorporate them into their world of ideas. The basic example is that of a simple instruction, except that cultural instructions are more abstract. Rather than saying, very concretely, "Press the left button when you see the red light," people are acquiring worldviews, such as "Eat less, and that will make you live longer." With consistent messages about how mind and body work, our culture can change our beliefs and change the way we behave.

The nature and existence of free will have long been topics of considerable controversy. However, in the last few decades, scientists have joined the fray and claimed to have evidence that free will is an illusion (Crick, 1994; Wegner, 2003). But what might be the effects of these anti–free will viewpoints if they became cultural priors?

Some philosophers believe that undermining people's belief in free will would be disastrous (e.g., Smilansky, 2002). Even if the experience of free will is an illusion, our belief in it is needed for us to have the commitment to behave morally. If we don't believe in free will, we will revert to our selfish and animalistic urges. Others have a different view (e.g., Caruso, 2014), suggesting that abandoning our belief in free will make us less willing to apply harsh punishments to others.

There have been several experimental tests of these ideas, and, at first glance, both predictions have been confirmed. To reduce their belief in free will, people are presented with statements such as "Most rational people now recognize that free will is an illusion," citing Nobel Prize winner Francis Crick as an authority and example of a rational person (Crick, 1994). Bad effects were observed in these early studies. People who came to doubt the existence of free will showed increased aggression and reduced helping behavior (Baumeister, Masicampo, and Dewall, 2009). They were also more likely to cheat in exams (Vohs and Schooler, 2008). On the other hand, people became more prosocial by recommending lighter prison sentences (Shariff et al., 2014).

But should we trust these results that purport that it is easy to change the strong cultural prior of belief in free will? Perhaps we should have been suspicious that both prosocial and antisocial effects can be conjured up in the lab by priming people about the existence of free will. Indeed, subsequent studies have failed to replicate these results (Genschow et al., 2021). Perhaps it is just as well that attempts to manipulate beliefs about free will are very fragile. If believing in free will is a cultural prior, then it may be similar to believing in justice and truth. Our commitment to justice and the associated legal systems enhances social cohesion through the resolution of conflicts. But we don't usually think about the value of justice in this way.

We are committed to justice for its own sake (Rawls, 2020, 416). This high-level prior has been fixed in most of us from earliest infancy. At a lower level, the behavior that is consistent with this high-level prior has become automatic and engrained. Behaving according to the tenets of justice has become habitual for us. Our commitment to justice would not be so robust if we always had to think in detail about the psychological origins of this commitment (Rawls, 2020, 451). This combination of a high-level prior and a low-level habit is difficult to dislodge.[3]

We suggest that same principles apply when we consider responsibility and free will. Linked to our vivid experience of being on control of our actions (the sense of free will), we have a commitment to accepting responsibility for our actions. From childhood onward, we have learned that if we want to overcome our urge to be selfish, we must exert self-control (pull yourself together; think about others). Free will is key to such control (Nahmias et al., 2005).

We believe that it is this intentional, top-down control that enables us to behave in a moral fashion and it emerges through cultural learning. True, constantly exerting top-down control over our actions is very effortful. In the long term, however, this effort can be minimized by developing habits that are consistent with our high-level priors. As in the case of our belief in justice, the combination of the higher-level prior of being responsible for our actions and the lower-level habit of acting accordingly is remarkably robust.

Groups and Their Limitations—Some Problems with Diversity

Through cooperation in groups, we can achieve much more than we can as individuals. So why does human cooperation fail so often? We can point to a couple of reasons. While mixed experience, perspectives, and skills are a boon in most collaborations, mixed ability is a drawback. As discussed in chapter 15, groups have an equality bias, which means that the less competent member can have undue influence on a joint decision when a better decision could have been made by the most competent individual. However, it is hard to assess the ability of individuals and match them up. We prefer to use confidence as a marker of ability. Unfortunately, when part of the Machiavelli strand, this marker is easily subverted, and therefore it can often be misleading.

One problem is that we often form groups that are not sufficiently diverse. A major theme in this book has been that most problems can be solved better when we work in groups that are diverse. As we mentioned in chapter 15, the diversity we have in mind has little to do with visible differences in appearance, age, gender, or ethnicity. Instead, it is about usually invisible differences in perspectives, experiences, and skills (Sulik,

3. This idea of justice was foreshadowed in David Hume's *A Treatise of Human Nature*.

Bahrami, and Deroy, 2021). By comparing different approaches, we can find better solutions to a problem. By exposing our own mind to other minds, we can allow our priors to open up, and even change.

Another major problem we see is a lack of mutual understanding. Clearly, optimal cooperation requires that knowledge is shared. So one approach for reducing failure is to find strategies for better information sharing. Formal education plays a major role in creating shared cultural knowledge, and so does informal gossip (chapter 9). Gossip spreads rapidly within groups of similar people, and sometimes it can create networks between diverse people.

The internet and the almost global adoption of English as the preferred language has given us the ability to create ever-larger and more diverse groups and has made communication faster and more fluent. Can we expect even greater advantages? Unfortunately, no. When people are too widely connected, some advantages, such as common ground and stability, are lost. At first sight, we might expect information sharing to become optimal when everyone interacts with everyone else. The problem arises because people are limited in their ability to reciprocate with a large number of social partners. There are psychological limits on our capacity to process information. These limits create an attentional bottleneck. This is especially true for the explicit metacognitive processes on which the transmission of cultural norms depends (Cowan, 2001). When there is too much information, selection must come into play to overcome the bottleneck. But this selection prioritizes information that is consistent with current beliefs, that is negative, that is social, and that is predictive (Hills, 2018).

The consequences can be calamitous. Selection for belief-consistent information drives increasingly polarized views. Selection for negative information amplifies emotional responses. Selection for social information (i.e., finding out what other people are doing) drives herding behavior (Raafat, Chater, and Frith, 2009). This in turn impairs objective assessments and reduces exploration for solutions to hard problems. Selection for predictive patterns drives overfitting. Excessive information sampling will inevitably turn up spurious correlations and false positive results. Because of the pervasive human need for everything to make sense, these false predictions are especially attractive. These are exactly the problems that have emerged through widespread use of the internet, where everyone is connected to everyone else.

Are there forms of connectivity that can overcome this problem? One possibility is the small-world network. Such networks occupy the middle ground between networks where the connections are completely regular and networks where the connections are completely random (Watts and Strogatz, 1998). Small-world networks can be observed throughout nature, including commercial organizations and professional groups such as scientists (Uzzi, Amaral, and Reed-Tsochas, 2007). These networks typically consist of a small number of subgroups (cliques) with many interactions between the partners within these cliques, while interaction between the cliques happens at a much lower rate (figure 19.2). Interestingly, such networks can optimize information sharing and diminish the occurrence of information bubbles and polarization (Momennejad, Duker, and Coman, 2019). One type of diversity is particularly important for such networks. This is a blend of less sociable people, who interact *within* the cliques, and of more sociable people, who also interact *between* the cliques. Such a mix can attain and maintain the structure.

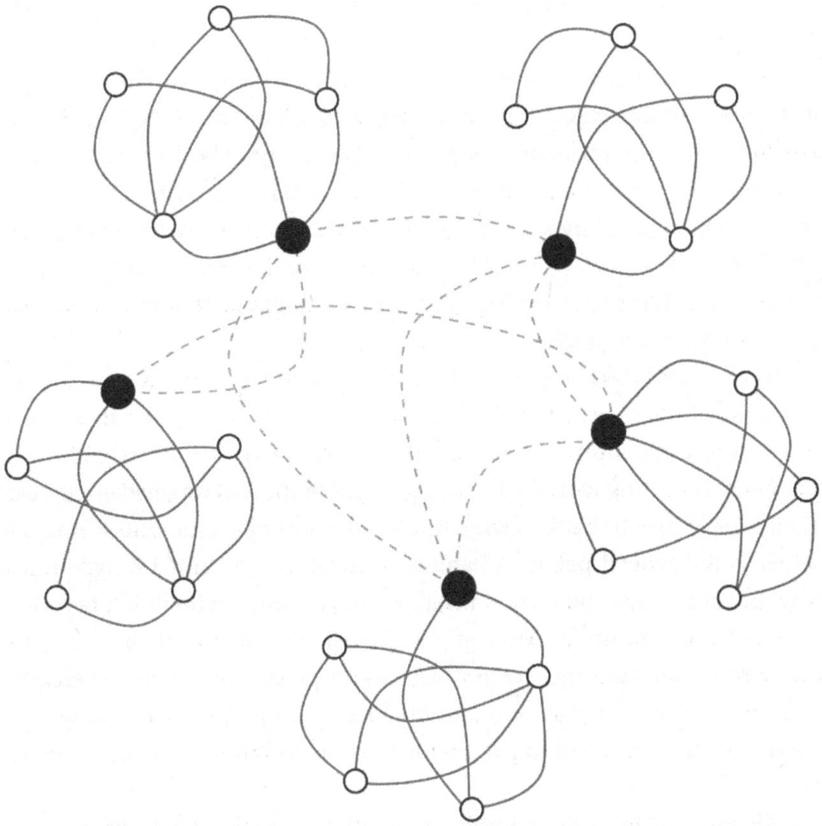

Figure 19.2
Diversity in a small-world network: there are many connections within the subgroups (solid lines) and fewer connections between the subgroups (dotted lines) The black dots mark the more sociable people, who interact with other subgroups. Redrawn from figure 2c in Shanahan (2012). Copyright 2012, with permission from the Royal Society.

The Way Ahead?

One might think that the creation of groups that promote altruistic and cooperative behavior and make their members proud to belong to them would benefit everyone. But unfortunately, the creation of any group, even a morally admirable group, always leads to the simultaneous creation of outgroups. Members of outgroups are treated very differently, and usually very badly compared to the ingroup. As discussed in chapter 8, we prefer to interact with members of our ingroup, not least because we have shared memories and conventions. It is easier to communicate with them and to predict their behavior. We avoid outgroup members and will not cooperate with them. We find ways to justify this behavior via various false portrayals of the outgroup, racial prejudice being one glaring example.

One solution is to enlarge the ingroup by bringing in outgroup members, with globalization as an extreme case. With more travel and the spread of the World Wide Web, we have all become members of very much larger groups. Has this reduced prejudice and increased cooperation and altruism? Unfortunately, no. When groups compete, the group containing the most cooperative individuals is likely to prevail, but when individuals compete within groups, the uncooperative (i.e., selfish) people are likely to do best (Sober and Wilson, 1998). This pattern resembles one that has a long history in theories about the rise and fall of civilizations.

Such developments seem especially likely to occur in large groups such as nations or the even larger entities created by globalization. There are many ways in which people can be selfish in this context. A shocking possibility is that free riders join together to form ingroups competing with the rest of the population. This was evident, for example, in the behavior of various bankers who developed instruments for transferring risk from themselves to the general public. A banking scandal in the United Kingdom involved the misselling of payment protection insurance to customers who didn't need it or who would never be able to use it (McGagh, 2012). Salespeople and their bosses, the new ingroup, were maximizing their commissions and profits, all at the expense of their customers, who were treated as an outgroup. The law caught up with this misconduct, and as a result, the banks had to pay billions in compensation. Hurray for our system of justice![4]

Tax avoidance, in its various forms, is another example of one group exploiting another. This practice makes more money for individuals or companies, while they still

4. But more work remains to be done. Banks still manage to keep their profits while sharing the losses with the rest of us because they are 'too big to fail."

benefit from the infrastructure paid for by the taxes of those who continue to pay them. A similar story can be told about vaccination (Fine, Eames, and Heymann, 2011). A selfish free rider, who wishes to avoid the possible risks of vaccination, can take advantage of the herd immunity created when the majority of the population continues to be vaccinated. But if enough people adopt this strategy, herd immunity is lost, and diseases that were almost eradicated, such as measles and smallpox, start to spread again.

In response to the development of new, selfish groups, other groups arise in competition. It is as if the modules in the small-world network start to break away from one another. One example of a competing ingroup is the Occupy movement, which arose in response to the banking crisis and to inequality more generally, as indicated by the slogan "We are the 99%." This movement began in Wall Street in 2011 and spread all over the world. Its major goals were to end tax evasion and tighten banking regulations. At the opposite extreme is the Tea Party movement, which also arose partly in response to the banking crisis of 2008. The goals of this movement were to reduce government spending (e.g., by not bailing out banks), reducing financial regulations, and reducing taxes.

These are just two examples of the many new ingroups that have appeared in last few decades, and which have little interaction with each other. They are manifestations of the increased polarization affecting society and especially politics. These groups become polarized since they no longer share memories and norms. Without common ground, communication between groups becomes too difficult and is therefore avoided. This is especially true for communications (e.g., tweets) concerning moral norms (Brady et al., 2017).

There are, as yet, few experimental studies that suggest that opposing groups can be brought together and their norms realigned. However, we would like to end with a positive story, and we are delighted that Joaquin Navajas and colleagues (2019) have begun to explore this important problem in a promising set of experiments. They recruited large numbers of volunteers and divided them into groups consisting of just three individuals. These small groups were instructed to discuss highly controversial topics, such as abortion, with the aim to reach a consensus. As might be expected with such topics, the groups often included people with extreme positions, believing, for example, that abortion was either always permissible or always wrong.

Typically, people at these extremes were more confident in their views than people in the middle. In spite of these extreme views, about 50 percent of the triads were able to reach a consensus. In some cases, there was also evidence of depolarization (i.e., individual views became less extreme after the discussion). What was special about the groups in which consensus and depolarization were achieved? Whether they reached

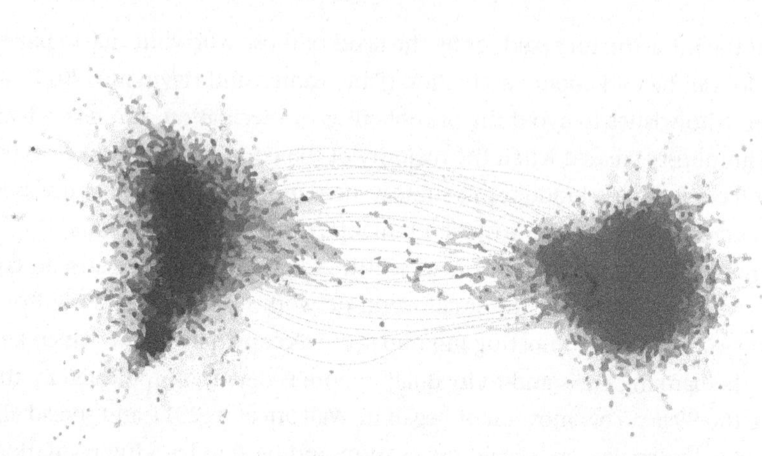

Figure 19.3
Traffic between two polarized groups on Twitter: Here, we see lots of traffic within groups and very little traffic between groups. Redrawn from figure 3 in Brady et al. (2017).

consensus depended on *the person in the middle*. If this person held their middle position with high confidence, then they could act as mediators, engaging with both sides and bringing them together. If this result is validated by future studies, we must learn to feel and express more confidence in our more moderate beliefs.

Of course, in practice, we would want depolarizing effects to spill over beyond groups of three. Perhaps we can rely on the social influence resulting from our human wish to be part of a wider group? We would predict that if enough of these small groups depolarize, so that this becomes the position of the majority, then those who remain at the extremes will start to conform. But there are many situations where depolarization may not be what we want to achieve. Rather than bringing people together from the extremes, we may want to eliminate one of the extremes altogether. This is the case when a group has acquired traditions that are widely considered undesirable.[5] Examples would be the acceptance of honor killings and forced marriages. History tells us that group traditions can be changed, but it does not tell us how this happens.

There Is Much More to Do

We are aware that we have not always managed to answer the questions that we have posed in this book, let alone those we haven't. However, we hope that we have convinced

5. But who decides what is undesirable?

you that the methods of cognitive neuroscience can provide insights that can help us to solve some of the hardest problems that we encounter in our social lives. We need more research to tell us how groups change their values; how moral values, such as equality and fairness, can become normative; and how norms can be upheld that make it easier for human beings of all kinds to get along with each other.

Our aim in this book has been to write about what we find strange and fascinating about our social nature. Even though the work we have reviewed will rapidly be superseded, our hope is that our basic framework will remain useful for somewhat longer. This framework should point to the gaps in our knowledge, as well as where further research is most likely to be fruitful. It might also reveal where conceptual confusion is hindering progress.

Our framework is derived from information processing, and we used evidence from cognitive neuroscience to apply this framework to processing social information in conscious (Machiavelli thread) and unconscious (Zombie thread) forms. We have gone down many avenues and talked about social creatures ranging from fruit flies to humans. We believe that humans share the Zombie thread with other animals, but are virtually unique in using the Machiavelli thread, for good or ill. We have shown that we can learn, not only from direct experience, but also from observing others and from listening to instructions. We have commented on our prior beliefs about how our mind works and on how these priors can be modified, whether by discussions with other people, or by long-term trends in our culture more generally.

We hope that our passion for a neuroscience of social cognition will spread via contagion to our readers. We believe that this is currently one of the most exciting areas where psychology meets neuroscience. And here, we mean psychology in the broadest sense. When we talk about culture, for example, we hope to excite rather than annoy those who consider psychology to be part of the social sciences and humanities.

When considering so many areas, our treatment is bound to be somewhat superficial. We have stepped back a bit to look at the big picture, which has allowed us a glimpse of three worlds: the world of objects, the world of agents, and the world of ideas. The fact that, as social beings, humans can all navigate these worlds is rather wonderful. And this makes us optimistic that there will be ways to preserve these worlds and perhaps even make us better navigators.

Epilogue

20 Facing a Pandemic Can Bring Out the Good in Us

Shortly after we began to revise the final versions of the chapters in this book, the COVID-19 virus started spreading through the world. One striking result of this crisis was that science suddenly took center stage. Instead of disparaging the experts, some politicians started to appear on television flanked by scientists. Presumably, these politicians hoped that this would increase their own credibility. The most obvious scientific input relevant to the crisis concerned the nature of the virus and mathematical modeling of how the epidemic was likely to spread. However, once we had this information, the major problem we were faced with was how to minimize the effects of the pandemic. And this problem is essentially about human behavior. We knew that the virus was highly contagious. But how could we stop its spread? We needed to reduce contact between people by instituting lockdowns. To achieve this change in behavior, it is necessary to know something about how groups of people react when under severe threat.

A long-standing and popular idea is that imminent danger brings out the worst in us. When we believe that we are faced with a severe threat, we abandon the social niceties and, being naturally competitive, fall back into "brutishness and misery" (Hobbes, 1651, chapter XIII). Indeed, whenever there is some mass disaster, such as a fire in a crowded disco, we often read reports that the deaths occurred because the crowd panicked and people were crushed to death, fighting with each other to get through the exit. In anticipation of such behavior, the police and the army are often prepared for violent public disturbances.

These beliefs about human nature had an impact on the strategies adopted to limit the spread of the virus. Governments needed to impose rules to safeguard citizens, accompanied by appropriate sanctions if these rules were disobeyed. However, the governments of some nations, such as the United Kingdom, did not trust the public to obey the rules. This was probably rooted in the idea that a crowd is incapable of behaving sensibly. They firmly expected people to rebel when asked to stay at home and keep their distance from each other. As a result, the necessary measures were delayed.

Michael Bang Petersen (2021) argues that a better strategy is to trust the public with hard truths. This was the case in Denmark, where there is a high level of mutual trust between the government and the people. The Danish government was not afraid to present the hard truths about the virus. As a result, lockdowns could be imposed early, preventing many deaths and laying the groundwork for a high vaccination rate.

But how do people actually behave during the crisis? As we conclude in several of the chapters in this book, our natural state is not selfish competition. If anything, it is the opposite. In one of the economic games that reveal so much about selfishness and altruism, "dictators" can give away as little or as much of their money as they like and get no disadvantage from choosing the option of keeping all the money to themselves. Even so, the majority give some money away (Forsythe et al., 1994). It seems that most of us always take some account of the welfare of others. We conclude that the majority of us are cooperative (prosocial) rather than competitive (van Lange et al., 1997). We are naturally empathic and have an aversion to unfair behavior (Contreras-Huerta et al., 2020). We align with others, feeling their pain and imitating their actions (see chapters 3 and 4).

Furthermore, prosocial behavior is not something we have to work hard to achieve by suppressing our "naturally" selfish and competitive impulses. For most of us, most of the time, this behavior occurs automatically. We are influenced by the feelings (Morelli and Lieberman, 2013) and actions of others (Belot, Crawford, and Heyes, 2013). And, for most of us, our behavior becomes more prosocial when our cognitive capacity is filled by having to think about something else (cognitive load; see Haruno, Kimura, and Frith, 2014; Rand, 2016).

But will a threat, such as the inexorable spread of the COVID-19 virus, change our priorities so that we will think only about how we shall escape the danger, not about the welfare of others? What happened is quite contrary to the long-standing belief that, faced with danger, people (especially crowds) will panic and revert to selfish and irrational behavior (Le Bon, 1895). However, if the observations that we have mentioned here are correct, then under such pressure, we become more prosocial rather than more selfish.

It turns out that emergency situations do not typically give rise to collective panic (Dezecache, 2015). For example, at a Who concert in 1979 attended by about 18,000 people, 11 deaths occurred. The press reported that the deaths were caused by panic. But statements collected afterward (from police officers, employees, and private security guards) did not report competition between people as they tried to reach safety. Rather, people tried to prevent others from being crushed (Johnson, 2014). Likewise, numerous stories about behavior after the July 7, 2005, bomb attacks on the London public transportation system mentioned the help that the victims gave to each other. Selfish behavior was rarely reported (Drury, Cocking, and Reicher, 2009).

But how reliable are these statements given after the event? To get around this problem, Dezecache, Grèzes, and Dahl (2017) studied the response to the mild threat associated with entering the Haunted House ride in a fairground from photographs taken at the time of the ride. At these moments of apparent danger, people held onto each other, seeking affiliation and comfort. Just as in natural disaster, people turn toward their loved ones and form clusters of familiar individuals before trying to escape.

It is not just that people seek comfort in the face of danger and threat. They also want to help others, especially their family members. It can be difficult to get people to evacuate because they wait for all their friends to get together before leaving (Sime, 1983). Even people with no previous ties will get together and experience a shared social identity as a result of the common threat. "In disasters, people are more likely to be killed by compassion than competition. They often tarry to help friends or family members" (Drury and Reicher, 2010, 59)

This urge to help others when we are all threatened has been evident in people's responses to the spread of COVID-19. Throughout the world, health workers have exposed themselves to danger by continuing to do their jobs, some even coming out of retirement to do so. When the UK government asked for volunteers to help the National Health Service (NHS), 750,000 people signed up, three times as many as were expected (Butler, 2020).

But this urge to help and to be with our families and friends is very problematic when we are asked to practice social distancing and go into isolation (Dezecache, Frith, and Deroy, 2020). If our friends are ill, we want to comfort and help them. If they seem perfectly well, we don't see why they should be isolated from us. Of course, most of us recognized and respected the reasons for going into lockdown. And even if we didn't, most of us, at least in the United Kingdom, believed that we should follow the advice of government scientists.

Given our desire for social affiliation, the experience of lockdown is hard. However, for many of us, it was not as hard as it could have been. This is because, in our times, physical distancing need not eliminate social interactions. We can keep in touch with our friends via the telephone and email. And even more recent technological developments have allowed us to have face-to-face meetings via the internet. The number of such meetings dramatically increased after social distancing was introduced.

The day that the lockdown was announced in the United Kingdom, the remote conferencing app Zoom was downloaded 2.13 million times around the world, up from 56,000 a day two months earlier. At the same time, there was a dramatic rise in the number of informal mutual aid "good neighbor" organizations. These are local groups that keep volunteers in touch via social media. Shortly after the beginning of the lockdown in Britain, there were more than 4,300 such groups, connecting an estimated 3 million people (Neate, 2020). In our case, we were isolated at home, but we instituted teaching and chatting sessions with our grandchildren. And at work, our weekly seminars continued online and included more participants who now attended from all over the world. As a result of the lockdown, the number and scope of our social interactions actually increased.

Humans are very good at adapting to changing circumstances. We have a remarkable ability to develop new tools and to discover novel and unexpected ways to use old tools. And this is especially true for tools concerned with social interactions. Under the threat of the virus, we have developed all sorts of clever new ways for interacting at a distance.

Facing a Pandemic Can Bring Out the Worst in Us

In this book, we have claimed that there is a strong tendency to behave in a prosocial manner. However, as we have also emphasized, there is a dark side to this tendency. The social solidarity that we share with others seems to come at the expense of defining an outgroup from which we strenuously distinguish ourselves (see chapter 8). And when we inevitably start looking for someone to blame for our troubles. At the beginning of the pandemic, people started referring to COVID-19 as the "Chinese virus." Then there was the "Kent variant." This is not a new stratagem. We always look for some outgroup to blame. When syphilis spread through Europe and beyond in the fifteenth century, the Germans blamed the French, calling it the "French disease." Meanwhile, the French blamed the Italians; the Poles blamed the Russians; the Persians blamed the Turks; the Muslims blamed the Hindus; and the Japanese blamed the Portuguese. To combat this blame game, new variants of the coronavirus are being identified by letters, such as delta and omicron.

The special status given to ingroups was also visible in the behavior of national governments during the pandemic. In many countries, dramatically large sums of money were provided to help when their residents suddenly lost their livelihoods, and equally large sums were made available to develop and buy vaccines. But will the same governments extend their altruism toward other, poorer nations? Everyone agrees that COVID-19 is a global issue, following the motto that nobody is safe until everyone is safe. But making vaccines available to everyone has so far been only a lofty ideal, with few practical consequences.

The pandemic has amplified the divide between rich and poor and added to the pressures from climate change, habitat erosion, and resource depletion. The poor have suffered far more than the rich from the effects of the virus. But there is little sign that the rich are helping the poor. We fear that the risk of violent conflict due to lack of resources is likely to increase.

But what about the increased interconnectedness that has been made possible by social media? As many of us would agree, this connectedness has made the periods of lockdown more bearable. We can continue to interact with our families, colleagues, and friends. But the increase in connectedness also created unexpected problems. New ingroups have been created that focused on the virus and the strategies for coping with its effects. Vaccines were vilified. Lockdowns were said to be pointless. Following the science was considered incompatible with freedom.

There is a problem: When everyone is connected to everyone else, information overload is created, and selection must come into play to overcome this bottleneck. Selection means ignoring information, which can be calamitous when in fact we are searching for information. Attempts to simplify and make sense of the picture inevitably turn up spurious correlations and false positives. False predictions with a highly emotional tone are especially attractive. And then the selection for belief-consistent information creates groups with increasingly polarized views.

All these processes are accompanied by the increase in epistemic vigilance elicited by stress. Any information supplied by a rival group is automatically classified as untrustworthy. A friend of ours tried to convert an antivaxxer by sending a clearly written and well-reasoned argument in which all the sources of evidence were fully specified. The response was that, since the assertion was published in the *Guardian* newspaper, it was obviously untrue.

The groups that emerged, such as antivaxxers, can gain support from social media, and this gives us a glimpse of both the stickiness and the malleability of beliefs and behavior in the world of ideas. Luckily, antivaxxers remain in a minority, and many of them regret their views when they find themselves in the hospital with COVID. People who consistently refused to wear face masks and those who continued to gather in groups were more difficult to persuade. This might be because cause and effect (spreading the virus and catching the illness) tend to be well apart in time and space.

One of the major claims that we have made throughout this book is that there are distinctly different kinds of cognitive process that are arranged in a hierarchy. At the top, the processes involve reasoning. We referred to them as the "Machiavelli thread," without implying that this always results in antisocial actions. It is through these processes that we can provide argued justifications, and it is these processes that most clearly distinguish cultural groups. For example, ignoring norms can be readily justified in the Machiavelli thread (e.g., "I never wore masks before and never got ill. Others I know who got ill did wear masks. So why should I wear a mask now?"). These arguments become widely promoted and can become part of a subgroup culture. People who hesitate to be vaccinated are strongly impressed by claims about possible side effects and highly skeptical of statements to the contrary by experts. We note that there has been a long history of distrust in vaccination, which has proved extremely difficult to overcome. Antivaxxers are not just hesitant about getting vaccinated; they embrace their label as a form of social identity and have effectively become a new cultural group (Motta et al., 2021).

In time, we presume, the virus will be overcome, and these subgroups will dissolve. But it is likely that distrust of governments and of experts will last longer than COVID.

. . . and the Future?

The impact of COVID-19 has been worldwide, and many believe that there are even worse threats facing us right now, so that the world will never be the same. But what will this new world be like? The answer depends on the nature of human social cognition.

One thing that we are fairly confident about is that people will continue to use social media, such as Zoom, in the new ways developed during the lockdown. And there will be further developments in remote interactions, including the creation of novel signals for controlling the flow of speech during group meetings and the creation of new conventions for greetings and farewells. In consequence, remote meetings will become less cognitively taxing and even more popular. These developments have the potential to increase the diversity of interacting groups. This would lead to more efficient social networks. If so, this development should reduce polarization and allow information to be shared more widely. At the same time, group decision-making would improve.

We mentioned earlier that one of the requirements for successful group decision-making is that everyone should have the same goals. The external threat engendered by COVID-19 certainly helped to align our goals. Helping the vulnerable and rewarding those who give the most help became priorities. Our hope is that these will become defining characteristics of the sort of people we are, and thus help solve our existential crisis. This new norm will be guarded, because we will start to feel regret and shame if we fail to uphold these principles (Schaumberg and Skowronek, 2022).

COVID-19 and the associated lockdowns have affected both low- and high-level processes. Bottom-up effects emerge from our natural desires to be with others and to help them. Top-down effects arise when people begin to talk to each other about the new norms that are developing and whether to adopt them. At a higher level still, top-down effects can be applied when governments start hearing what people are saying. Unfortunately, it is at this highest level that justifications can be developed and used to inhibit our natural inclinations to help others. Justifications can also be used to take advantage of our inherently negative attitudes toward outgroups. But fortunately, the Machiavelli thread can also come up with justifications that increase cooperation. What happens in the future critically depends upon the public discourse that occurs between individuals and between groups. What kinds of justifications will emerge? What sorts of values will be promoted? We can't tell right now because there are so many other factors and new events that are bound to come into play.

One major factor relevant to a global change of values in our society is the continuum that goes from cheating (or free riding) to an active society of voluntary social work (Tausch and Heshmati, 2009). Many societies at present seem to be placed near the cheating end of this continuum, and this may be because of an uncertainty about who counts as "our society" (i.e., people like us). This question was easy to answer when population levels were low and families and tribes defined whom you belonged to, and who your enemy was (Henrich, 2020). It seems that if a population becomes too big, many trusted mechanisms that regulate social behavior may start to collapse. This is a huge problem, and the uncertainty that it creates has brought into sharp relief the indisputable difficulty of coping with genuine crises such as climate change, risk of species extinction, wars, and collapsing justice systems. However, precisely because of our experience with the COVID-19 pandemic, we have, for a moment, been able to experience more directly the better nature of our social cognition. With any luck, this experience may reset our social values and facilitate actions that benefit all of us, and indeed the whole living world.

Acknowledgments

This book would not exist were it not for the Jean Nicod Prize, which we were fortunate to receive in 2014. One of the conditions of accepting this prize was to publish a book on the content of our Jean Nicod lectures at the École Normale Supérieure. It was clear to us at the time that we would need to read up much more on the burgeoning research in social neuroscience, but we did not realize just how rapidly this field was developing and that it would take years to absorb the evidence we needed to sort out our own rather sketchy thoughts.

Something happened on the way to our lectures. We habitually passed through the Rue Dante, where there are many shops selling comics. It was a wonderland for us because the standards of these comics were extremely high, with a wide range of artistic expression and an equally wide range of content. In short, we were inspired to produce a book on the content of the lectures in the form of a graphic novel. Our son, Alex Frith, the author of nonfiction books for children and a lifelong comics fan, agreed to be in charge, and found the artist Daniel Locke, who was excited to collaborate with us. That book was published in 2022. All the time that we had fun collaborating on the comic, we kept in mind that we would present the scientific evidence in a proper academic book. And here it is.

As everyone knows, scientific evidence is always work in progress, and we have tried to keep up with developments, bearing in mind the crucial importance of independent replication of study results. This gave us lots of opportunity to scrutinize our own beliefs and biases. We are quite pleased that each of us brought a somewhat different set of biases to the project because our reading of the literature convinced us that diversity and discussion are vital to broadening our thoughts, and different perspectives on the same problem are more likely to achieve good solutions. We quickly agreed on one thing: psychology is always social. It is other people who form the Umwelt that is our natural habitat. While writing, we came to agree on another thing: explicit metacognition is our human superpower. It creates the interface between culture and brain.

How did we manage the collaboration? Roughly, it turned out that Chris provided the structure, and Uta did the coloring in. Our big picture was of three worlds: the worlds of objects, agents, and ideas. Our guiding framework is an information-processing hierarchy. To allow both bottom-up and top-down dynamics, it is held in place by prior expectations on the one side, and by current evidence on the other. This framework necessitated frequent mentions of unconscious or implicit processes, which are widely shared between humans and other animals, and conscious or explicit processes, which in their richest form may be uniquely human. The coloring here refers to these processes as "Zombie" and "Machiavelli" threads.

We had many silent collaborators in our friends and colleagues, whose work we tried to understand better. Luckily, not all of them stayed silent—we asked them to read chapters in the making and to give honest criticism. We benefited greatly from their advice, and this made us deeply grateful for their willingness to read repeated drafts and the generosity and kindness behind their responses.

Alex and Martin Frith read our introduction and said, "It'll do." Ros Ridley commented on our stories about learning and checked our pictures of the brain. Antonia Hamilton cast a critical eye on our ideas concerning learning from others and mirroring actions. Natalie Sebanz gave valuable comments on joint action, while Mattia Gallotti commented on the We-mode.

Sarah-Jayne Blakemore, Essi Viding, and Francesca Happé provided suggestions for our thoughts on mentalizing, communication, ingroups and outgroups, and the dark side. Patricia Lockwood and Matthew Apps made substantial comments concerning computational models and mentalizing. Nichola Raihani gave much needed critique on what we wanted to say about cooperation and competition as well as about reputation and teaching.

Barry C. Smith provided important suggestions for improving our ideas about communication. Johannes Schulz played a vital role in developing our ideas about detecting agents. Nick Shea, Steven Fleming, and Dan Bang were crucial to critiquing our ideas about computational models and metacognition. Celia Heyes fortunately approved of our concepts about culture. Alex Reuben helped us to remove jargon and gave valuable advice on the presentation of the figures, expertly drawn by Jamie Ball.

We are most grateful to Frederique de Vignemont, the organizer of the Jean Nicod Prize, and all those on the prize jury for having selected us and thereby bringing about this book. And finally, we must thank Philip Laughlin from MIT Press, who waited so patiently and sent a yearly nudge to keep us going. We are also very grateful to two anonymous reviewers, whose critical comments were so useful that we were persuaded to embark on a thorough revision.

There is something deeply gratifying to be able to collaborate on processes that in turn enable and hinder collaboration. We are in awe that our multilayered and conflicting needs and goals can be held together and put to work at all. We have benefited enormously from a lifetime of collaboration with the members of some incredibly inspiring (as well as nurturing) academic groups. And it is these groups that created our understanding of the major themes in this book: *cognition* from the Institute of Cognitive Neuroscience at University College London, *neuroscience* from the Functional Imaging Lab (aka the Wellcome Centre for Human NeuroImaging), *social interaction* from the Interacting Minds Centre at Aarhus University, and *explicit metacognition* from the Institute of Philosophy at the University of London. If this book has any value, it has emerged from our being embedded in these wonderfully competent and diverse groups.

June 22, 2022

References

Abell, F., Happé, F., and Frith, U. 2000. Do triangles play tricks? Attribution of mental states to animated shapes in normal and abnormal development. *Cognitive Development* 15, 1–16.

Adank, P., Hagoort, P., and Bekkering, H. 2010. Imitation improves language comprehension. *Psychological Science* 21, 1903–1909.

Addabbo, M., Vacaru, S. V., Meyer, M., and Hunnius, S. 2020. "Something in the way you move": Infants are sensitive to emotions conveyed in action kinematics. *Developmental Science* 23, e12873.

Albright, T. D. 2017. Why eyewitnesses fail. *Proceedings of the National Academy of Sciences* 114, 7758–7764.

Alós-Ferrer, C., García-Segarra, J., and Ritschel, A. 2021. Generous with individuals and selfish to the masses. *Nature Human Behaviour* 6, 88–96.

Alpizar, F., Carlsson, F., and Johansson-Stenman, O. 2008. Anonymity, reciprocity, and conformity: Evidence from voluntary contributions to a national park in Costa Rica. *Journal of Public Economics* 92, 1047–1060.

Altay, S., Majima, Y., and Mercier, H. 2020. It's my idea! Reputation management and idea appropriation. *Evolution and Human Behavior* 41, 235–243.

Ambrosini, E., Blomberg, O., Mandrigin, A., and Costantini, M. 2014. Social exclusion modulates pre-reflective interpersonal body representation. *Psychological Research* 78, 28–36.

Anderson, C., Hildreth, J. A. D., and Howland, L. 2015. Is the desire for status a fundamental human motive? A review of the empirical literature. *Psychological Bulletin* 141, 574–601.

Andreoni, J., and Rao, J. M. 2011. The power of asking: How communication affects selfishness, empathy, and altruism. *Journal of Public Economics* 95, 513–520.

Andrews, J. L., Ahmed, S. P., and Blakemore, S.-J. 2021. Navigating the social environment in adolescence: The role of social brain development. *Biological Psychiatry* 89, 109–118.

Apperly, I. A., and Butterfill, S. A. 2009. Do humans have two systems to track beliefs and belief-like states? *Psychological Review* 116, 953–970.

Aquino, T. G., Minxha, J., Dunne, S., Ross, I. B., Mamelak, A. N., Rutishauser, U., and Doherty, J. P. 2020. Value-related neuronal responses in the human amygdala during observational learning. *Journal of Neuroscience* 40, 4761–4772.

Arora, A., Schurz, M., and Perner, J. 2017. Systematic comparison of brain imaging meta-analyses of ToM with vPT. *BioMed Research International* 6875850, https://doi.org/10.1155/2017/6875850.

Aslin, R. N. 2017. Statistical learning: A powerful mechanism that operates by mere exposure. *WIREs Cognitive Science* 8, e1373, https://doi.org/10.1002/wcs.1373.

Astuti, R. 2015. Implicit and explicit theory of mind. *Anthropology of This Century* 13, http://aotcpress.com/articles/implicit-explicit-theory-mind/.

Atlas, L. Y., Doll, B. B., Li, J., Daw, N. D., and Phelps, E. A. 2016. Instructed knowledge shapes feedback-driven aversive learning in striatum and orbitofrontal cortex, but not the amygdala. *eLife* 5, e15192, https://doi.org/10.7554/eLife.15192.

Aucouturier, J.-J., and Canonne, C. 2017. Musical friends and foes: The social cognition of affiliation and control in improvised interactions. *Cognition* 161, 94–108.

Auger, S. D., and Maguire, E. A. 2018. Retrosplenial cortex indexes stability beyond the spatial domain. *Journal of Neuroscience* 38, 1472–1481.

Aumann, R. J. 1987. Correlated equilibrium as an expression of Bayesian rationality. *Econometrica* 55, 1–18.

Axelrod, R. 1986. An evolutionary approach to norms. *American Political Science Review* 80, 1095–1111.

Axelrod, R., and Hamilton, W. D. 1981. The evolution of cooperation. *Science* 211, 1390–1396.

Bahrami, B., Olsen, K., Latham, P. E., Roepstorff, A., Rees, G., and Frith, C. D. 2010. Optimally interacting minds. *Science* 329, 1081–1085.

Ballerini, M., Cabibbo, N., Candelier, R., Cavagna, A., Cisbani, E., Giardina, I., et al. 2008. Interaction ruling animal collective behavior depends on topological rather than metric distance: Evidence from a field study. *Proceedings of the National Academy of Sciences* 105, 1232–1237.

Balliet, D. 2010. Communication and cooperation in social dilemmas: A meta-analytic review. *Journal of Conflict Resolution* 54, 39–57.

Banakou, D., Hanumanthu, P. D., and Slater, M. 2016. Virtual embodiment of White people in a Black virtual body leads to a sustained reduction in their implicit racial bias. *Frontiers in Human Neuroscience* 29, https://doi.org/10.3389/fnhum.2016.00601.

Bandura, A. 1965. Vicarious processes: A case of no-trial learning. In *Advances in Experimental Social Psychology*, ed. L. Berkowitz, 1–55. Academic Press.

Bang, D., Aitchison, L., Moran, R., Herce Castanon, S., Rafiee, B., Mahmoodi, A., et al. 2017. Confidence matching in group decision-making. *Nature Human Behaviour* 1, 0117, https://doi.org/10.1038/s41562-017-0117.

Bang, D., and Frith, C. D. 2017. Making better decisions in groups. *Royal Society Open Science* 4, http://doi.org/10.1098/rsos.170193.

Bang, D., Fusaroli, R., Tylén, K., Olsen, K., Latham, P. E., Lau, J. Y., et al. 2014. Does interaction matter? Testing whether a confidence heuristic can replace interaction in collective decision-making. *Consciousness and Cognition* 26, 13–23.

Barack, D. L., and Krakauer, J. W. 2021. *Nature Reviews Neuroscience* 22, 359–371

Bardi, L., Gheza, D., and Brass, M. 2017. TPJ-M1 interaction in the control of shared representations: New insights from tDCS and TMS combined. *Neuroimage* 146, 734–740.

Barlow, H. B. 1989. Unsupervised learning. *Neural Computation* 1, 295–311.

Baron-Cohen, S., Leslie, A. M., and Frith, U. 1985. Does the autistic child have a "theory of mind"? *Cognition* 21, 37–46.

Bastiaansen, J. A. C. J., Thioux, M., and Keysers, C. 2009. Evidence for mirror systems in emotions. *Philosophical Transactions of the Royal Society B: Biological Sciences* 364, 2391–2404.

Batchelor, T. P., and Briffa, M. 2011. Fight tactics in wood ants: Individuals in smaller groups fight harder but die faster. *Proceedings of the Royal Society B: Biological Sciences* 278, 3243–3250.

Baumeister, R. F., Bratslavsky, E., Muraven, M., and Tice, D. M. 1998. Ego depletion: Is the active self a limited resource? *Journal of Personality and Social Psychology* 74, 1252–1265.

Baumeister, R. F., Masicampo, E. J., and Dewall, C. N. 2009. Prosocial benefits of feeling free: Disbelief in free will increases aggression and reduces helpfulness. *Personality and Social Psychology Bulletin* 35, 260–268.

Baumgartner, T., Gotte, L., Gugler, R., and Fehr, E. 2011. The mentalizing network orchestrates the impact of parochial altruism on social norm enforcement. *Human Brain Mapping* 33, 1452–1469.

Bavelas, J. B., Black, A., Lemery, C. R., and Mullett, J. 1986. I show how you feel—Motor mimicry as a communicative act. *Journal of Personality and Social Psychology* 50, 322–329.

Bayes, T. 1763/1958. Studies in the history of probability and statistics: IX. Thomas Bayes' essay towards solving a problem in the doctrine of chances. *Biometrika* 45, 296–315.

Bayliss, A. P., and Tipper, S. P. 2006. Predictive gaze cues and personality judgments: Should eye trust you? *Psychological Science* 17, 514–520.

Beck, D. M., Rees, G., Frith, C. D., and Lavie, N. 2001. Neural correlates of change detection and change blindness. *Nature Neuroscience* 4, 645–650.

Beck, S. R., McColgan, K. L. T., Robinson, E. J., and Rowley, M. G. 2011. Imagining what might be: Why children underestimate uncertainty. *Journal of Experimental Child Psychology* 110, 603–610.

Beck, S. R., Robinson, E. J., Carroll, D. J., and Apperly, I. A. 2006. Children's thinking about counterfactuals and future hypotheticals as possibilities. *Child Development* 77, 413–426.

Beersma, B., and Van Kleef, G. A. 2012. Why people gossip: An empirical analysis of social motives, antecedents, and consequences. *Journal of Applied Social Psychology* 42, 2640–2670.

Behrens, T. E., Hunt, L. T., Woolrich, M. W., and Rushworth, M. F. 2008. Associative learning of social value. *Nature* 456, 245–249.

Beilock, S. L., Carr, T. H., MacMahon, C., and Starkes, J. L. 2002. When paying attention becomes counterproductive: Impact of divided versus skill-focused attention on novice and experienced performance of sensorimotor skills. *Journal of Experimental Psychology-Applied* 8, 6–16.

Belot, M., Crawford, V. P., and Heyes, C. 2013. Players of matching pennies automatically imitate opponents' gestures against strong incentives. *Proceedings of the National Academy of Sciences* 110, 2763–2768.

Berdahl, A., Torney, C. J., Ioannou, C. C., Faria, J. J., and Couzin, I. D. 2013. Emergent sensing of complex environments by mobile animal groups. *Science* 339, 574–576.

Berg, J., Dickhaut, J., and McCabe, K. 1995. Trust, reciprocity, and social history. *Games and Economic Behavior* 10, 122–142.

Bernard, S., Proust, J., and Clément, F. 2015. Four- to six-year-old children's sensitivity to reliability versus consensus in the endorsement of object labels. *Child Development* 86, 1112–1124.

Bernhardt, B. C., and Singer, T. 2012. The neural basis of empathy. *Annual Review of Neuroscience* 35, 1–23.

Berridge, K. C., and Kringelbach, M. L. 2015. Pleasure systems in the brain. *Neuron* 86, 646–664.

Bian, L., and Baillargeon, R. 2022. When are similar individuals a group? Early reasoning about similarity and in-group support. *Psychological Science* 33, 752–764.

Bikhchandani, S., Hirshleifer, D., and Welch, I. 1992. A theory of fads, fashion, custom, and cultural change as informational cascades. *Journal of Political Economy* 100, 992–1026.

Birch, J. 2020. The search for invertebrate consciousness. *Noûs* 56, 1–21.

Biro, D., Sumpter, D. J. T., Meade, J., and Guilford, T. 2006. From compromise to leadership in pigeon homing. *Current Biology* 16, 2123–2128.

Blair, R. J., Jones, L., Clark, F., and Smith, M. 1997. The psychopathic individual: A lack of responsiveness to distress cues? *Psychophysiology* 34, 192–198.

Blakemore, S.-J. 2018. *Inventing Ourselves: The Secret Life of the Teenage Brain.* Doubleday.

Blakemore, S.-J., den Ouden, H., Choudhury, S., and Frith, C. 2007. Adolescent development of the neural circuitry for thinking about intentions. *Social Cognitive and Affective Neuroscience* 2, 130–139.

Blakemore, S. J., Bristow, D., Bird, G., Frith, C., and Ward, J. 2005. Somatosensory activations during the observation of touch and a case of vision-touch synaesthesia. *Brain* 128, 1571–1583.

Bloom, P. 2000. *How Children Learn the Meaning of Words.* MIT Press.

Bloom, P. 2002. Mindreading, communication and the learning of names for things. *Mind & Language* 17, 37–54.

Bloom, P. 2017. *Against Empathy: The Case for Rational Compassion*. Random House.

Bobzien, S. 2006. Moral responsibility and moral development in Epicurus' philosophy. In *The Virtuous Life in Greek Ethics*, ed. B. Reis, 206–299. Cambridge University Press.

Böckler, A., Knoblich, G., and Sebanz, N. 2011. Giving a helping hand: Effects of joint attention on mental rotation of body parts. *Experimental Brain Research* 211, 531–545.

Boebinger, D. L., Norman-Haignere, S. V., McDermott, J. H., and Kanwisher, N. 2020. Cortical music selectivity does not require musical training. *bioRxiv*, 2020.2001.2010.902189.

Bohn, M., Kachel, G., and Tomasello, M. 2019. Young children spontaneously recreate core properties of language in a new modality. *Proceedings of the National Academy of Sciences* 116, 26072–26077.

Boissin, E., Caparos, S., Raoelison, M., and De Neys, W. 2021. From bias to sound intuiting: Boosting correct intuitive reasoning. *Cognition* 211, 104645.

Bok, S. 1978. *Lying: Moral Choice in Public and Private Life*. Vintage Books.

Bond, C. F., and Robinson, M. 1988. The evolution of deception. *Journal of Nonverbal Behavior* 12, 295–307.

Bond, C. F., Jr., and DePaulo, B. M. 2006. Accuracy of deception judgments. *Personality and Social Psychology Review* 10, 214–234.

Bongelli, R., Riccioni, I., Burro, R., and Zuczkowski, A. 2019. Writers' uncertainty in scientific and popular biomedical articles: A comparative analysis of the British Medical Journal and Discover Magazine. *PLOS One* 14, e0221933.

Bonnefon, J.-F., Hopfensitz, A., and De Neys, W. 2017. Can we detect cooperators by looking at their face? *Current Directions in Psychological Science* 26, 276–281.

Boschin, E. A., Piekema, C., and Buckley, M. J. 2015. Essential functions of primate frontopolar cortex in cognition. *Proceedings of the National Academy of Sciences* 112, E1020–E1027.

Botvinick, M., and Cohen, J. 1998. Rubber hands "feel" touch that eyes see. *Nature* 391, 756.

Boyd, R., Richerson, P. J., and Henrich, J. 2011. The cultural niche: Why social learning is essential for human adaptation. *Proceedings of the National Academy of Sciences* 108, 10918.

Bozdag, E. 2013. Bias in algorithmic filtering and personalization. *Ethics and Information Technology* 15, 209–227.

Bradley, N. 1970. Colour blindness: Notes on its developmental and clinical significance. *International Journal of Psycho-Analysis* 51, 59–70.

Brady, W. J., Wills, J. A., Jost, J. T., Tucker, J. A., and Van Bavel, J. J. 2017. Emotion shapes the diffusion of moralized content in social networks. *Proceedings of the National Academy of Sciences* 114, 7313–7318.

Brahm, F., and Poblete, J. 2021. The evolution of productive organizations. *Nature Human Behaviour* 5, 39–48.

Brandner, P., Güroğlu, B., van de Groep, S., Spaans, J. P., and Crone, E. A. 2021. Happy for us not them: Differences in neural activation in a vicarious reward task between family and strangers during adolescent development. *Developmental Cognitive Neuroscience* 51, 100985.

Brass, M., and Heyes, C. 2005. Imitation: Is cognitive neuroscience solving the correspondence problem? *Trends in Cognitive Sciences* 9, 489–495.

Brass, M., Derrfuss, J., Matthes-von Cramon, G., and von Cramon, D. Y. 2003. Imitative response tendencies in patients with frontal brain lesions. *Neuropsychology* 17, 265–271.

Brault, S., Bideau, B., Kulpa, R., and Craig, C. M. 2012. Detecting deception in movement: The case of the side-step in rugby. *PLOS One* 7, e37494, https://doi.org/10.1371/journal.pone.0037494.

Braun, D. A., Ortega, P. A., and Wolpert, D. M. 2011. Motor coordination: When two have to act as one. *Experimental Brain Research* 211, 631–641.

Brembs, B. 2011. Towards a scientific concept of free will as a biological trait: Spontaneous actions and decision-making in invertebrates. *Proceedings of the Royal Society B: Biological Sciences* 278, 930–939.

Brembs, B., Lorenzetti, F. D., Reyes, F. D., Baxter, D. A., and Byrne, J. H. 2002. Operant reward learning in *Aplysia*: Neuronal correlates and mechanisms. *Science* 296, 1706–1709.

Bressan, P., and Kramer, P. 2015. Human kin detection. *Wiley Interdisciplinary Reviews: Cognitive Science* 6, 299–311.

Bressler, S. L., and Seth, A. K. 2011. Wiener–Granger causality: A well established methodology. *NeuroImage* 58, 323–329.

Brethel-Haurwitz, K. M., Cardinale, E. M., Vekaria, K. M., Robertson, E. L., Walitt, B., VanMeter, J. W., and Marsh, A. A. 2018. Extraordinary altruists exhibit enhanced self–other overlap in neural responses to distress. *Psychological Science* 29, 1631–1641.

Brewer, R., Cook, R., and Bird, G. 2016. Alexithymia: A general deficit of interoception. *Royal Society Open Science* 3, 150664, https://doi.org/10.1098/rsos.150664.

Brodsky, P., and Waterfall, H. 2007. Characterizing motherese: On the computational structure of child-directed language. *Proceedings of the Annual Meeting of the Cognitive Science Society* 29, 833–838.

Brown, D. E. 1991. *Human Universals*. Temple University Press.

Bryant, G. A., Fessler, D. M. T., Fusaroli, R., Clint, E., Aarøe, L., Apicella, C. L., et al. 2016. Detecting affiliation in colaughter across 24 societies. *Proceedings of the National Academy of Sciences* 113, 4682–4687.

Bryant, G. A., Wang, C. S., and Fusaroli, R. 2020. Recognizing affiliation in colaughter and cospeech. *Royal Society Open Science* 7, 201092, http://doi.org/10.1098/rsos.201092.

Bshary, R., and Grutter, A. S. 2006. Image scoring and cooperation in a cleaner fish mutualism. *Nature* 441, 975–978.

Bshary, R., and Würth, M. 2001. Cleaner fish Labroides dimidiatus manipulate client reef fish by providing tactile stimulation. *Proceedings of the Royal Society of London Series B: Biological Sciences* 268, 1495–1501.

Buckholtz, J. W. 2015. Social norms, self-control, and the value of antisocial behavior. *Current Opinion in Behavioral Sciences* 3, 122–129.

Burgess, N. 2006. Spatial memory: How egocentric and allocentric combine. *Trends in Cognitive Sciences* 10, 551–557.

Burnham, D., Kitamura, C., and Vollmer-Conna, U. 2002. What's new, pussycat? On talking to babies and animals. *Science* 296, 1435.

Buskell, A. (2017). What are cultural attractors? *Biology & Philosophy* 32, 377–394.

Butler, P. (2020). A million volunteer to help NHS and others during Covid-19 outbreak. *The Guardian,* April 13.

Buttelmann, D., Zmyj, N., Daum, M., and Carpenter, M. 2013. Selective imitation of in-group over out-group members in 14-month-old infants. *Child Development* 84, 422–428.

Butterworth, G., and Morissette, P. 1996. Onset of pointing and the acquisition of language in infancy. *Journal of Reproductive and Infant Psychology* 14, 219–231.

Byrne, R., and Whiten, A., eds. 1989. *Machiavellian Intelligence: Social Expertise and the Evolution of Intellect in Monkeys, Apes, and Humans.* Oxford University Press.

Campbell-Meiklejohn, D., Simonsen, A., Frith, C. D., and Daw, N. D. 2017. Independent neural computation of value from other people's confidence. *Journal of Neuroscience* 37, 673–684.

Cardellicchio, P., Sinigaglia, C., and Costantini, M. 2011. The space of affordances: A TMS study. *Neuropsychologia* 49, 1369–1372.

Cardellicchio, P., Sinigaglia, C., and Costantini, M. 2013. Grasping affordances with the other's hand: A TMS study. *Social Cognitive and Affective Neuroscience* 8, 455–459.

Carlebach, N., and Yeung, N. 2020. Subjective confidence acts as an internal cost-benefit factor when choosing between tasks. *Journal of Experimental Psychology: Human Perception and Performance* 46, 729–748.

Carpenter, M., Nagell, K., and Tomasello, M. 1998. Social cognition, joint attention, and communicative competence from 9 to 15 months of age. *Monographs of the Society for Research in Child Development* 63, 1–143.

Carpenter, M., Uebel, J., and Tomasello, M. 2013. Being mimicked increases prosocial behavior in 18-month-old infants. *Child Development* 84, 1511–1518.

Carr, E. W., Winkielman, P., and Oveis, C. 2014. Transforming the mirror: Power fundamentally changes facial responding to emotional expressions. *Journal of Experimental Psychology: General* 143, 997–1003.

Carruthers, P. 2009. How we know our own minds: The relationship between mindreading and metacognition. *Behavioral and Brain Sciences* 32, 121–138.

Carruthers, P. 2011. *The Opacity of Mind: An Integrative Theory of Self-Knowledge.* Oxford University Press.

Carter, E. C., and McCullough, M. E. 2014. Publication bias and the limited strength model of self-control: Has the evidence for ego depletion been overestimated? *Frontiers in Psychology* 5, 823, https://doi.org/10.3389/fpsyg.2014.0082.

Caruso, G. D. 2014. (Un) just deserts: The dark side of moral responsibility. *Southwest Philosophy Review* 30, 27–38.

Caspar, E. A., Christensen, J. F., Cleeremans, A., and Haggard, P. 2016. Coercion changes the sense of agency in the human brain. *Current Biology* 26, 585–592.

Caspers, S., Zilles, K., Laird, A. R., and Eickhoff, S. B. 2010. ALE meta-analysis of action observation and imitation in the human brain. *NeuroImage* 50, 1148–1167.

Castelli, F., Frith, C., Happe, F., and Frith, U. 2002. Autism, Asperger syndrome and brain mechanisms for the attribution of mental states to animated shapes. *Brain* 125, 1839–1849.

Castelli, F., Happe, F., Frith, U., and Frith, C. 2000. Movement and mind: A functional imaging study of perception and interpretation of complex intentional movement patterns. *Neuroimage* 12, 314–325.

Catania 2010. Born knowing: Tentacled snakes innately predict future prey behavior. *PLOS One* 5, e10953, https://doi.org/10.1371/journal.pone.0010953.106.

Chambon, V., and Haggard, P. 2012. Sense of control depends on fluency of action selection, not motor performance. *Cognition* 125, 441–451.

Chang, L. J., Doll, B. B., van't Wout, M., Frank, M. J., and Sanfey, A. G. 2010. Seeing is believing: Trustworthiness as a dynamic belief. *Cognitive Psychology* 61, 87–105.

Charlesworth, T. E. S., Kurdi, B., and Banaji, M. R. 2019. Children's implicit attitude acquisition: Evaluative statements succeed, repeated pairings fail. *Developmental Science* 23, e12911, https://doi.org/10.1111/desc.12911.

Charpentier, C. J., Iigaya, K., and Doherty, J. P. 2020. A neuro-computational account of arbitration between choice imitation and goal emulation during human observational learning. *Neuron*106, 687–699.

Chartrand, T. L., and Bargh, J. A. 1999. The chameleon effect: The perception-behavior link and social interaction. *Journal of Personality and Social Psychology* 76, 893–910.

Choi, Y.-j., Mou, Y., and Luo, Y. 2018. How do 3-month-old infants attribute preferences to a human agent? *Journal of Experimental Child Psychology* 172, 96–106.

Chrastil, E. R. 2018. Heterogeneity in human retrosplenial cortex: A review of function and connectivity. *Behavioral Neuroscience* 132, 317–338.

Christiansen, M. H., and Chater, N. 2008. Language as shaped by the brain. *Behavioral and Brain Sciences* 31, 489–509.

Cicero, M. T. 1913. *De Officiis*. Harvard University Press.

Cikara, M. 2015. Intergroup Schadenfreude: Motivating participation in collective violence. *Current Opinion in Behavioral Sciences* 3, 12–17.

Cikara, M., Botvinick, M. M., and Fiske, S. T. 2011. Us versus them: Social identity shapes neural responses to intergroup competition and harm. *Psychological Science* 22, 306–313.

Cisek, P. 2019. Resynthesizing behavior through phylogenetic refinement. *Attention, Perception, & Psychophysics* 81, 2265–2287.

Clark, A. 2013. The many faces of precision (replies to commentaries on "Whatever next? Neural prediction, situated agents, and the future of cognitive science"). *Frontiers in Psychology* 4, 270, https://doi.org/10.3389/fpsyg.2013.0027.

Clark, A. 2015. *Surfing Uncertainty: Prediction, Action, and the Embodied Mind*. Oxford University Press.

Clark, E. V., and Wong, A. D.-W. 2002. Pragmatic directions about language use: Offers of words and relations. *Language in Society* 31, 181–212.

Clark, H. H., and Brennan, S. E. 1991. Grounding in communication. *Perspectives on Socially Shared Cognition* 13, 127–149.

Clayton, N. S., Dally, J. M., and Emery, N. J. 2007. Social cognition by food-caching corvids. The Western scrub-jay as a natural psychologist. *Philosophical Transactions of the Royal Society B* 362, 507–522.

Clements, W. A., and Perner, J. 1994. Implicit understanding of false beliefs. *Cognitive Development* 9, 377–395.

Cogsdill, E. J., Todorov, A. T., Spelke, E. S., and Banaji, M. R. 2014. Inferring character from faces: A developmental study. *Psychological Science* 25, 1132–1139.

Cohen, J. D., McClure, S. M., and Yu, A. J. 2007. Should I stay or should I go? How the human brain manages the trade-off between exploitation and exploration. *Philosophical Transactions of the Royal Society B* 362, 933–942.

Cole, M. W., Laurent, P., and Stocco, A. 2013. Rapid instructed task learning: A new window into the human brain's unique capacity for flexible cognitive control. *Cognitive, Affective, & Behavioral Neuroscience* 13, 1–22. doi: 10.3758/s13415-13012-10125-13417.

Colman, A. M., and Gold, N. 2018. Team reasoning: Solving the puzzle of coordination. *Psychonomic Bulletin & Review* 25, 1770–1783.

Coman, A., Momennejad, I., Drach, R. D., and Geana, A. 2016. Mnemonic convergence in social networks: The emergent properties of cognition at a collective level. *Proceedings of the National Academy of Sciences* 113, 8171–8176.

Contreras-Huerta, L. S., Lockwood, P. L., Bird, G., Apps, M. A. J., and Crockett, M. J. 2020. Prosocial behavior is associated with transdiagnostic markers of affective sensitivity in multiple domains. *Emotion* 22, 820–835.

Cook, R., Bird, G., Lunser, G., Huck, S., and Heyes, C. 2012. Automatic imitation in a strategic context: Players of rock-paper-scissors imitate opponents' gestures. *Proceedings of the Royal Society B-Biological Sciences* 279, 780–786.

Coricelli, G., and Nagel, R. 2009. Neural correlates of depth of strategic reasoning in medial prefrontal cortex. *Proceedings of the National Academy of Sciences* 106, 9163–9168.

Couzin, I. D. 2009. Collective cognition in animal groups. *Trends in Cognitive Sciences* 13, 36–43.

Cowan, N. 2001. The magical number 4 in short-term memory: A reconsideration of mental storage capacity. *Behavioral and Brain Sciences* 24, 87–114; discussion 114–185.

Craig, A. D. 2002. How do you feel? Interoception: The sense of the physiological condition of the body. *Nature Reviews Neuroscience* 3, 655–666.

Crapse, T. B., and Sommer, M. A. 2008. Corollary discharge across the animal kingdom. *Nature Reviews Neuroscience* 9, 587–600.

Crawford, V.P., Costa-Gomes, M.A., and Iriberri, N. 2013. Structural models of nonequilibrium strategic thinking: Theory, evidence, and applications. *Journal of Economic Literature* 51, 5–62.

Crick, F. 1994. *The Astonishing Hypothesis: The Scientific Search for the Soul*. Scribner.

Crockett, M. J., Kurth-Nelson, Z., Siegel, J. Z., Dayan, P., and Dolan, R. J. 2014. Harm to others outweighs harm to self in moral decision making. *Proceedings of the National Academy of Sciences* 111, 17320–17325.

Cross, E. S., Ramsey, R., Liepelt, R., Prinz, W., and Hamilton, A. F. d. C. 2016. The shaping of social perception by stimulus and knowledge cues to human animacy. *Philosophical Transactions of the Royal Society B: Biological Sciences* 371, 20150075, http://doi.org/10.1098/rstb.2015.0075.

Croy, I., Olgun, S., Mueller, L., Schmidt, A., Muench, M., Gisselmann, G., et al. 2016. Spezifische Anosmie als Prinzip olfaktorischer Wahrnehmung [Specific anosmia as a principle of olfactory perception]. *HNO* 64, 292–295.

Csibra, G. 2008. Goal attribution to inanimate agents by 6.5-month-old infants. *Cognition* 107, 705–717.

Csibra, G., and Gergely, G. 2009. Natural pedagogy. *Trends in Cognitive Sciences* 13, 148–153.

Csibra, G., and Gergely, G. 2011. Natural pedagogy as evolutionary adaptation. *Philosophical Transactions of the Royal Society B: Biological Sciences* 366, 1149–1157.

Csibra, G., Gergely, G., Biro, S., Koos, O., and Brockbank, M. 1999. Goal attribution without agency cues: The perception of "pure reason" in infancy. *Cognition* 72, 237–267.

Cunningham, W. A., Johnson, M. K., Raye, C. L., Gatenby, J. C., Gore, J. C., and Banaji, M. R. 2004. Separable neural components in the processing of black and white faces. *Psychological Science* 15, 806–813.

Curtin, C. M., Barrett, H. C., Bolyanatz, A., Crittenden, A. N., Fessler, D. M. T., Fitzpatrick, S., et al. 2020. Kinship intensity and the use of mental states in moral judgment across societies. *Evolution and Human Behavior* 41, 415–429.

Danchin, E., Nöbel, S., Pocheville, A., Dagaeff, A.-C., Demay, L., Alphand, M., et al. 2018. Cultural flies: Conformist social learning in fruitflies predicts long-lasting mate-choice traditions. *Science* 362, 1025.

Davidov, M., Paz, Y., Roth-Hanania, R., Uzefovsky, F., Orlitsky, T., Mankuta, D., and Zahn-Waxler, C. 2020. Caring babies: Concern for others in distress during infancy. *Developmental Science* 24, e13016, https://doi.org/10.1111/desc.13016.

Davidson, D. 1969. How is weakness of the will possible? In *Moral Concepts*, ed. J. Feinberg. Oxford University Press.

Davidson, D. 1991. Three varieties of knowledge. In *A. J. Ayer: Memorial Essays*, ed. A. P. Griffiths, 153–166. Cambridge University Press.

Davidson, M. J., Macdonald, J., and Yeung, N. 2021. Alpha oscillations and stimulus-evoked activity dissociate metacognitive reports of attention, visibility and confidence in a rapid visual detection task. *Journal of Vision* 22, https://doi.org/10.1167/jov.22.10.20.

Davila Ross, M., J Owren, M., and Zimmermann, E. 2009. Reconstructing the evolution of laughter in great apes and humans. *Current Biology* 19, 1106–1111.

de Bruxelles, S. 2009. Sleepwalker Brian Thomas admits killing wife while fighting intruders in nightmare. *The Times*, https://www.thetimes.co.uk/article/sleepwalker-brian-thomas-admits-killing-wife-while-fighting-intruders-in-nightmare-sv79ljzk7lh.

Decety, J., Bartal, I. B., Uzefovsky, F., and Knafo-Noam, A. 2016. Empathy as a driver of prosocial behaviour: Highly conserved neurobehavioural mechanisms across species. *Philosophical Transactions of the Royal Society B* 371, 20150077, http://doi.org/10.1098/rstb.2015.0077.

Decety, J., Echols, S., and Correll, J. 2009. The lame game: The effect of responsibility and social stigma on empathy for pain. *Journal of Cognitive Neuroscience* 22, 985–997.

Decker, J. H., Otto, A. R., Daw, N. D., and Hartley, C. A. 2016. From creatures of habit to goal-directed learners: Tracking the developmental emergence of model-based reinforcement learning. *Psychological Science* 27, 848–858.

De Dreu, C. K., Greer, L. L., Van Kleef, G. A., Shalvi, S., and Handgraaf, M. J. 2011. Oxytocin promotes human ethnocentrism. *Proceedings of the National Academy of Sciences* 108, 1262–1266.

Dehaene, S. 2009. *Reading in the Brain*. Penguin Books.

Dehaene, S., and Cohen, L. 2011. The unique role of the visual word form area in reading. *Trends in Cognitive Sciences* 15, 254–262.

Dehaene, S., and Dehaene-Lambertz, G. 2016. Is the brain prewired for letters? *Nature Neuroscience* 19, 1192–1193.

Dehaene, S., Naccache, L., Cohen, L., Bihan, D. L., Mangin, J. F., Poline, J. B., and Riviere, D. 2001. Cerebral mechanisms of word masking and unconscious repetition priming. *Nature Neuroscience* 4, 752–758.

de Klerk, C., Albiston, H., Bulgarelli, C., Southgate, V., and Hamilton, A. 2020. Observing third-party ostracism enhances facial mimicry in 30-month-olds. *Journal of Experimental Child Psychology* 196, 104862.

de Klerk, C. C. J. M., Bulgarelli, C., Hamilton, A., and Southgate, V. 2019. Selective facial mimicry of native over foreign speakers in preverbal infants. *Journal of Experimental Child Psychology* 183, 33–47.

de Klerk, C. C. J. M., Hamilton, A. F. d. C., and Southgate, V. 2018. Eye contact modulates facial mimicry in 4-month-old infants: An EMG and fNIRS study. *Cortex* 106, 93–103.

Delgado, M. R., Frank, R. H., and Phelps, E. A. 2005. Perceptions of moral character modulate the neural systems of reward during the trust game. *Nature Neuroscience* 8, 1611–1618.

De Martino, B., O'Doherty, J. P., Ray, D., Bossaerts, P., and Camerer, C. 2013. In the mind of the market: Theory of mind biases value computation during financial bubbles. *Neuron* 79, 1222–1231.

Dennett, D. C. 1971. Intentional systems. *Journal of Philosophy* 68, 87–106.

Dennett, D. C. 1987. *The Intentional Stance*. MIT Press.

Desender, K., Boldt, A., and Yeung, N. 2018. Subjective confidence predicts information seeking in decision making. *Psychological Science* 29, 761–778.

Desimone, R., and Duncan, J. 1995. Neural mechanisms of selective visual attention. *Annual Review of Neuroscience* 18, 193–222.

Devaine, M., Hollard, G., and Daunizeau, J. 2014a. The social Bayesian brain: Does mentalizing make a difference when we learn? *PLOS Computational Biology* 10, e1003992, https://doi.org/10.1371/journal.pcbi.1003992.

Devaine, M., Hollard, G., and Daunizeau, J. 2014b. Theory of mind: Did evolution fool us? *PLOS One* 9, e87619, https://doi.org/10.1371/journal.pone.0087619.

de Vignemont, F. 2014. Shared body representations and the "Whose" system. *Neuropsychologia* 55, 128–136.

de Vignemont, F., and Jacob, P. 2012. What is it like to feel another's pain? *Philosophy of Science* 79, 295–316.

de Vignemont, F., and Jacob, P. 2016. Beyond empathy for pain. *Philosophy of Science* 83, 434–445.

de Vignemont, F., and Singer, T. 2006. The empathic brain: How, when and why? *Trends in Cognitive Sciences* 10, 435–441.

de Villiers, J. 2007. The interface of language and theory of mind. *Lingua International Review of General Linguistics (Revue internationale de Linguistique Generale)* 117, 1858–1878.

Dezecache, G. 2015. Human collective reactions to threat. *WIREs Cognitive Science* 6, 209–219.

Dezecache, G., Frith, C. D., and Deroy, O. 2020. Pandemics and the great evolutionary mismatch. *Current Biology* 30, R417–R429.

Dezecache, G., Grèzes, J., and Dahl, C. D. 2017. The nature and distribution of affiliative behaviour during exposure to mild threat. *Royal Society Open Science* 4, 170265, http://doi.org/10.1098/rsos.170265.

Dezecache, G., Jacob, P., and Grèzes, J. 2015. Emotional contagion: Its scope and limits. *Trends in Cognitive Sciences* 19, 297–299.

Diggle, S. P., Gardner, A., West, S. A., and Griffin, A. S. 2007. Evolutionary theory of bacterial quorum sensing: When is a signal not a signal? *Philosophical Transactions of the Royal Society B: Biological Sciences* 362, 1241–1249.

Di Giorgio, E., Lunghi, M., Vallortigara, G., and Simion, F. 2021. Newborns' sensitivity to speed changes as a building block for animacy perception. *Scientific Reports–UK* 11, 542, https://doi.org/10.1038/s41598-020-79451-3.

Dijkstra, P. D., Maguire, S. M., Harris, R. M., Rodriguez, A. A., DeAngelis, R. S., Flores, S. A., and Hofmann, H. A. 2017. The melanocortin system regulates body pigmentation and social behaviour in a colour polymorphic cichlid fish. *Proceedings of the Royal Society B: Biological Sciences* 284, 20162838.

Dimberg, U., Thunberg, M., and Elmehed, K. 2000. Unconscious facial reactions to emotional facial expressions. *Psychological Science* 11, 86–89.

Dinstein, I., Hasson, U., Rubin, N., and Heeger, D. J. 2007. Brain areas selective for both observed and executed movements. *Journal of Neurophysiology* 98, 1415–1427.

Di Stefano, A., Scatà, M., La Corte, A., Liò, P., Catania, E., Guardo, E., and Pagano, S. 2015. Quantifying the role of homophily in human cooperation using multiplex evolutionary game theory. *PLOS One* 10, e0140646, https://doi.org/10.1371/journal.pone.014064.

Ditrich, L., and Sassenberg, K. 2016. It's either you or me! Impact of deviations on social exclusion and leaving. *Group Processes & Intergroup Relations* 19, 630–652.

Doebel, S., and Munakata, Y. 2018. Group influences on engaging self-control: Children delay gratification and value it more when their in-group delays and their out-group doesn't. *Psychological Science* 29, 738–748.

Dogge, M., Schaap, M., Custers, R., Wegner, D. M., and Aarts, H. 2012. When moving without volition: Implied self-causation enhances binding strength between involuntary actions and effects. *Consciousness and Cognition* 21, 501–506.

Domenici, P., Booth, D., Blagburn, J. M., and Bacon, J. P. 2008. Cockroaches keep predators guessing by using preferred escape trajectories. *Current Biology* 18, 1792–1796.

Drury, J., and Reicher, S. D. 2010. Crowd control. *Scientific American Mind* 21, 58–65.

Drury, J., Cocking, C., and Reicher, S. 2009. The nature of collective resilience: Survivor reactions to the 200 London bombings. *International Journal of Mass Emergencies and Disasters* 27, 66–95.

Dumontheil, I., Apperly, I. A., and Blakemore, S.-J. 2010. Online usage of theory of mind continues to develop in late adolescence. *Developmental Science* 13, 331–338.

Dunbar, R. I. M. 2004. Gossip in evolutionary perspective. *Review of General Psychology* 8, 100–110.

Dunham, Y. 2018. Mere membership. *Trends in Cognitive Sciences* 22, 780–793.

Dunham, Y., Baron, A. S., and Carey, S. 2011. Consequences of "minimal" group affiliations in children. *Child Development* 82, 793–811.

Durkheim, É. 1912. *The Elementary Forms of Religious Life*. Translated by Karen Fields. 1995, New York, Free Press.

Easterbrook, G. 2010. Brett Favre, felled by his fatal flaw. In *ESPN (Entertainment and Sports Programming Network)*, http://www.espn.com/espn/page2/story?sportCat=nfl&page=easterbrook/100126.

Eaton, R. C., Bombardieri, R. A., and Meyer, D. L. 1977. The Mauthner-initiated startle response in teleost fish. *Journal of Experimental Biology* 66, 65–81.

Echterhoff, G., Higgins, E. T., and Levine, J. M. 2009. Shared reality: Experiencing commonality with others' inner states about the world. *Perspectives on Psychological Science* 4, 496–521.

Edwards, K., and Low, J. 2017. Reaction time profiles of adults' action prediction reveal two mindreading systems. *Cognition* 160, 1–16.

Eerkens, J. W., and Lipo, C. P. 2005. Cultural transmission, copying errors, and the generation of variation in material culture and the archaeological record. *Journal of Anthropological Archaeology* 24, 316–334.

Egyed, K., Kiraly, I., and Gergely, G. 2013. Communicating shared knowledge in infancy. *Psychological Science* 24, 1348–1353.

Eigsti, I.-M., and Irvine, C. A. 2021. Verbal mediation of theory of mind in verbal adolescents with autism spectrum disorder. *Language Acquisition*, 28, 195–213.

Eisenberger, N. I., and Lieberman, M. D. 2004. Why rejection hurts: A common neural alarm system for physical and social pain. *Trends in Cognitive Sciences* 8, 294–300.

Elekes, F., Bródy, G., Halász, E., and Király, I. 2016. Enhanced encoding of the co-actor's target stimuli during a shared non-motor task. *Quarterly Journal of Experimental Psychology* 69, 2376–2389.

El Kaddouri, R., Bardi, L., De Bremaeker, D., Brass, M., and Wiersema, J. R. 2020. Measuring spontaneous mentalizing with a ball detection task: Putting the attention-check hypothesis by Phillips and colleagues (2015) to the test. *Psychological Research (Psychologische Forschung)* 84, 1749–1757.

Ellemers, N., Spears, R., and Doosje, B. 2002. Self and social identity. *Annual Review of Psychology* 53, 161–186.

El Zein, M., and Bahrami, B. 2020. Joining a group diverts regret and responsibility away from the individual. *Proceedings of the Royal Society B: Biological Sciences* 287, 20192251, http://doi.org/10.1098/rspb.2019.2251.

Engelmann, J. M., and Rapp, D. J. 2018. The influence of reputational concerns on children's prosociality. *Current Opinion in Psychology* 20, 92–95.

Engelmann, J. M., Herrmann, E., and Tomasello, M. 2016. Preschoolers affect others' reputations through prosocial gossip. *British Journal of Developmental Psychology* 34, 447–460.

Epley, N., Morewedge, C. K., and Keysar, B. 2004. Perspective taking in children and adults: Equivalent egocentrism but differential correction. *Journal of Experimental Social Psychology* 40, 760–768.

Ereira, S., Dolan, R. J., and Kurth-Nelson, Z. 2018. Agent-specific learning signals for self–other distinction during mentalising. *PLOS Biology* 16, e2004752, https://doi.org/10.1371/journal.pbio.2004752.

Ereira, S., Hauser, T. U., Moran, R., Story, G. W., Dolan, R. J., and Kurth-Nelson, Z. 2020. Social training reconfigures prediction errors to shape self-other boundaries. *Nature Communications* 11, 3030, https://doi.org/10.1038/s41467-020-16856-8.

Evans, J. S. B. T., and Stanovich, K. E. 2013. Dual-process theories of higher cognition: Advancing the debate. *Perspectives on Psychological Science* 8, 223–241.

Fadiga, L., Fogassi, L., Pavesi, G., and Rizzolatti, G. 1995. Motor facilitation during action observation: A magnetic stimulation study. *Journal of Neurophysiology* 73, 2608–2611.

Falck-Ytter, T., Gredebäck, G., and von Hofsten, C. 2006. Infants predict other people's action goals. *Nature Neuroscience* 9, 878–879.

Faria, J. J., Krause, S., and Krause, J. 2010. Collective behavior in road crossing pedestrians: The role of social information. *Behavioral Ecology* 21, 1236–1242.

Farmer, H., Ciaunica, A., and Hamilton Antonia, F. d. C. 2018. The functions of imitative behaviour in humans. *Mind & Language* 33, 378–396.

Fehr, E. 2004. Human behaviour: Don't lose your reputation. *Nature* 432, 449–450.

Fehr, E., and Gächter, S. 2002. Altruistic punishment in humans. *Nature* 415, 137–140.

Fehr, E., Bernhard, H., and Rockenbach, B. 2008. Egalitarianism in young children. *Nature* 454, 1079–1083.

Fehr, E., Goette, L., and Zehnder, C. 2009. A behavioral account of the labor market: The role of fairness concerns. *Annual Review of Economics* 1, 355–384.

Feinberg, M., Willer, R., and Schultz, M. 2014. Gossip and ostracism promote cooperation in groups. *Psychological Science* 25, 656–664.

Feinman, S., Roberts, D., Hsieh, K. F., Sawyer, D., and Swanson, K. 1992. A critical review of social referencing in infancy. In *Social Referencing and the Social Construction of Reality in Infancy*, ed. S. Feinman, 15–54. Plenum Press.

Feinstein, J. S., Buzza, C., Hurlemann, R., Follmer, R. L., Dahdaleh, N. S., Coryell, W. H., et al. 2013. Fear and panic in humans with bilateral amygdala damage. *Nature Neuroscience* 16, 270–272.

Ferguson, B., and Waxman, S. R. 2016. What the [beep]? Six-month-olds link novel communicative signals to meaning. *Cognition* 146, 185–189.

Filiz-Ozbay, E., and Ozbay, E. Y. 2007. Auctions with anticipated regret: Theory and experiment. *American Economic Review* 97, 1407–1418.

Fine, P., Eames, K., and Heymann, D. L. 2011. "Herd immunity": A rough guide. *Clinical Infectious Diseases* 52, 911–916.

Fleming, S. M. 2021. *Know Thyself: The Science of Self-Awareness*. Basic Books.

Fleming, S. M., and Daw, N. D. 2017. Self-evaluation of decision-making: A general Bayesian framework for metacognitive computation. *Psychological Review* 124, 91–114.

Fleming, S. M., Maniscalco, B., Ko, Y., Amendi, N., Ro, T., and Lau, H. 2015. Action-specific disruption of perceptual confidence. *Psychological Science* 26, 89–98.

Fleming, S. M., Weil, R. S., Nagy, Z., Dolan, R. J., and Rees, G. 2010. Relating introspective accuracy to individual differences in brain structure. *Science* 329, 1541–1543.

Fletcher, P. C., Happe, F., Frith, U., Baker, S. C., Dolan, R. J., Frackowiak, R. S., and Frith, C. D. 1995. Other minds in the brain: A functional imaging study of "theory of mind" in story comprehension. *Cognition* 57, 109–128.

Flom, R., and Johnson, S. 2011. The effects of adults' affective expression and direction of visual gaze on 12-month-olds' visual preferences for an object following a 5-minute, 1-day, or 1-month delay. *British Journal of Developmental Psychology* 29, 64–85.

Flombaum, J. I., and Santos, L. R. 2005. Rhesus monkeys attribute perceptions to others. *Current Biology* 15, 447–452.

Fogarty, L., and Creanza, N. 2017. The niche construction of cultural complexity: Interactions between innovations, population size and the environment. *Philosophical Transactions of the Royal Society B: Biological Sciences* 372, 20160428, http://doi.org/10.1098/rstb.2016.0428.

Folke, T., Jacobsen, C., Fleming, S. M., and De Martino, B. 2016. Explicit representation of confidence informs future value-based decisions. *Nature Human Behaviour* 1, 0002, https://doi.org/10.1038/s41562-016-0002.

Forgács, B., Parise, E., Csibra, G., Gergely, G., Jacquey, L., and Gervain, J. 2018. Fourteen-month-old infants track the language comprehension of communicative partners. *Developmental Science* 22, e12751, https://doi.org/10.1111/desc.12751.

Forgeot d'Arc, B., Devaine, M., and Daunizeau, J. 2020. Social behavioural adaptation in Autism. *PLOS Computational Biology* 16, e1007700, https://doi.org/10.1371/journal.pcbi.1007700.

Forsythe, R., Horowitz, J. L., Savin, N. E., and Sefton, M. 1994. Fairness in simple bargaining experiments. *Games and Economic Behavior* 6, 347–369.

Foster, E. K. 2004. Research on gossip: Taxonomy, methods, and future directions. *Review of General Psychology* 8, 78–99.

Foster, K. R., and Kokko, H. 2009. The evolution of superstitious and superstition-like behaviour. *Proceedings Biological Sciences* 276, 31–37.

Fouragnan, E., Chierchia, G., Greiner, S., Neveu, R., Avesani, P., and Coricelli, G. 2013. Reputational priors magnify striatal responses to violations of trust. *Journal of Neuroscience* 33, 3602–3611.

Fox, S. G., and Walters, H. A. 1986. The impact of general versus specific expert testimony and eyewitness confidence upon mock juror judgment. *Law and Human Behavior* 10, 215–228.

Francis, D. D., Champagne, F. C., and Meaney, M. J. 2000. Variations in maternal behaviour are associated with differences in oxytocin receptor levels in the rat. *Journal of Neuroendocrinology* 12, 1145–1148.

Freeman, J. B., Pauker, K., and Sanchez, D. T. 2016. A perceptual pathway to bias: Interracial exposure reduces abrupt shifts in real-time race perception that predict mixed-race bias. *Psychological Science* 27, 502–517.

Freundlieb, M., Kovács, Á. M., and Sebanz, N. 2016. When do humans spontaneously adopt another's visuospatial perspective? *Journal of Experimental Psychology: Human Perception and Performance* 42, 401–412.

Fried, C. 1978. *Right and Wrong*. Volume 153. Harvard University Press.

Friston, K. 2005. A theory of cortical responses. *Philosophical Transactions of the Royal Society B* 360, 815–836.

Friston, K. 2008. Hierarchical models in the brain. *PLOS Computational Biology* 4, e1000211, https://doi.org/10.1371/journal.pcbi.1000211.

Friston, K., and Frith, C. 2015. A duet for one. *Consciousness and Cognition* 36, 390–405.

Friston, K., Kilner, J., and Harrison, L. 2006. A free energy principle for the brain. *Journal of Physiology* 100, 70–87.

Friston, K., Rigoli, F., Ognibene, D., Mathys, C., Fitzgerald, T., and Pezzulo, G. 2015. Active inference and epistemic value. *Journal of Cognitive Neuroscience* 6, 187–214.

Frith, C. 1995. Consciousness is for other people. *Behavioral and Brain sciences* 18, 682–683.

Frith, C. 2002. How can we share experiences? Comment from Chris Frith. *Trends in Cognitive Sciences* 6, 374.

Frith, C. D. 1992. *The Cognitive Neuropsychology of Schizophrenia.* Lawrence Erlbaum Associates.

Frith, C. D. 2000. The role of dorsolateral prefrontal cortex in the selection of action as revealed by functional imaging. In *Control* of Cognitive Processes, eds. S. Monsell and J. Driver, 549–565. MIT Press.

Frith, C. D. 2007. *Making up the Mind: How the Brain Creates Our Mental World.* Blackwell.

Frith, C. D. 2014. How the brain creates culture. *Nova Acta Leopoldina* 120, 3–26.

Frith, C. D. 2021. The neural basis of consciousness. *Psychological Medicine* 51, 550–562.

Frith, C. D., and Frith, U. 1991. Elective affinities in schizophrenia and childhood autism. In *Social Psychiatry: Theory, Methodology, and Practice*, ed. P. Bebbington, 65–88. Transactions Publishers.

Frith, C. D., and Frith, U. 1999. Interacting minds—a biological basis. *Science* 286, 1692–1695.

Frith, C. D., and Metzinger, T. 2016. What's the use of consciousness? In *Where's the Action? The Pragmatic Turn in Cognitive Science*, eds. A. K. Engel, K. Friston, and D. Kragic, 193–214. MIT Press.

Frith, U. 1979. Reading by eye and writing by ear. In *Processing of Visible Language*, eds. P. A. Kolers, M. E. Wrolstad, and H. Bouma, 379–390, Springer.

Frith, U. 1989. *Autism: Explaining the Enigma.* Blackwell.

Frith, U., and Frith, C. D. 1980. Relationships between reading and spelling. In *Orthography, Reading and Dyslexia,* eds. J. Kavanagh, and R. Venezky, 287–295. University Park Press.

Frith, U., and Frith, C. 2010. The social brain: Allowing humans to boldly go where no other species has been. *Philosophical Transactions of the Royal Society B* 365, 165–176.

Frith, U., Morton, J., and Leslie, A. M. 1991. The cognitive basis of a biological disorder: Autism. *Trends in Neurosciences* 14, 433–438.

Fu, F., Nowak, M. A., Christakis, N. A., and Fowler, J. H. 2012. The evolution of homophily. *Scientific Reports–UK* 2, 845, https://doi.org/10.1038/srep00845.

Fusaroli, R., Bahrami, B., Olsen, K., Roepstorff, A., Rees, G., Frith, C., and Tylén, K. 2012. Coming to terms: Quantifying the benefits of linguistic coordination. *Psychological Science* 23, 931–939.

Gagnepain, P., Vallée, T., Heiden, S., Decorde, M., Gauvain, J.-L., Laurent, A., et al. 2020. Collective memory shapes the organization of individual memories in the medial prefrontal cortex. *Nature Human Behaviour* 4, 189–200.

Galantucci, B. 2005. An experimental study of the emergence of human communication systems. *Cognitive Science* 29, 737–767.

Galef, B. G., Jr., and Giraldeau, L. A. 2001. Social influences on foraging in vertebrates: Causal mechanisms and adaptive functions. *Animal Behaviour* 61, 3–15.

Gallagher, H. L., Jack, A. I., Roepstorff, A., and Frith, C. D. 2002. Imaging the intentional stance in a competitive game. *Neuroimage* 16, 814–821.

Gallotti, M., and Frith, C. D. 2013. Social cognition in the we-mode. *Trends in Cognitive Sciences* 17, 160–165.

Gallotti, M., Fairhurst, M., and Frith, C. 2017. Alignment in social interactions. *Consciousness and Cognition* 48, 253–261.

Galton, F. 1907. Vox populi. *Nature* 75, 450–451.

Gambetta, D. 2009. *Codes of the Underworld: How Criminals Communicate.* Princeton University Press.

Gao, T., Scholl, B. J., and McCarthy, G. 2012. Dissociating the detection of intentionality from animacy in the right posterior superior temporal sulcus. *Journal of Neuroscience* 32, 14276–14280.

Garrod, S., and Pickering, M. J. 2009. Joint action, interactive alignment, and dialog. *Topics in Cognitive Science* 1, 292–304.

Gawronski, B., and Quinn, K. A. 2013. Guilty by mere similarity: Assimilative effects of facial resemblance on automatic evaluation. *Journal of Experimental Social Psychology* 49, 120–125.

Genschow, O., Cracco, E., Schneider, J., Protzko, J., Wisniewski, D., Brass, M., and Schooler, J. 2022. Manipulating belief in free will and its downstream consequences: A meta-analysis. *Personality and Social Psychology Review*, in press. https://doi.org/10.1177/10888683221087527.

Germar, M., and Mojzisch, A. 2019. Learning of social norms can lead to a persistent perceptual bias: A diffusion model approach. *Journal of Experimental Social Psychology* 84, 130801, https://doi.org/10.1016/j.jesp.2019.03.012.

Gershman, S. J. 2019. The generative adversarial brain. *Frontiers in Artificial Intelligence* 2, https://doi.org/10.3389/frai.2019.00018.

Gershman, S. J., Markman, A. B., and Otto, A. R. 2014. Retrospective revaluation in sequential decision making: A tale of two systems. *Journal of Experimental Psychology–General* 143, 182–194.

Ghetti, S., and Angelini, L. 2008. The development of recollection and familiarity in childhood and adolescence: Evidence from the dual-process signal detection model. *Child Development* 79, 339–358.

Gibson, J. J. 1979. *The Ecological Approach to Visual Perception.* Houghton Mifflin.

Gilead, M., Liberman, N., and Maril, A. 2014. From mind to matter: Neural correlates of abstract and concrete mindsets. *Social Cognitive and Affective Neuroscience* 9, 638–645.

Gilead, M., and Ochsner, K. N. 2021. *The Neural Basis of Mentalizing.* Springer.

Gino, F. 2015. Understanding ordinary unethical behavior: Why people who value morality act immorally. *Current Opinion in Behavioral Sciences* 3, 107–111.

Gintis, H. 2010. Social norms as choreography. *Politics, Philosophy & Economics* 9, 251–264.

Giudice, M. D., Manera, V., and Keysers, C. 2009. Programmed to learn? The ontogeny of mirror neurons. *Developmental Science* 12, 350–363.

Godfrey-Smith, P. 2012. Darwinism and cultural change. *Philosophical Transactions of the Royal Society B: Biological Sciences* 367, 2160–2170.

Godfrey-Smith, P. 2005. Folk psychology as a model. *Philosophers' Imprint* 5, 1–16.

Goffman, E. 2008. *Behavior in Public Places*. Simon and Schuster.

Goldsmith, D. J., and Baxter, L. A. 1996. Constituting relationships in talk A taxonomy of speech events in social and personal relationships. *Human Communication Research* 23, 87–114.

Golkar, A., Castro, V., and Olsson, A. 2015. Social learning of fear and safety is determined by the demonstrator's racial group. *Biology Letters* 11, 2014847, http://doi.org/10.1098/rsbl.2014.0817.

Goodale, M. A., and Milner, A. D. 2004. *Sight Unseen*. Oxford University Press.

Goodale, M. A., Milner, A. D., Jakobson, L. S., and Carey, D. P. 1991. A neurological dissociation between perceiving objects and grasping them. *Nature* 349, 154–156.

Gough, P. B. 1972. One second of reading. In *Language by Ear and by Eye*, eds. J. F. Kavanagh, and I. G. Mattingly, 291–320, MIT Press.

Grammer, K., Schiefenhövel, W., Schleidt, M., Lorenz, B., and Eibl-Eibesfeldt, I. 1988. Patterns on the face: The eyebrow flash in crosscultural comparison. *Ethology* 77, 279–299.

Greenwald, A. G., and Banaji, M. R. 1995. Implicit social cognition—attitudes, self-esteem, and stereotypes. *Psychological Review* 102, 4–27.

Greenwald, A. G., McGhee, D. E., and Schwartz, J. L. 1998. Measuring individual differences in implicit cognition: The Implicit Association Test. *Journal of Personality and Social Psychology* 74, 1464–1480.

Grice, H. P. 1989. *Studies in the Way of Words*. Harvard University Press.

Grosenick, L., Clement, T. S., and Fernald, R. D. 2007. Fish can infer social rank by observation alone. *Nature* 445, 429–432.

Grossenbacher, P. G., and Lovelace, C. T. 2001. Mechanisms of synesthesia: Cognitive and physiological constraints. *Trends in Cognitive Sciences* 5, 36–41.

Grosse Wiesmann, C., Schreiber, J., Singer, T., Steinbeis, N., and Friederici, A. D. 2017. White matter maturation is associated with the emergence of Theory of Mind in early childhood. *Nature Communications* 8, 14692.

Grossman, E., Donnelly, M., Price, R., Pickens, D., Morgan, V., Neighbor, G., and Blake, R. 2000. Brain areas involved in perception of biological motion. *Journal of Cognitive Neuroscience* 12, 711–720.

Grunbaum, D. 1998. Schooling as a strategy for taxis in a noisy environment. *Evolutionary Ecology* 12, 503–522.

Grutter, A. 1996. Parasite removal rates by the cleaner wrasse Labroides dimidiatus. *Marine Ecology Progress Series* 130, 61–70.

Grutter, A. S., and Bshary, R. 2003. Cleaner wrasse prefer client mucus: Support for partner control mechanisms in cleaning interactions. *Proceedings of the Royal Society of London Series B-Biological Sciences* 270, S242–S244.

Gu, X., Wang, X., Hula, A., Wang, S., Xu, S., Lohrenz, T. M., et al. 2015. Necessary, yet dissociable contributions of the insular and ventromedial prefrontal cortices to norm adaptation: Computational and lesion evidence in humans. *Journal of Neuroscience* 35, 467–473.

Guggenmos, M., Wilbertz, G., Hebart, M. N., and Sterzer, P. 2016. Mesolimbic confidence signals guide perceptual learning in the absence of external feedback. *eLife* 5, e13388, https://doi.org/10.7554/eLife.13388.

Gürerk, Ö., Irlenbusch, B., and Rockenbach, B. 2006. The competitive advantage of sanctioning institutions. *Science* 312, 108–111.

Hagenmuller, F., Rössler, W., Wittwer, A., and Haker, H. 2014. Juicy lemons for measuring basic empathic resonance. *Psychiatry Research* 219, 391–396.

Haggard, P., Clark, S., and Kalogeras, J. 2002. Voluntary action and conscious awareness. *Nature Neuroscience* 5, 382–385.

Hahn, M., Jurafsky, D., and Futrell, R. 2020. Universals of word order reflect optimization of grammars for efficient communication. *Proceedings of the National Academy of Sciences* 117, 2347–2353.

Haldane, J. B. S. 1964. A defense of beanbag genetics. *Perspectives in Biology and Medicine* 7, 343–360.

Hale, J., and Hamilton, A. F. D. C. 2016. Testing the relationship between mimicry, trust and rapport in virtual reality conversations. *Scientific Reports-UK* 6, 35295, https://doi.org/10.1038/srep35295.

Hall, K., and Brosnan, S. F. 2017. Cooperation and deception in primates. *Infant Behavior and Development* 48, 38–44.

Hamilton, A. F. 2008. Emulation and mimicry for social interaction: A theoretical approach to imitation in autism. *Quarterly Journal of Experimental Psychology (Colchester)* 61, 101–115.

Hamilton, A. F. D. 2015. Cognitive underpinnings of social interaction. *Quarterly Journal of Experimental Psychology* 68, 417–432.

Hamilton, W. D. 1964. The genetical evolution of social behaviour. II. *Journal of Theoretical Biology* 7, 17–52.

Hamlin, J. K., Mahajan, N., Liberman, Z., and Wynn, K. 2013. Not like me=bad: Infants prefer those who harm dissimilar others. *Psychological Science* 24, 589–594.

Hampton, A. N., and O'Doherty, J. P. 2007. Decoding the neural substrates of reward-related decision making with functional MRI. *Proceedings of the National Academy of Sciences* 104, 1377–1382.

Hampton, A. N., Bossaerts, P., and O'Doherty, J. P. 2008. Neural correlates of mentalizing-related computations during strategic interactions in humans. *Proceedings of the National Academy of Sciences* 105, 6741–6746.

Happé, F. G. E. 1994. An advanced test of theory of mind: Understanding of story characters' thoughts and feelings by able autistic, mentally handicapped, and normal children and adults. *Journal of Autism and Developmental Disorders* 24, 129–154.

Happé, F. G. E. 1995. The role of age and verbal ability in the theory of mind task performance of subjects with autism. *Child Development* 66, 843–855.

Hardwick, R. M., Caspers, S., Eickhoff, S. B., and Swinnen, S. P. 2018. Neural correlates of action: Comparing meta-analyses of imagery, observation, and execution. *Neuroscience & Biobehavioral Reviews* 94, 31–44.

Harlow, H. F. 1949. The formation of learning sets. *Psychological Review* 56, 51–65.

Harris, L. T., and Fiske, S. T. 2006. Dehumanizing the lowest of the low: Neuroimaging responses to extreme out-groups. *Psychological Science* 17, 847–853.

Hartley, T., Maguire, E. A., Spiers, H. J., and Burgess, N. 2003. The well-worn route and the path less traveled: Distinct neural bases of route following and wayfinding in humans. *Neuron* 37, 877–888.

Haruno, M., and Frith, C. D. 2010. Activity in the amygdala elicited by unfair divisions predicts social value orientation. *Nature Neuroscience* 13, 160–161.

Haruno, M., Kimura, M., and Frith, C. D. 2014. Activity in the nucleus accumbens and amygdala underlies individual differences in prosocial and individualistic economic choices. *Journal of Cognitive Neuroscience* 26, 1861–1870.

Hassabis, D., Kumaran, D., Summerfield, C., and Botvinick, M. 2017. Neuroscience-inspired artificial intelligence. *Neuron* 95, 245–258.

Hasson, U., and Frith, C. D. 2016. Mirroring and beyond: Coupled dynamics as a generalized framework for modelling social interactions. *Philosophical Transactions of the Royal Society B* 371, 20150366, http://doi.org/10.1098/rstb.2015.0366.

Healey, P. G. T., Mills, G. J., Eshghi, A., and Howes, C. 2018. Running repairs: Coordinating meaning in dialogue. *Topics in Cognitive Science* 10, 367–388.

Hebb, D. O. 1949. *The Organization of Behavior: A Neuropsychological Theory*. Wiley.

Heider, F., and Simmel, M. 1944. An experimental study of apparent behavior. *American Journal of Psychology* 57, 243–249.

Heilbron, M., and Meyniel, F. 2019. Confidence resets reveal hierarchical adaptive nature of learning in humans. *PLoS Comput Biol* 15, e1006972, https://doi.org/10.1371/journal.pcbi.1006972.

Hein, G., Silani, G., Preuschoff, K., Batson, C. D., and Singer, T. 2010. Neural responses to ingroup and outgroup members' suffering predict individual differences in costly helping. *Neuron* 68, 149–160.

Hembacher, E., and Ghetti, S. 2014. Don't look at my answer: Subjective uncertainty underlies preschoolers' exclusion of their least accurate memories. *Psychological Science* 25, 1768–1776.

Henderson, A. M. E., and Woodward, A. L. 2012. Nine-month-old infants generalize object labels, but not object preferences across individuals. *Developmental Science* 15, 641–652.

Henrich, J. 2004. Demography and cultural evolution: How adaptive cultural processes can produce maladaptive losses—The Tasmanian case. *American Antiquity* 69, 197–214.

Henrich, J. 2018. *The Secret of Our Success: How Culture Is Driving Human Evolution, Domesticating Our Species, and Making Us Smarter*. Princeton University Press.

Henrich, J. 2020. *The Weirdest People in the World: How the West Became Psychologically Peculiar and Particularly Prosperous*. Penguin Books Ltd. (Allen Lane).

Henry, O. 1906. The gift of the magi in *The Four Million*. McClure, Phillips & Co. http://webhome.auburn.edu/~vestmon/Gift_of_the_Magi.html.

Hertz, U., Bell, V., and Raihani, N. J. 2021. Trusting and learning from others: Immediate and long-term effects of learning from observation and advice. *Proceedings of the Royal Society, series B* 288, 20211414, http://doi.org/10.1098/rspb.2021.1414.

Hertz, U., Palminteri, S., Brunetti, S., Olesen, C., Frith, C. D., and Bahrami, B. 2017. Neural computations underpinning the strategic management of influence in advice giving. *Nature Communications* 8, 2191, https://doi.org/10.1038/s41467-017-02314-5.

Heyes, C. 2011. Automatic imitation. *Psychological Bulletin* 137, 463–483.

Heyes, C. 2014. Submentalizing: I am not really reading your mind. *Perspectives on Psychological Science* 9, 131–143.

Heyes, C. 2018. *Cognitive Gadgets: The Cultural Evolution of Thinking*. Harvard University Press.

Heyes, C. M., and Frith, C. D. 2014. The cultural evolution of mind reading. *Science* 344, 1243091, https://doi.org/10.1126/science.1243091.

Heyes, C., Bang, D., Shea, N., Frith, C. D., and Fleming, S. M. 2020. Knowing ourselves together: The cultural origins of metacognition. *Trends in Cognitive Sciences* 24, 349–362.

Hickok, G. 2009. Eight problems for the mirror neuron theory of action understanding in monkeys and humans. *Journal of Cognitive Neuroscience* 21, 1229–1243.

Hillebrandt, H., Friston, K. J., and Blakemore, S.-J. 2014. Effective connectivity during animacy perception—dynamic causal modelling of Human Connectome Project data. *Scientific Reports* 4, 6240, https://doi.org/10.1038/srep06240.

Hills, T. T. 2018. The dark side of information proliferation. *Perspectives on Psychological Science* 14, 323–330.

Hirst, W., and Echterhoff, G. 2012. Remembering in conversations: The social sharing and reshaping of memories. *Annual Review of Psychology* 63, 55–79.

Hirst, W., Yamashiro, J. K., and Coman, A. 201. Collective memory from a psychological perspective. *Trends in Cognitive Sciences* 22, 438–451.

Hobbes, T. 1651. *Leviathan, or The Matter, Forme, & Power of a Common-Wealth.* Andrew Crooke.

Hoehl, S., Keupp, S., Schleihauf, H., McGuigan, N., Buttelmann, D., and Whiten, A. 2019. "Over-imitation": A review and appraisal of a decade of research. *Developmental Review* 51, 90–108.

Hohwy, J. 2013. *The Predictive Mind.* Oxford University Press.

Hojat, M., Gonnella, J. S., Nasca, T. J., Mangione, S., Vergare, M., and Magee, M. 2002. Physician empathy: Definition, components, measurement, and relationship to gender and specialty. *American Journal of Psychiatry* 159, 1563–1569.

Holmes, W. G., and Sherman, P. W. 1983. Kin recognition in animals: The prevalence of nepotism among animals raises basic questions about how and why they distinguish relatives from unrelated individuals. *American Scientist* 71, 46–55.

Hoppitt, W., and Laland, K. N. 2013. *Social Learning: An Introduction to Mechanisms, Methods, and Models*: Princeton University Press.

Hoppitt, W. J. E., Brown, G. R., Kendal, R., Rendell, L., Thornton, A., Webster, M. M., and Laland, K. N. 2008. Lessons from animal teaching. *Trends in Ecology and Evolution* 23, 486–493.

Horner, V., and Whiten, A. 2005. Causal knowledge and imitation/emulation switching in chimpanzees (Pan troglodytes) and children (Homo sapiens). *Animal Cognition* 8, 164–181.

Hsu, Y. K., and Cheung, H. 2013. Two mentalizing capacities and the understanding of two types of lie telling in children. *Developmental Psychology* 49, 1650–1659.

Hu, J., Lucas, C. G., Griffiths, T. L., and Xu, F. 2015. Preschoolers' understanding of graded preferences. *Cognitive Development* 36, 93–102.

Huber, L., Range, F., Voelkl, B., Szucsich, A., Viranyi, Z., and Miklosi, A. 2009. The evolution of imitation: What do the capacities of non-human animals tell us about the mechanisms of imitation? *Philosophical Transactions of the Royal Society B* 364, 2299–2309.

Hurks, P. P. 2012. Does instruction in semantic clustering and switching enhance verbal fluency in children? *The Clinical Neuropsychologist* 26, 1019–1037.

Iriki, A., Tanaka, M., and Iwamura, Y. 1996. Coding of modified body schema during tool use by macaque postcentral neurones. *Neuroreport* 7, 2325–2330.

Izuma, K., Saito, D. N., and Sadato, N. 2010. The roles of the medial prefrontal cortex and striatum in reputation processing. *Society for Neuroscience* 5, 133–147.

Jack, A. I., Dawson, A. J., Begany, K. L., Leckie, R. L., Barry, K. P., Ciccia, A. H., and Snyder, A. Z. 2013. fMRI reveals reciprocal inhibition between social and physical cognitive domains. *NeuroImage* 66, 385–401.

Jacob, P., and Jeannerod, M. 2003. *Ways of Seeing: The Scope and Limits of Visual Cognition.* Oxford University Press.

Jacob, P., and Jeannerod, M. 2005. The motor theory of social cognition: A critique. *Trends in Cognitive Sciences* 9, 21–25.

Jaeger, B., Oud, B., Williams, T., Krumhuber, E. G., Fehr, E., and Engelmann, J. B. 2022. Can people detect the trustworthiness of strangers based on their facial appearance? *Evolution and Human Behavior* 43, 296–303.

Jamali, M., Grannan, B. L., Fedorenko, E., Saxe, R., Báez-Mendoza, R., and Williams, Z. M. 2021. Single-neuronal predictions of others' beliefs in humans. *Nature* 591, 610–614.

James, W. 1890. *The Principles of Psychology*. Vol. II. Henry Holt.

James, W. 1904. A world of pure experience II (the conterminousness of different minds). *Journal of Philosophy, Psychology and Scientific Methods* 1, 561–570.

Jara-Ettinger, J., Gweon, H., Schulz, L. E., and Tenenbaum, J. B. 2016. The naïve utility calculus: Computational principles underlying commonsense psychology. *Trends in Cognitive Sciences* 20, 589–604.

Jaswal, V. K., Croft, A. C., Setia, A. R., and Cole, C. A. 2010. Young children have a specific, highly robust bias to trust testimony. *Psychological Science* 21, 1541–1547.

Jerez-Fernandez, A., Angulo, A. N., and Oppenheimer, D. M. 2013. Show me the numbers: Precision as a cue to others' confidence. *Psychological Science* 25, 633–635.

Jessen, S., and Grossmann, T. 2016. Neural and behavioral evidence for infants' sensitivity to the trustworthiness of faces. *Journal of Cognitive Neuroscience* 28, 1728–1736.

Job, V., Dweck, C. S., and Walton, G. M. 2010. Ego depletion—is it all in your head? Implicit theories about willpower affect self-regulation. *Psychological Science* 21, 1686–1693.

Johnson, M. H., Dziurawiec, S., Ellis, H., and Morton, J. 1991. Newborns' preferential tracking of face-like stimuli and its subsequent decline. *Cognition* 40, 1–19.

Johnson, N. R. 2014. Panic at "The Who Concert Stampede": An empirical assessment. *Social Problems* 34, 362–373.

Johnson, S. C. 2003. Detecting agents. *Philosophical Transactions of the Royal Society of London Series B-Biological Sciences* 358, 549–559.

Jolly, E., and Chang, L. J. 2021. Gossip drives vicarious learning and facilitates social connection. *Current Biology* 31, 1–11.

Jordan, G., Deeb, S. S., Bosten, J. M., and Mollon, J. D. 2010. The dimensionality of color vision in carriers of anomalous trichromacy. *Journal of Vision* 10, 12–12.

Jordan, J. J., Sommers, R., Bloom, P., and Rand, D. G. 2017. Why do we hate hypocrites? Evidence for a theory of false signaling. *Psychological Science* 28, 356–368.

Jouravlev, O., Schwartz, R., Ayyash, D., Mineroff, Z., Gibson, E., and Fedorenko, E. 2018. Tracking colisteners' knowledge states during language comprehension. *Psychological Science* 30, 3–19.

Joyce, M. K. P., García-Cabezas, M. Á., John, Y. J., and Barbas, H. 2020. Serial prefrontal pathways are positioned to balance cognition and emotion in primates. *Journal of Neuroscience* 40, 8306–8328.

Jueptner, M., Stephan, K. M., Frith, C. D., Brooks, D. J., Frackowiak, R. S. J., and Passingham, R. E. 1997. Anatomy of motor learning. 1. Frontal cortex and attention to action. *Journal of Neurophysiology* 77, 1313–1324.

Kahneman, D. 2011. *Thinking, Fast and Slow*. Farrar, Straus and Giroux.

Kalra, P. B., Gabrieli, J. D. E., and Finn, A. S. 2019. Evidence of stable individual differences in implicit learning. *Cognition* 190, 199–211.

Kampe, K. K., Frith, C. D., and Frith, U. 2003. "Hey John": Signals conveying communicative intention toward the self activate brain regions associated with "mentalizing," regardless of modality. *Journal of Neuroscience* 23, 5258–5263.

Kampis, D., Kármán, P., Csibra, G., Southgate, V., and Hernik, M. 2021. A two-lab direct replication attempt of Southgate, Senju and Csibra (2007). *Royal Society Open Science* 8, 210190, http://doi.org/10.1098/rsos.210190.

Kandel, S., Orliaguet, J.-P., and Viviani, P. 2000. Perceptual anticipation in handwriting: The role of implicit motor competence. *Perception & Psychophysics* 62, 706–716.

Kang, P., Burke, C. J., Tobler, P. N., and Hein, G. 2021. Why we learn less from observing outgroups. *Journal of Neuroscience* 41, 144–152.

Kanizsa, G., and Vicario, G. 1968. The perception of intentional reaction. In *Experimental Research on Perception*, eds. G. Kanizsa and G. Vicario, 71–126. University of Trieste.

Karpus, J., Krüger, A., Verba, J. T., Bahrami, B., and Deroy, O. 2021. Algorithm exploitation: Humans are keen to exploit benevolent AI. *iScience* 24, 102679, https://doi.org/10.1016/j.isci.2021.102679.

Kashtelyan, V., Lichtenberg, N. T., Chen, M. L., Cheer, J. F., and Roesch, M. R. 2014. Observation of reward delivery to a conspecific modulates dopamine release in ventral striatum. *Current Biology* 24, 2564–2568.

Kavanagh, L. C., Suhler, C. L., Churchland, P. S., and Winkielman, P. 2011. When it's an error to mirror. *Psychological Science* 22, 1274–1276.

Keltner, D., and Buswell, B. N. 1997. Embarrassment: Its distinct form and appeasement functions. *Psychological Bulletin* 122, 250–270.

Kendal, R., Hopper, L. M., Whiten, A., Brosnan, S. F., Lambeth, S. P., Schapiro, S. J., and Hoppitt, W. 2015. Chimpanzees copy dominant and knowledgeable individuals: Implications for cultural diversity. *Evolution and Human Behavior* 36, 65–72.

Kent, S. 1994. *Sherman Kent and the Board of National Estimates: Collected Essays*. History Staff, Center for the Study of Intelligence, Central Intelligence Agency, University of Michigan Library.

Kerr, N. L., and Tindale, R. S. 2004. Group performance and decision making. *Annual Review of Psychology* 55, 623–655.

Keysers, C., and Gazzola, V. 2009. Expanding the mirror: Vicarious activity for actions, emotions, and sensations. *Current Opinion in Neurobiology* 19, 666–671.

Kidd, C., Palmeri, H., and Aslin, R. N. 2013. Rational snacking: Young children's decision-making on the marshmallow task is moderated by beliefs about environmental reliability. *Cognition* 126, 109–114.

Kihlstrom, J. F. 1987. The cognitive unconscious. *Science* 237, 1445–1452.

Kihlstrom, J. F. 2002. Demand characteristics in the laboratory and the clinic: Conversations and collaborations with subjects and patients. *Prevention & Treatment* 5, https://doi.org/10.1037/1522-3736.5.1.536c.

Kilner, J. M., Friston, K. J., and Frith, C. D. 2007. Predictive coding: An account of the mirror neuron system. *Cognitive Process* 8, 159–166.

Kilner, J. M., Neal, A., Weiskopf, N., Friston, K. J., and Frith, C. D. 2009. Evidence of mirror neurons in human inferior frontal gyrus. *Journal of Neuroscience* 29, 10153–10159.

Kim, A. J., Fitzgerald, J. K., and Maimon, G. 2015. Cellular evidence for efference copy in Drosophila visuomotor processing. *Nature Neuroscience* 18, 1247–1255.

King-Casas, B., Tomlin, D., Anen, C., Camerer, C. F., Quartz, S. R., and Montague, P. R. 2005. Getting to know you: Reputation and trust in a two-person economic exchange. *Science* 308, 78–83.

Klein, J. T., Deaner, R. O., and Platt, M. L. 2008. Neural correlates of social target value in macaque parietal cortex. *Current Biology* 18, 419–424.

Kline, M. A. 2015. How to learn about teaching: An evolutionary framework for the study of teaching behavior in humans and other animals. *Behavioral and Brain Sciences* 38, E31, https://doi.org/10.1017/S0140525X14000090.

Klink, P. C., Jentgens, P., and Lorteije, J. A. M. 2014. Priority maps explain the roles of value, attention, and salience in goal-oriented behavior. *Journal of Neuroscience* 34, 13867–13869.

Kloo, D., Sodian, B., Kristen-Antonow, S., Kim, S., and Paulus, M. 2021. Knowing minds: Linking early perspective taking and later metacognitive insight. *British Journal of Developmental Psychology* 39, 39–53.

Knoblich, G., and Sebanz, N. 2008. Evolving intentions for social interaction: from entrainment to joint action. *Philosophical Transactions of the Royal Society B* 363, 2021–2031.

Kobayashi, H., and Kohshima, S. 2001. Unique morphology of the human eye and its adaptive meaning: Comparative studies on external morphology of the primate eye. *Journal of Human Evolution* 40, 419–435.

Kochukhova, O., and Gredebäck, G. 2010. Preverbal infants anticipate that food will be brought to the mouth: An eye tracking study of manual feeding and flying spoons. *Child Development* 81, 1729–1738.

Koenig, M. A., and Harris, P. L. 2005. Preschoolers mistrust ignorant and inaccurate speakers. *Child Development* 76, 1261–1277.

Köksal, Ö., Sodian, B., and Legare, C. H. 2021. Young children's metacognitive awareness of confounded evidence. *Journal of Experimental Child Psychology* 205, 105080, https://doi.org/10.1016/j.jecp.2020.105080.

Konvalinka, I., and Roepstorff, A. 2012. The two-brain approach: How can mutually interacting brains teach us something about social interaction? *Frontiers in Human Neuroscience* 6, 215, https://doi.org/10.3389/fnhum.2012.00215.

Konvalinka, I., Bauer, M., Stahlhut, C., Hansen, L. K., Roepstorff, A., and Frith, C. D. 2014. Frontal alpha oscillations distinguish leaders from followers: Multivariate decoding of mutually interacting brains. *Neuroimage* 94, 79–88.

Konvalinka, I., Vuust, P., Roepstorff, A., and Frith, C. D. 2010. Follow you, follow me: Continuous mutual prediction and adaptation in joint tapping. *Quarterly Journal of Experimental Psychology (Colchester)* 63, 2220–2230.

Koriat, A. 2012. When are two heads better than one and why? *Science* 336, 360–362.

Kourtis, D., Woźniak, M., Sebanz, N., and Knoblich, G. 2019. Evidence for we-representations during joint action planning. *Neuropsychologia* 131, 73–83.

Kovács, Á. M., Téglás, E., and Endress, A. D. 2010. The social sense: Susceptibility to others' beliefs in human infants and adults. *Science* 330, 1830–1834.

Kragness, H. E., Johnson, E. K., and Cirelli, L. K. 2021. The song, not the singer: Infants prefer to listen to familiar songs, regardless of singer identity. *Developmental Science* 25, e13147, https://doi.org/10.1111/desc.13149.

Kruger, J., and Dunning, D. 1999. Unskilled and unaware of it: How difficulties in recognizing one's own incompetence lead to inflated self-assessments. *Journal of Personality and Social Psychology* 77, 1121–1134.

Krupenye, C., and Call, J. 2019. Theory of mind in animals: Current and future directions. *WIREs Cognitive Science* 10, e1503, https://doi.org/10.1002/wcs.1503.

Kuhl, P. K., Tsao, F.-M., and Liu, H.-M. 2003. Foreign-language experience in infancy: Effects of short-term exposure and social interaction on phonetic learning. *Proceedings of the National Academy of Sciences* 100, 9096–9101.

Kulesza, W., Dolinski, D., and Wicher, P. 2016. Knowing that you mimic me: The link between mimicry, awareness and liking. *Social Influence* 11, 68–74.

Kulke, L., and Hinrichs, M. A. B. 2021. Implicit theory of mind under realistic social circumstances measured with mobile eye-tracking. *Scientific Reports–UK* 11, 1215.

Kulke, L., von Duhn, B., Schneider, D., and Rakoczy, H. 2018. Is implicit theory of mind a real and robust phenomenon? Results from a systematic replication study. *Psychological Science* 29, 888–900.

Kunst-Wilson, W. R., and Zajonc, R. B. 1980. Affective discrimination of stimuli that cannot be recognized. *Science* 207, 557–558.

Kutas, M., and Federmeier, K. D. 2011. Thirty years and counting: Finding meaning in the N400 component of the event related brain potential (ERP). *Annual Review of Psychology* 62, 621–647.

Kuzmics, C. 2018. The game theory of everyday life—the secret handshake. Graz Economics Blog. Graz Economics Club.

Lafer-Sousa, R., Hermann, K. L., and Conway, B. R. 2015. Striking individual differences in color perception uncovered by "the dress" photograph. *Current Biology* 25, R545–R546.

Lakin, J. L., and Chartrand, T. L. 2003. Using nonconscious behavioral mimicry to create affiliation and rapport. *Psychological Science* 14, 334–339.

Lakin, J. L., Chartrand, T. L., and Arkin, R. M. 2008. I am too just like you—Nonconscious mimicry as an automatic behavioral response to social exclusion. *Psychological Science* 19, 816–822.

Laland, K. N. 2004. Social learning strategies. *Learning & Behavior* 32, 4–14.

Lambie, J. A., and Marcel, A. J. 2002. Consciousness and the varieties of emotion experience: A theoretical framework. *Psychological Review* 109, 219–259.

Lany, J., and Saffran, J. R. 2010. From statistics to meaning: Infants' acquisition of lexical categories. *Psychological Science* 21, 284–291.

Lathem, E. C. 1966. *Interviews with Robert Frost.* Jonathan Cape.

Lau, H., and Rosenthal, D. 2011. Empirical support for higher-order theories of conscious awareness. *Trends in Cognitive Sciences* 15, 365–373.

Lausic, D., Tennebaum, G., Eccles, D., Jeong, A., and Johnson, T. 2009. Intrateam communication and performance in doubles tennis. *Research Quarterly for Exercise and Sport* 80, 281–290.

Lea, R., and Taylor, M. 2010. Historian Orlando Figes admits posting Amazon reviews that trashed rivals. *The Guardian,* April 23, https://www.theguardian.com/books/2010/apr/23/historian-orlando-figes-amazon-reviews-rivals.

Le Bon, G. 1895. *Psychologie des foules.* Félix Alcan.

Lebreton, M., Kawa, S., Forgeot d'Arc, B., Daunizeau, J., and Pessiglione, M. 2012. Your goal is mine: Unraveling mimetic desires in the human brain. *Journal of Neuroscience* 32, 7146–7157.

LeDoux, J. E., and Brown, R. 2017. A higher-order theory of emotional consciousness. *Proceedings of the National Academy of Sciences* 114, E2016–E2025.

Lee, B. P. H. 2001. Mutual knowledge, background knowledge and shared beliefs: Their roles in establishing common ground. *Journal of Pragmatics* 33, 21–44.

Lee, D., McGreevy, B. P., and Barraclough, D. J. 2005. Learning and decision making in monkeys during a rock-paper-scissors game. *Brain Research. Cognitive Brain Research* 25, 416–430.

Legare, C. H. 2017. Cumulative cultural learning: Development and diversity. *Proceedings of the National Academy of Sciences* 114, 7877–7883.

Legros, S., and Cislaghi, B. 2019. Mapping the social-norms literature: An overview of reviews. *Perspectives on Psychological Science* 15, 62–80.

Lengersdorff, L. L., Wagner, I. C., Lockwood, P. L., and Lamm, C. 2020. When implicit prosociality trumps selfishness: The neural valuation system underpins more optimal choices when learning to avoid harm to others than to oneself. *Journal of Neuroscience* 40, 7286–7299.

Lepage, M., Ghaffar, O., Nyberg, L., and Tulving, E. 2000. Prefrontal cortex and episodic memory retrieval mode. *Proceedings of the National Academy of Sciences* 97, 506–511.

Leslie, A. M. 1987. Pretense and representation: The origins of "theory of mind." *Psychological Review* 94, 412–426.

Leslie, A. M. 1994. ToMM, ToBY, and agency: Core architecture and domain specificity. In *Mapping the Mind: Domain Specificity in Cognition and Culture*, eds. L. A. Hirschfeld and S. A. Gelman, 119–148. Cambridge University Press.

Leung, A., Tunkel, A., and Yurovsky, D. 2021. Parents fine-tune their speech to children's vocabulary knowledge. *Psychological Science* 32, 957–984.

Levy, D. J., and Glimcher, P. W. 2012. The root of all value: A neural common currency for choice. *Current Opinion in Neurobiology* 22, 1027–1038.

Lewis, D. K. 1973. Causation. *Journal of Philosophy* 70, 556–567.

Lewis, H. M., and Laland, K. N. 2012. Transmission fidelity is the key to the build-up of cumulative culture. *Philosophical Transactions of the Royal Society of London Series B, Biological Sciences* 367, 2171–2180.

Lewis, M. B. 2016. Arguing that black is white: Racial categorization of mixed-race faces. *Perception* 45, 505–514.

Lhermitte, F., Pillon, B., and Serdaru, M. 1986. Human autonomy and the frontal lobes. Part I: Imitation and utilization behavior: A neuropsychological study of 75 patients. *Annals of Neurology* 19, 326–334.

Li, J., Delgado, M. R., and Phelps, E. A. 2011. How instructed knowledge modulates the neural systems of reward learning. *Proceedings of the National Academy of Sciences* 108, 55–60.

Li, L., Britvan, B., and Tomasello, M. 2021. Young children conform more to norms than to preferences. *PLOS One* 16, e0251228, https://doi.org/10.1371/journal.pone.0251228.

Liang, Z. S., Nguyen, T., Mattila, H. R., Rodriguez-Zas, S. L., Seeley, T. D., and Robinson, G. E. 2012. Molecular determinants of scouting behavior in honey bees. *Science* 335, 1225–1228.

Liebal, K., Behne, T., Carpenter, M., and Tomasello, M. 2009. Infants use shared experience to interpret pointing gestures. *Developmental Science* 12, 264–271.

Lin, C., Adolphs, R., and Alvarez, R. M. 2018. Inferring whether officials are corruptible from looking at their faces. *Psychological Science* 29, 1807–1823.

Lin, L. C., Qu, Y., and Telzer, E. H. 2018. Intergroup social influence on emotion processing in the brain. *Proceedings of the National Academy of Sciences* 115, 10630–10635.

Lindström, B., Golkar, A., Jangard, S., Tobler, P. N., and Olsson, A. 2019. Social threat learning transfers to decision making in humans. *Proceedings of the National Academy of Sciences* 116, 4732–4737.

Lindström, B., Haaker, J., and Olsson, A. 2018. A common neural network differentially mediates direct and social fear learning. *NeuroImage* 167, 121–129.

Liu, S., and Spelke, E. S. 2017. Six-month-old infants expect agents to minimize the cost of their actions. *Cognition* 160, 35–42.

Lloyd, J. E. 1986. Firefly communication and deception: "Oh, what a tangled web." In *Deception: Perspectives on Human and Nonhuman Deceit*, eds. R. W. Mitchell and N. S. Thompson, 113–128. State University of New York Press.

Lockwood, P. L., Apps, M. A. J., and Chang, S. W. C. 2020. Is there a "social" brain? Implementations and algorithms. *Trends in Cognitive Sciences* 24, 802–813.

Lockwood, P. L., Apps, M. A. J., Valton, V., Viding, E., and Roiser, J. P. 2016. Neurocomputational mechanisms of prosocial learning and links to empathy. *Proceedings of the National Academy of Sciences* 113, 9763.

Logan, G. D., and Crump, M. J. 2010. Cognitive illusions of authorship reveal hierarchical error detection in skilled typists. *Science* 330, 683–686.

Lohrenz, T., McCabe, K., Camerer, C. F., and Montague, P. R. 2007. Neural signature of fictive learning signals in a sequential investment task. *Proceedings of the National Academy of Sciences* 104, 9493–9498.

Lovett, L. 2005. The Popeye principle: Selling child health in the first nutrition crisis. *Journal of Health Politics, Policy and Law* 30, 803–838.

Luce, R. D., and Raiffa, H. 1989. *Games and Decisions: Introduction and Critical Survey*. Dover Publications.

Luria, A. R. (2012). *Higher Cortical Functions in Man*. Springer Science & Business Media.

Lyons, D. E., Damrosch, D. H., Lin, J. K., Macris, D. M., and Keil, F. C. 2011. The scope and limits of overimitation in the transmission of artefact culture. *Proceedings of the National Academy of Sciences* 366, 1158–1167.

Lyons, D. E., Young, A. G., and Keil, F. C. 2007. The hidden structure of overimitation. *Proceedings of the National Academy of Sciences* 104, 19751–19756.

Ma, F., Zeng, D., Xu, F., Compton, B. J., and Heyman, G. D. 2020. Delay of gratification as reputation management. *Psychological Science* 31, 1174–1182.

MacAskill, W. 2015. *Doing Good Better: Effective Altruism and a Radical New Way to Make a Difference*. Guardian Faber Publishing.

Machiavelli, N. 2008 (orig. 1532). *The Prince*. Hackett Publishing.

Mahmoodi, A., Bang, D., Olsen, K., Zhao, Y. A., Shi, Z., Broberg, K., et al. 2015. Equality bias impairs collective decision-making across cultures. *Proceedings of the National Academy of Sciences* 112, 3835–3840.

Maister, L., Sebanz, N., Knoblich, G., and Tsakiris, M. 2013. Experiencing ownership over a dark-skinned body reduces implicit racial bias. *Cognition* 128, 170–178.

Malafouris, L. 2013. *How Things Shape the Mind*. MIT Press.

Maren, S. 2001. Neurobiology of Pavlovian fear conditioning. *Annual Review of Neuroscience* 24, 897–931.

Marsh, A. A. 2019. The caring continuum: Evolved hormonal and proximal mechanisms explain prosocial and antisocial extremes. *Annual Review of Psychology* 70, 347–371.

Marsh, A. A., Finger, E. C., Fowler, K. A., Adalio, C. J., Jurkowitz, I. T. N., Schechter, J. C., et al. 2013. Empathic responsiveness in amygdala and anterior cingulate cortex in youths with psychopathic traits. *Journal of Child Psychology and Psychiatry* 54, 900–910.

Marsh, A. A., Stoycos, S. A., Brethel-Haurwitz, K. M., Robinson, P., VanMeter, J. W., and Cardinale, E. M. 2014a. Neural and cognitive characteristics of extraordinary altruists. *Proceedings of the National Academy of Sciences* 111, 15036.

Marsh, L. E., Mullett, T. L., Ropar, D., and Hamilton, A. F. d. C. 2014b. Responses to irrational actions in action observation and mentalising networks of the human brain. *NeuroImage* 103, 81–90.

Marshall, J., and McAuliffe, K., 2022. Children as assessors and agents of third-party punishment. *Nature Reviews Psychology* 1, 334–344.

Mascaro, O., and Csibra, G. 2012. Representation of stable social dominance relations by human infants. *Proceedings of the National Academy of Sciences* 109, 6862–6867.

Mascaro, O., and Morin, O. 2015. Epistemology for beginners: Two- to five-year-old children's representation of falsity. *PLOS One* 10, e0140658, https://doi.org/10.1371/journal.pone.0140658.

Mascaro, O., and Sperber, D. 2009. The moral, epistemic, and mindreading components of children's vigilance towards deception. *Cognition* 112, 367–380.

McAuliffe, K., and Thornton, A. 2015. The psychology of cooperation in animals: An ecological approach. *Journal of Zoology* 295, 23–35.

McClung, J. S., Placì, S., Bangerter, A., Clément, F., and Bshary, R. 2017. The language of cooperation: Shared intentionality drives variation in helping as a function of group membership. *Proceedings of the Royal Society B: Biological Sciences* 284, 20171682, http://doi.org/10.1098/rspb.2017.1682.

McCormack, T., O'Connor, E., Cherry, J., Beck, S. R., and Feeney, A. 2019. Experiencing regret about a choice helps children learn to delay gratification. *Journal of Experimental Child Psychology* 179, 162–175.

McCulloch, W. S., and Pitts, W. H. 1943. A logical calculus of the ideas immanent in nervous activity. *Bulletin of Mathematical Biophysics* 5, 115–133.

McDougall, M. 1926. "Imitation, play and habit." In *An Introduction to Social Psychology* (revised edition), 332–358. John W. Luce & Co.

McGagh, M. 2012. PPI becomes most complained about product ever. *Citywire*, https://citywire.com/funds-insider/news/ppi-becomes-most-complained-about-product-ever/a590862.

McGeer, V. 2007. The regulative dimension of folk psychology. In *Folk Psychology Re-Assessed*, eds. D. Hutto and M. Ratcliffe, 137–156. Springer Netherlands.

McGuigan, N., Makinson, J., and Whiten, A. 2011. From over-imitation to super-copying: Adults imitate causally irrelevant aspects of tool use with higher fidelity than young children. *British Journal of Psychology* 102, 1–18.

McKee, S. P., and Westhe, G. 1978. Improvement in vernier acuity with practice. *Perception & Psychophysics* 24, 258–262.

McLoughlin, N., and Over, H. 2017. Young children are more likely to spontaneously attribute mental states to members of their own group. *Psychological Science* 28, 1503–1509.

McLoughlin, N., and Over, H. 2018. Encouraging children to mentalise about a perceived outgroup increases prosocial behaviour towards outgroup members. *Developmental Science* 22, e12774, https://doi.org/10.1111/desc.12774.

McManus, R. M., Kleiman-Weiner, M., and Young, L. 2020. What we owe to family: The impact of special obligations on moral judgment. *Psychological Science*, 31, 227–242.

McNally, L., and Jackson, A. L. 2013. Cooperation creates selection for tactical deception. *Proceedings of the Royal Society B: Biological Sciences* 280, 20130699, http://doi.org/10.1098/rspb.2013.0699.

McNamara, J. M., and Barta, Z. 2020. Behavioural flexibility and reputation formation. *Proceedings of the Royal Society B: Biological Sciences* 287, 20201758, http://doi.org/10.1098/rspb.2020.1758.

Mercier, H. 2016. The argumentative theory: Predictions and empirical evidence. *Trends in Cognitive Sciences* 20, 689–700.

Mercier, H., and Sperber, D. 2011. Why do humans reason? Arguments for an argumentative theory. *Behavioral and Brain Sciences* 34, 57–74; discussion 74–111.

Meriau, K., Wartenburger, I., Kazzer, P., Prehn, K., Lammers, C. H., van der Meer, E., et al. 2006. A neural network reflecting individual differences in cognitive processing of emotions during perceptual decision making. *NeuroImage* 33, 1016–1027.

Mesoudi, A. 2016. Cultural evolution: Integrating psychology, evolution and culture. *Current Opinion in Psychology* 7, 17–22.

Michael, J., Sebanz, N., and Knoblich, G. 2015. The sense of commitment: A minimal approach. *Frontiers in Psychology* 6, 1968, http://doi.org/10.3389/fpsyg.2015.01968.

Michotte, A. E. 1963. *The Perception of Causality*. Methuen.

Miele, D. B., Wager, T. D., Mitchell, J. P., and Metcalfe, J. 2011. Dissociating neural correlates of action monitoring and metacognition of agency. *Journal of Cognitive Neuroscience* 23, 3620–3636.

Milinski, M., and Rockenbach, B. 2007. Spying on others evolves. *Science* 317, 464–465.

Milinski, M., Semmann, D., and Krambeck, H. J. 2002. Reputation helps solve the "tragedy of the commons." *Nature* 415, 424–426.

Miller, K. J., Shenhav, A., and Ludvig, E. A. 2019. Habits without values. *Psychological Review* 126, 292–311.

Mills, C. M. 2013. Knowing when to doubt: Developing a critical stance when learning from others. *Developmental Psychology* 49, 404–418.

Mills, K. L., Lalonde, F., Clasen, L. S., Giedd, J. N., and Blakemore, S.-J. 2014. Developmental changes in the structure of the social brain in late childhood and adolescence. *Social Cognitive and Affective Neuroscience* 9, 123–131.

Milner, B. 1963. Effects of different brain lesions on card sorting. *Archives of Neurology* 9, 90–100.

Mineka, S., and Ohman, A. 2002. Phobias and preparedness: The selective, automatic, and encapsulated nature of fear. *Biological Psychiatry* 52, 927–937.

Mirza, M. B., Adams, R. A., Friston, K., and Parr, T. 2019. Introducing a Bayesian model of selective attention based on active inference. *Scientific Reports* 9, 13915, https://doi.org/10.1038/s41598-019-50138-8.

Misch, A., Over, H., and Carpenter, M. 2016. I won't tell: Young children show loyalty to their group by keeping group secrets. *Journal of Experimental Child Psychology* 142, 96–106.

Mischel, W. 2015. *The Marshmallow Test: Why Self-Control Is the Engine of Success*. Little, Brown and Company.

Misyak, J., Noguchi, T., and Chater, N. 2016. Instantaneous conventions: The emergence of flexible communicative signals. *Psychological Science* 27, 1550–1561.

Mitani, J. C., Watts, D. P., and Amsler, S. J. 2010. Lethal intergroup aggression leads to territorial expansion in wild chimpanzees. *Current Biology* 20, R507–R508.

Mitchell, A. S., Czajkowski, R., Zhang, N., Jeffery, K., and Nelson, A. J. D. 2018. Retrosplenial cortex and its role in spatial cognition. *Brain and Neuroscience Advances* 2, 1–13.

Mitchell, K. J. 2018. *Innate: How the Wiring of Our Brains Shapes Who We Are*. Princeton University Press.

Miyamoto, Y. R., Wang, S., and Smith, M. A. 2020. Implicit adaptation compensates for erratic explicit strategy in human motor learning. *Nature Neuroscience* 23, 443–455.

Mizutani, A., Chahl, J. S., and Srinivasan, M. V. 2003. Insect behaviour: Motion camouflage in dragonflies. *Nature* 423, 604.

Mobbs, D., Yu, R., Meyer, M., Passamonti, L., Seymour, B., Calder, A., Schweizer, S., Frith, C., and Dalgleish, T. 2009. A key role for similarity in vicarious reward. *Science* 324, 900.

Moessnang, C., Otto, K., Bilek, E., Schäfer, A., Baumeister, S., Hohmann, S., et al. 2017. Differential responses of the dorsomedial prefrontal cortex and right posterior superior temporal sulcus to spontaneous mentalizing. *Human Brain Mapping* 38, 3791–3803.

Moll, H., and Khalulyan, A. 2016. "Not see, not hear, not speak": Preschoolers think they cannot perceive or address others without reciprocity. *Journal of Cognition and Development* 18, 152–162.

Moltz, H. 1962. The fixed action pattern: Empirical properties and theoretical implications. In *Recent Advances in Biological Psychiatry*, ed. J. Wortis, 69–85. Springer.

Momennejad, I., Duker, A., and Coman, A. 2019. Bridge ties bind collective memories. *Nature Communications* 10, 1578, https://doi.org/10.1038/s41467-019-09452-y.

Moore, J., and Haggard, P. 2008. Awareness of action: Inference and prediction. *Consciousness and Cognition* 17, 136–144.

Moore, R., Liebal, K., and Tomasello, M. 2013. Three-year-olds understand communicative intentions without language, gestures, or gaze. *Interaction Studies* 14, 62–80.

Moran, R., Keramati, M., and Dolan, R. J. 2021. Model based planners reflect on their model-free propensities. *PLOS Computational Biology* 17, e1008552.

Morelli, S., and Lieberman, M. 2013. *Frontiers in Human Neuroscience* 7, 160, https://doi.org/10.3389/fnhum.2013.00160.

Morelli, S. A., Sacchet, M. D., and Zaki, J. (2015). Common and distinct neural correlates of personal and vicarious reward: A quantitative meta-analysis. *NeuroImage* 112, 244–253.

Morris, J. S., Frith, C. D., Perrett, D. I., Rowland, D., Young, A. W., Calder, A. J., and Dolan, R. J. 1996. A differential neural response in the human amygdala to fearful and happy facial expressions. *Nature* 383, 812–815.

Morton, J., and Johnson, M. H. 1991. CONSPEC and CONLERN: A two-process theory of infant face recognition. *Psychological Review* 98, 164–181.

Moshman, D., and Geil, M. 1998. Collaborative reasoning: Evidence for collective rationality. *Thinking & Reasoning* 4, 231–248.

Motta, M., Callaghan, T., Sylvester, S., and Lunz-Trujillo, K. 2021. Identifying the prevalence, correlates, and policy consequences of anti-vaccine social identity. *Politics, Groups, and Identities*, https://doi.org/10.1080/21565503.2021.1932528.

Mullally, S. L., and Maguire, E. A. 2013. Memory, imagination, and predicting the future: A common brain mechanism? *The Neuroscientist* 20, 220–234.

Müller, C. A., and Cant, M. A. 2010. Imitation and traditions in wild banded mongooses. *Current Biology* 20, 1171–1175.

Muro, C., Escobedo, R., Spector, L., and Coppinger, R. P. 2011. Wolf-pack (Canis lupus) hunting strategies emerge from simple rules in computational simulations. *Behavioural Processes* 88, 192–197.

Murray, J., Theakston, A., and Wells, A. 2016. Can the attention training technique turn one marshmallow into two? Improving children's ability to delay gratification. *Behaviour Research and Therapy* 77, 34–39.

Naber, M., Pashkam, M. V., and Nakayama, K. 2013. Unintended imitation affects success in a competitive game. *Proceedings of the National Academy of Sciences* 110, 20046–20050.

Nagy, M., Akos, Z., Biro, D., and Vicsek, T. 2010. Hierarchical group dynamics in pigeon flocks. *Nature* 464, 890–893.

Nahmias, E., Morris, S., Nadelhoffer, T., and Turner, J. 2005. Surveying freedom: Folk intuitions about free will and moral responsibility. *Philosophical Psychology* 18, 561–584.

Nakahara, K., Hayashi, T., Konishi, S., and Miyashita, Y. 2002. Functional MRI of macaque monkeys performing a cognitive set-shifting task. *Science* 295, 1532.

Naughtin, C. K., Horne, K., Schneider, D., Venini, D., York, A., and Dux, P. E. 2017. Do implicit and explicit belief processing share neural substrates? *Human Brain Mapping* 38, 4760–4772.

Navajas, J., Heduan, F. Á., Garrido, J. M., Gonzalez, P. A., Garbulsky, G., Ariely, D., and Sigman, M. 2019. Reaching consensus in polarized moral debates. *Current Biology* 29, 4124–4129.

Neate, R. 2020. Zoom booms as demand for video-conferencing tech grows. *The Guardian*, March 31.

Nesse, R. M. 2004. Natural selection and the elusiveness of happiness. *Philosophical Transactions of the Royal Society of London Series B: Biological Sciences* 359, 1333–1347.

Neves, L., Cordeiro, C., Scott, S. K., Castro, S. L., and Lima, C. F. 2018. High emotional contagion and empathy are associated with enhanced detection of emotional authenticity in laughter. *Quarterly Journal of Experimental Psychology* 71, 2355–2363.

Newman-Norlund, S. E., Noordzij, M. L., Newman-Norlund, R. D., Volman, I. A., Ruiter, J. P., Hagoort, P., and Toni, I. 2009. Recipient design in tacit communication. *Cognition* 111, 46–54.

Nichols, P. M. 1998. FILM; In a social mirror, the faces of apes. *New York Times*. https://www.nytimes.com/1998/08/30/movies/film-in-a-social-mirror-the-faces-of-apes.html.

Nielsen, M., and Blank, C. 2011. Imitation in young children: When who gets copied is more important than what gets copied. *Developmental Psychology* 47, 1050–1053.

Nielsen, R. H., Vuust, P., and Wallentin, M. 2015. Perception of animacy from the motion of a single sound object. *Perception* 44, 183–197.

Nölle, J., Staib, M., Fusaroli, R., and Tylén, K. 2018. The emergence of systematicity: How environmental and communicative factors shape a novel communication system. *Cognition* 181, 93–104.

Norman, D. A., and Shallice, T. 1986. Attention to action: Willed and automatic control of behavior. In *Consciousness and self regulation: Advances in research*, eds. R. J. Davidson, G. E. Schwartz, and D. Shapiro, 1–18. Plenum Press.

Northoff, G., and Bermpohl, F. 2004. Cortical midline structures and the self. *Trends in Cognitive Sciences* 8, 102–107.

Nowak, M. A., and Sigmund, K. 1998a. The dynamics of indirect reciprocity. *Journal of Theoretical Biology* 194, 561–574.

Nowak, M. A., and Sigmund, K. 1998b. Evolution of indirect reciprocity by image scoring. *Nature* 393, 573–577.

Oaksford, M., and Chater, N. 2019. New paradigms in the psychology of reasoning. *Annual Review of Psychology* 71, 305–330.

Oberman, L. M., and Ramachandran, V. S. 2008. Reflections on the mirror neuron system: Their evolutionary functions beyond motor representation. In *Mirror Neuron Systems*, ed. J. A. Pineda, 39–59. Springer.

Ochsner, K. N., Knierim, K., Ludlow, D. H., Hanelin, J., Ramachandran, T., Glover, G., and Mackey, S. C. 2004. Reflecting upon feelings: An fMRI study of neural systems supporting the attribution of emotion to self and other. *Journal of Cognitive Neuroscience* 16, 1746–1772.

Ochsner, K. N., Silvers, J. A., and Buhle, J. T. 2012. Functional imaging studies of emotion regulation: A synthetic review and evolving model of the cognitive control of emotion. *Annals of the New York Academy of Sciences* 1251, E1–E24.

O'Doherty, J. P., Hampton, A., and Kim, H. 2007. Model-based fMRI and its application to reward learning and decision making. *Annals of the New York Academy of Sciences* 1104, 35–53.

Oja, E. 1982. Simplified neuron model as a principal component analyzer. *Journal of Mathematical Biology* 15, 267–273.

Olsson, A., and Phelps, E. A. 2004. Learned fear of "unseen" faces after Pavlovian, observational, and instructed fear. *Psychological Science* 15, 822–828.

Olsson, A., Knapska, E., and Lindström, B. 2020. The neural and computational systems of social learning. *Nature Reviews Neuroscience* 21, 197–212.

Olsson, A., Nearing, K. I., and Phelps, E. A. 2007. Learning fears by observing others: The neural systems of social fear transmission. *Social Cognitive and Affective Neuroscience* 2, 3–11.

O'Nions, E., Lima, C. F., Scott, S. K., Roberts, R., McCrory, E. J., and Viding, E. 2017. Reduced laughter contagion in boys at risk for psychopathy. *Current Biology* 27, 3049–3055.

Onishi, K. H., and Baillargeon, R. 2005. Do 15-month-old infants understand false beliefs? *Science* 308, 255–258.

Onishi, K. H., Baillargeon, R., and Leslie, A. M. 2007. 15-month-old infants detect violations in pretend scenarios. *Acta Psychologica* 124, 106–128.

Oostenbroek, J., and Over, H. 2015. Young children contrast their behavior to that of out-group. *Journal of Experimental Child Psychology* 139, 234–241.

Oosterhof, N. N., and Todorov, A. 2008. The functional basis of face evaluation. *Proceedings of the National Academy of Sciences* 105, 11087–11092.

Over, H., and Carpenter, M. 2009. Priming third-party ostracism increases affiliative imitation in children. *Developmental Science* 12, F1–F8.

Over, H., and Carpenter, M. 2013. The social side of imitation. *Child Development Perspectives* 7, 6–11.

Over, H., Carpenter, M., Spears, R., and Gattis, M. 2013. Children selectively trust individuals who have imitated them. *Social Development* 22, 215–224.

Palmer, S. E., Rosch, E., and Chase, P. (1981). Canonical perspective and the perception of objects. In *Attention and Peformance IX*, J. Long and A. D. Baddeley, eds., 135–151. Erlbaum.

Palser, E. R., Fotopoulou, A., and Kilner, J. M. 2018. Altering movement parameters disrupts metacognitive accuracy. *Consciousness and Cognition* 57, 33–40.

Panchanathan, K., and Boyd, R. 2004. Indirect reciprocity can stabilize cooperation without the second-order free rider problem. *Nature* 432, 499–502.

Panksepp, J. 2007. Neuroevolutionary sources of laughter and social joy: Modeling primal human laughter in laboratory rats. *Behavioural Brain Research* 182, 231–244.

Parish-Morris, J., Hennon, E. A., Hirsh-Pasek, K., Golinkoff, R. M., and Tager-Flusberg, H. 2007. Children with autism illuminate the role of social intention in word learning. *Child Development* 78, 1265–1287.

Parvin, D. E., McDougle, S. D., Taylor, J. A., and Ivry, R. B. 2018. Credit assignment in a motor decision making task is influenced by agency and not sensory prediction errors. *Journal of Neuroscience* 38, 4521–4530.

Patel, D., Fleming, S. M., and Kilner, J. M. 2012. Inferring subjective states through the observation of actions. *Proceedings of the Royal Society B: Biological Sciences* 279, 4853–4860.

Paulesu, E., Démonet, J. F., Fazio, F., McCrory, E., Chanoine, V., Brunswick, N., et al. 2001. Dyslexia: Cultural diversity and biological unity. *Science* 291, 2165–2167.

Paulesu, E., McCrory, E., Fazio, F., Menoncello, L., Brunswick, N., Cappa, S. F., et al. (2000). A cultural effect on brain function. *Nature Neuroscience* 3, 91–96.

Pavlov, I. P. 1927. *Conditioned Reflexes: An Investigation of the Physiological Activity of the Cerebral Cortex*. Oxford University Press.

Pearl, J., and Mackenzie, D. 2018. *The Book of Why: The New Science of Cause and Effect.* Basic Books.

Peer, M., Hayman, M., Tamir, B., and Arzy, S. 2021. Brain coding of social network structure. *Journal of Neuroscience* 41, 4897–4909.

Pellegrino, G. di, Fadiga, L., Fogassi, L., Gallese, V., and Rizzolatti, G. 1992. Understanding motor events: A neurophysiological study. *Experimental Brain Research* 91, 176–180.

Pelphrey, K. A., Singerman, J. D., Allison, T., and McCarthy, G. 2003. Brain activation evoked by perception of gaze shifts: The influence of context. *Neuropsychologia* 41, 156–170.

Perlstein, S., Waller, R., Wagner, N. J., and Saudino, K. J. 2021. Low social affiliation predicts increases in callous-unemotional behaviors in early childhood. *Journal of Child Psychology and Psychiatry* 63, 109–117.

Perner, J., and Roessler, J. 2012. From infants' to children's appreciation of belief. *Trends in Cognitive Sciences* 16, 519–525.

Perner, J., Rendl, B., and Garnham, A. 2007. Objects of desire, thought, and reality: Problems of anchoring discourse referents in development. *Mind & Language* 22, 475–513.

Pessiglione, M., Schmidt, L., Draganski, B., Kalisch, R., Lau, H., Dolan, R. J., and Frith, C. D. 2007. How the brain translates money into force: An fMRI study of subliminal motivation. *Science* 316, 904–906.

Peters, K., Jetten, J., Radova, D., and Austin, K. 2017. Gossiping about deviance: Evidence that deviance spurs the gossip that builds bonds. *Psychological Science* 28, 1610–1619.

Petersen, M. B. 2021. COVID lesson: Trust the public with hard truths. *Nature* 58, 237–237.

Pezzulo, G., Donnarumma, F., and Dindo, H. 2013. Human sensorimotor communication: A theory of signaling in online social interactions. *PLOS One* 8, e79876, https://doi.org/10.1371/journal.pone.0079876.

Phelps, E. A., O'Connor, K. J., Cunningham, W. A., Funayama, E. S., Gatenby, J. C., Gore, J. C., and Banaji, M. R. 2000. Performance on indirect measures of race evaluation predicts amygdala activation. *Journal of Cognitive Neuroscience* 12, 729–738.

Pickering, M. J., and Garrod, S. 2004. Toward a mechanistic psychology of dialogue. *Behavioral and Brain Sciences* 27, 169–190; discussion 190–226.

Pickering, M. J., and Garrod, S. 2007. Do people use language production to make predictions during comprehension? *Trends in Cognitive Sciences* 11, 105–110.

Pierrehumbert, J. B. 2006. The next toolkit. *Journal of Phonetics* 34, 516–530.

Pinker, S. 2002. *The Blank Slate: The Modern Denial of Human Nature*. Viking Penguin.

Pinker, S. 2011. *The Better Angels of Our Nature: The Decline of Violence in History and Its Causes*. Penguin.

Pinto, A., Oates, J., Grutter, A., and Bshary, R. 2011. Cleaner wrasses Labroides dimidiatus are more cooperative in the presence of an audience. *Current Biology* 21, 1140–1144.

Pisella, L., Grea, H., Tilikete, C., Vighetto, A., Desmurget, M., Rode, G., Boisson, D., and Rossetti, Y. 2000. An "automatic pilot" for the hand in human posterior parietal cortex: Toward reinterpreting optic ataxia. *Nature Neuroscience* 3, 729–736.

Pisor, A. C., and Surbeck, M. 2019. The evolution of intergroup tolerance in nonhuman primates and humans. *Evolutionary Anthropology: Issues, News, and Reviews* 28, 210–223.

Poulin-Dubois, D., Lepage, A., and Ferland, D. 1996. Infants' concept of animacy. *Cognitive Development* 11, 19–36.

Powell, L. J., and Spelke, E. S. 2013. Preverbal infants expect members of social groups to act alike. *Proceedings of the National Academy of Sciences* 110, E3965–E3972.

Powell, L. J., and Spelke, E. S. 2018. Human infants' understanding of social imitation: Inferences of affiliation from third party observations. *Cognition* 170, 31–48.

Pratt, S. C., and Sumpter, D. J. 2006. A tunable algorithm for collective decision-making. *Proceedings of the National Academy of Sciences* 103, 15906–15910.

Premack, D., and Woodruff, G. 1978. Does the chimpanzee have a theory of mind? *Behavioural and Brain Sciences* 4, 515–526.

Prinz, W. 1984. Modes of linkage between perception and action. In *Cognition and Motor Processes*, eds. W. Prinz and A. F. Sanders, 185–193. Springer.

Pronin, E., Gilovich, T., and Ross, L. 2004. Objectivity in the eye of the beholder: Divergent perceptions of bias in self versus others. *Psychological Review* 111, 781–799.

Proust, J. 2015. *The Representational Structure of Feelings*. T. Metzinger and J. M. Windt, eds. Open MIND, https://doi.org/10.15502/9783958570047.

Provine, R. R. 1989. Faces as releasers of contagious yawning: An approach to face detection using normal human subjects. *Bulletin of the Psychonomic Society* 27, 211–214.

Puce, A., and Perrett, D. 2003. Electrophysiology and brain imaging of biological motion. *Philosophical Transactions of the Royal Society B: Biological Sciences* 358, 435–445.

Pulvermüller, F., Tomasello, R., Henningsen-Schomers, M. R., and Wennekers, T. 2021. Biological constraints on neural network models of cognitive function. *Nature Reviews Neuroscience* 22, 488–502.

Purcell, B. A., and Kiani, R. 2016. Hierarchical decision processes that operate over distinct timescales underlie choice and changes in strategy. *Proceedings of the National Academy of Sciences* 113, E4531–E4540.

Qin, W., Zhao, L., Compton, B. J., Zheng, Y., Mao, H., Zheng, J., and Heyman, G. D. 2020. Overheard conversations can influence children's generosity. *Developmental Science* 24, e13068, https://doi.org/10.1111/desc.13068.

Quine, W. V. O. 2013. *Word and Object*. MIT Press.

Qureshi, A. W., Apperly, I. A., and Samson, D. 2010. Executive function is necessary for perspective selection, not Level-1 visual perspective calculation: Evidence from a dual-task study of adults. *Cognition* 117, 230–236.

Raafat, R. M., Chater, N., and Frith, C. 2009. Herding in humans. *Trends in Cognitive Sciences* 13, 420–428.

Rabbitt, P. M. 1966a. Error correction time without external error signals. *Nature* 212, 438.

Rabbitt, P. M. 1966b. Errors and error correction in choice-response tasks. *Journal of Experimental Psychology* 71, 264–272.

Radzevick, J. R., and Moore, D. A. 2011. Competing to be certain (but wrong): Market dynamics and excessive confidence in judgment. *Management Science* 57, 93–106.

Rafetseder, E., O'Brien, C., Leahy, B., and Perner, J. 2021. Extended difficulties with counterfactuals persist in reasoning with false beliefs: Evidence for teleology-in-perspective. *Journal of Experimental Child Psychology* 204, 105058, https://doi.org/10.1016/j.jecp.2020.105058.

Raihani, N. 2021. *The Social Instinct: How Cooperation Shaped the World*. Jonathan Cape.

Raihani, N. J. 2014. Hidden altruism in a real-world setting. *Biology Letters* 10, 20130884, http://doi.org/10.1098/rsbl.2013.0884.

Raihani, N. J., and Power, E. A. 2021. No good deed goes unpunished: The social costs of prosocial behaviour. *Evolutionary Human Sciences* 3, e40, https://doi.org/10.1017/ehs.2021.35.

Rakoczy, H., and Schmidt, M. F. H. 2013. The early ontogeny of social norms. *Child Development Perspectives* 7, 17–21.

Ramezanpour, H., and Thier, P. 2020. Decoding of the other's focus of attention by a temporal cortex module. *Proceedings of the National Academy of Sciences* 117, 2663–2670.

Ramsey, R., Kaplan, D. M., and Cross, E. S. 2021. Watch and learn: The cognitive neuroscience of learning from others' actions. *Trends in Neurosciences* 44, 478–491.

Rand, D. G. 2016. Cooperation, fast and slow: Meta-analytic evidence for a theory of social heuristics and self-interested deliberation. *Psychological Science* 27, 1171–1180.

Rand, D. G., Fudenberg, D., and Dreber, A. 2015. It's the thought that counts: The role of intentions in noisy repeated games. *Journal of Economic Behavior & Organization* 116, 481–499.

Rao, R. P. N., and Ballard, D. H. 1999. Predictive coding in the visual cortex: A functional interpretation of some extra-classical receptive-field effects. *Nature Neuroscience* 2, 79–87.

Rastle, K., Lally, C., Davis, M. H., and Taylor, J. S. H. 2021. The dramatic impact of explicit instruction on learning to read in a new writing system. *Psychological Science* 32, 471–484.

Rawls, J. 2020. *A Theory of Justice*. Harvard University Press.

Reber, R., Winkielman, P., and Schwarz, N. 1998. Effects of perceptual fluency on affective judgments. *Psychological Science* 9, 45–48.

Recanati, F. 2002. Does linguistic communication rest on inference? *Mind & Language* 17, 105–126.

Redcay, E., and Schilbach, L. 2019. Using second-person neuroscience to elucidate the mechanisms of social interaction. *Nature Reviews Neuroscience* 20, 495–505.

Reed, N., McLeod, P., and Dienes, Z. 2010. Implicit knowledge and motor skill: What people who know how to catch don't know. *Consciousness and Cognition* 19, 63–76.

Reiter, A. M. F., Moutoussis, M., Vanes, L., Kievit, R., Bullmore, E. T., Goodyer, I. M., et al. 2021. Preference uncertainty accounts for developmental effects on susceptibility to peer influence in adolescence. *Nature Communications* 12, 3823, https://doi.org/10.1038/s41467-021-23671-2.

Rendell, L., Boyd, R., Cownden, D., Enquist, M., Eriksson, K., Feldman, M. W., et al. 2010. Why copy others? Insights from the social learning strategies tournament. *Science* 328, 208–213.

Rendell, L., Fogarty, L., Hoppitt, W. J., Morgan, T. J., Webster, M. M., and Laland, K. N. 2011. Cognitive culture: Theoretical and empirical insights into social learning strategies. *Trends in Cognitive Sciences* 15, 68–76.

Reynolds, C. W. 1987. Flocks, herds, and schools: A distributed behavioral model. *Computer Graphics* 21, 25–34.

Richardson, D. C., and Dale, R. 2005. Looking to understand: The coupling between speakers' and listeners' eye movements and its relationship to discourse comprehension. *Cognitive Science* 29, 1045–1060.

Richardson, D. C., Dale, R., and Kirkham, N. Z. 2007. The art of conversation is coordination: Common ground and the coupling of eye movements during dialogue. *Psychological Science* 18, 407–413.

Richardson, H., and Saxe, R. 2020. Early signatures of and developmental change in brain regions for theory of mind. In *Neural Circuit and Cognitive Development, Second Edition*, eds. J. Rubenstein, P. Rakic, B. Chen, and K. Y. Kwan, 467–484. Academic Press.

Richerson, P. J., and Boyd, R. 2005. *Not by Genes Alone: How Culture Transformed Human Evolution*. University of Chicago Press.

Rieucau, G., and Giraldeau, L. A. 2011. Exploring the costs and benefits of social information use: An appraisal of current experimental evidence. *Philosophical Transactions of the Royal Society B: Biological Sciences* 366, 949–957.

Rilling, J. K., Sanfey, A. G., Aronson, J. A., Nystrom, L. E., and Cohen, J. D. 2004. The neural correlates of theory of mind within interpersonal interactions. *NeuroImage* 22, 1694–1703.

Rizzolatti, G., and Craighero, L. 2004. The mirror-neuron system. *Annual Review of Neuroscience* 27, 169–192.

Rizzolatti, G., and Destro, M. F. 2008. Mirror neurons. *Scholarpedia* 3, 2055, https://doi.org/10.4249/scholarpedia.2055.

Rizzolatti, G., and Fabbri-Destro, M. 2008. The mirror system and its role in social cognition. *Current Opinion in Neurobiology* 18(2), 179–184.

Rizzolatti, G., Camarda, R., Fogassi, L., Gentilucci, M., Luppino, G., and Matelli, M. 1988. Functional organization of inferior area 6 in the macaque monkey. *Experimental Brain Research* 71, 491–507.

Rizzolatti, G., Fadiga, L., Gallese, V., and Fogassi, L. 1996. Premotor cortex and the recognition of motor actions. *Brain Research* 3, 131–141.

Robalino, N., and Robson, A. 2012. The economic approach to "theory of mind." *Philosophical Transactions of the Royal Society B: Biological Sciences* 367, 2224–2233.

Robbins, J., and Rumsey, A. 2008. Introduction—cultural and linguistic anthropology and the opacity of other minds. *Anthropological Quarterly* 81, 407–420.

Roberts, G., Raihani, N., Bshary, R., Manrique, H. M., Farina, A., Samu, F., and Barclay, P. 2021. The benefits of being seen to help others: Indirect reciprocity and reputation-based partner choice. *Philosophical Transactions of the Royal Society B: Biological Sciences* 376, 20200290.

Roberts, I. D., Teoh, Y. Y., and Hutcherson, C. A. 2019. Oxytocin and the altruistic "Goldilocks zone." *Nature Neuroscience* 22, 510–512.

Roberts, S. O., Gelman, S. A., and Ho, A. K. 2017. So it is, so it shall be: Group regularities license children's prescriptive judgments. *Cognitive Science* 41, 576–600.

Robinson, E. J., Champion, H., and Mitchell, P. 1999. Children's ability to infer utterance veracity from speaker informedness. *Developmental Psychology* 35, 535–546.

Rockenbach, B., and Milinski, M. 2006. The efficient interaction of indirect reciprocity and costly punishment. *Nature* 444, 718–723.

Roelofs, K., Minelli, A., Mars, R. B., van Peer, J., and Toni, I. 2009. On the neural control of social emotional behavior. *Social Cognitive and Affective Neuroscience* 4, 50–58.

Roepstorff, A., and Frith, C. 2004. What's at the top in the top-down control of action? Script-sharing and "top-top" control of action in cognitive experiments. *Psychological Research* 68, 189–198.

Rolnick, J., and Parvizi, J. 2011. Automatisms: Bridging clinical neurology with criminal law. *Epilepsy & Behavior* 20, 423–427.

Roseberry, S., Hirsh-Pasek, K., and Golinkoff, R. M. 2014. Skype me! Socially contingent interactions help toddlers learn language. *Child Development* 85, 956–970.

Rosedahl, L. A., Serota, R., and Ashby, F. G. 2021. When instructions don't help: Knowing the optimal strategy facilitates rule-based but not information-integration category learning. *Journal of Experimental Psychology: Human Perception and Performance* 47, 1226–1236.

Rosenthal, D. M. 2008. Consciousness and its function. *Neuropsychologia* 46, 829–840.

Roumazeilles, L., Schurz, M., Lojkiewiez, M., Verhagen, L., Schüffelgen, U., Marche, K., et al. 2021. Social prediction modulates activity of macaque superior temporal cortex. *Science Advances* 7, eabh2392, https://doi.org/10.1126/sciadv.abh2.

Rozenblit, L., and Keil, F. 2002. The misunderstood limits of folk science: An illusion of explanatory depth. *Cognitive Science* 26, 521–562.

Ruff, C. C., Ugazio, G., and Fehr, E. 2013. Changing social norm compliance with noninvasive brain stimulation. *Science* 342, 482–484.

Rule, N. O., Krendl, A. C., Ivcevic, Z., and Ambady, N. 2013. Accuracy and consensus in judgments of trustworthiness from faces: Behavioral and neural correlates. *Journal of Personality and Social Psychology* 104, 409–426.

Rulkov, N. F., Sushchik, M. M., Tsimring, L. S., and Abarbanel, H. D. I. 1995. Generalized synchronization of chaos in directionally coupled chaotic systems. *Physical Review E* 51, 980–994.

Rumble, A. C., van Lange, P. A. M., and Parks, C. D. 2010. The benefits of empathy: When empathy may sustain cooperation in social dilemmas. *European Journal of Social Psychology* 40, 856–866.

Sabbagh, M. A., and Baldwin, D. A. 2001. Learning words from knowledgeable versus ignorant speakers: Links between preschoolers' theory of mind and semantic development. *Child Development* 72, 1054–1070.

Sacheli, L. M., Tidoni, E., Pavone, E. F., Aglioti, S. M., and Candidi, M. 2013. Kinematics fingerprints of leader and follower role-taking during cooperative joint actions. *Experimental Brain Research* 226, 473–486.

Saffran, J. R. 2003. Statistical language learning: Mechanisms and constraints. *Current Directions in Psychological Science* 12, 110–114.

Sakai, K., and Passingham, R. E. 2003. Prefrontal interactions reflect future task operations. *Nature Neuroscience* 6, 75–81.

Saldana, C., Kirby, S., Truswell, R., and Smith, K. 2019. Compositional hierarchical structure evolves through cultural transmission: An experimental study. *Journal of Language Evolution* 4, 83–107.

Samson, D., Apperly, I. A., Braithwaite, J. J., Andrews, B. J., and Bodley Scott, S. E. 2010. Seeing it their way: Evidence for rapid and involuntary computation of what other people see. *Journal of Experimental Psychology: Human Perception and Performance* 36, 1255–1266.

Sanborn, A. N., and Chater, N. 2016. Bayesian brains without probabilities. *Trends in Cognitive Sciences* 20, 883–893.

Sanborn, A. N., Mansinghka, V. K., and Griffiths, T. L. 2013. Reconciling intuitive physics and Newtonian mechanics for colliding objects. *Psychological Review* 120, 411–437.

Sarafyazd, M., and Jazayeri, M. 2019. Hierarchical reasoning by neural circuits in the frontal cortex. *Science* 364, eaav8911, https://doi.org/10.1126/science.aav8911.

Sarkissian, H., Chatterjee, A., De Brigard, F., Knobe, J., Nichols, S., and Sirker, S. 2010. Is belief in free will a cultural universal? *Mind & Language* 25, 346–358.

Sauciuc, G.-A., Zlakowska, J., Persson, T., Lenninger, S., and Alenkaer Madsen, E. 2020. *PLOS One* 15, e0232717, https://doi.org/10.1371/journal.pone.0232717.

Saxe, R., and Kanwisher, N. 2003. People thinking about thinking people: The role of the temporo-parietal junction in "theory of mind." *NeuroImage* 19, 1835–1842.

Saxe, R., Xiao, D. K., Kovacs, G., Perrett, D. I., and Kanwisher, N. 2004. A region of right posterior superior temporal sulcus responds to observed intentional actions. *Neuropsychologia* 42, 1435–1446.

Schafer, M., and Schiller, D. 2018. Navigating social space. *Neuron* 100, 476–489.

Schaller, G. B. 2009. *The Serengeti Lion: A Study of Predator-Prey Relations*. University of Chicago Press.

Schaumberg, R. L., and Skowronek, S. E. 2022. Shame broadcasts social norms: The positive social effects of shame on norm acquisition and normative behavior. *Psychological Science* 33, 1257–1277.

Schenk, T. 2006. An allocentric rather than perceptual deficit in patient D. F. *Nature Neuroscience* 9, 1369–1370.

Schlottmann, A., and Surian, L. 1999. Do 9-month-olds perceive causation-at-a-distance? *Perception* 28, 1105–1113.

Schmidt, M. F. H., Rakoczy, H., and Tomasello, M. 2012. Young children enforce social norms selectively depending on the violator's group affiliation. *Cognition* 124, 325–333.

Schneider, W., and Löffler, E. 2016. The development of metacognitive knowledge in children and adolescents. In *The Oxford Handbook of Metamemory*, eds. J. Dunlosky and S. K. Tauber, 491–518. Oxford University Press.

Scholl, B. J. 2005. Innateness and (Bayesian) visual perception: Reconciling nativism and development. In *The Innate Mind: Structure and Contents*, eds. P. Carruthers, S. Laurence, and S. Stich, 34–52. Oxford University Press.

Scholl, B. J., and Tremoulet, P. D. 2000. Perceptual causality and animacy. *Trends in Cognitive Sciences* 4, 299–309.

Schug, M. G., Shusterman, A., Barth, H., and Patalano, A. L. 2013. Minimal-group membership influences children's responses to novel experience with group members. *Developmental Science* 16, 47–55.

Schultz, J., and Bülthoff, H. H. 2013. Parametric animacy percept evoked by a single moving dot mimicking natural stimuli. *Journal of Vision* 13, 15.

Schultz, J., and Frith, C. D. (2022) Animacy and the prediction of behaviour. *Neuroscience and Biobehavioral Reviews* 140, 104766, https://doi.org/10.1016/j.neubiorev.2022.104766.

Schultz, W., Dayan, P., and Montague, P. R. 1997. A neural substrate of prediction and reward. *Science* 275, 1593–1599.

Schulze, C., and Hertwig, R. 2021. A description–experience gap in statistical intuitions: Of smart babies, risk-savvy chimps, intuitive statisticians, and stupid grown-ups. *Cognition* 210, 104580.

Schürmann, M., Hesse, M. D., Stephan, K. E., Saarela, M., Zilles, K., Hari, R., and Fink, G. R. 2005. Yearning to yawn: The neural basis of contagious yawning. *NeuroImage* 24, 1260–1264.

Schurz, M., Radua, J., Tholen, M. G., Maliske, L., Margulies, D. S., Mars, R. B., et al. 2021. Toward a hierarchical model of social cognition: A neuroimaging meta-analysis and integrative review of empathy and theory of mind. *Psychological Bulletin* 147, 293–327.

Scott-Phillips, T. C. 2008. Defining biological communication. *Journal of Evolutionary Biology* 21, 387–395.

Scott-Phillips, T. C., Blythe, R. A., Gardner, A., and West, S. A. 2012. How do communication systems emerge? *Proceedings of the Royal Society B: Biological Sciences* 279, 1943–1949.

Scott, S. K., Lavan, N., Chen, S., and McGettigan, C. 2014. The social life of laughter. *Trends in Cognitive Sciences* 18, 618–620.

Searle, J. R. 1995. *The Construction of Social Reality*. Simon and Schuster.

Sebanz, N., and Knoblich, G. 2021. Progress in joint-action research. *Current Directions in Psychological Science* 30, 138–143.

Sebanz, N., Bekkering, H., and Knoblich, G. 2006. Joint action: Bodies and minds moving together. *Trends in Cognitive Sciences* 10, 70–76.

Sebanz, N., Knoblich, G., and Prinz, W. 2003. Representing others' actions: Just like one's own? *Cognition* 88, B11–B21.

Seeley, T. D. 1983. Division of labor between scouts and recruits in honeybee foraging. *Behavioral Ecology and Sociobiology* 12, 253–259.

Seeley, T. D. 2010. *Honeybee Democracy*. Princeton University Press.

Seeley, T. D., and Buhrman, S. C. 2001. Nest-site selection in honey bees: How well do swarms implement the "best-of-N" decision rule? *Behavioral Ecology and Sociobiology* 49, 416–427.

Seemann, A. 2019. Reminiscing together: Joint experiences, epistemic groups, and sense of self. *Synthese* 196, 4813–4828.

Semendeferi, K., Armstrong, E., Schleicher, A., Zilles, K., and Van Hoesen, G. W. 2001. Prefrontal cortex in humans and apes: A comparative study of area 10. *American Journal of Biological Anthropology* 114, 224–241.

Senju, A., and Csibra, G. 2008. Gaze following in human infants depends on communicative signals. *Current Biology* 18, 668–671.

Senju, A., Southgate, V., White, S., and Frith, U. 2009. Mindblind eyes: An absence of spontaneous theory of mind in Asperger syndrome. *Science* 325, 883–885.

Shanahan, M. 2012. The brain's connective core and its role in animal cognition. *Philosophical Transactions of the Royal Society B: Biological Sciences* 367(1603): 2704–2714.

Shariff, A. F., Greene, J. D., Karremans, J. C., Luguri, J. B., Clark, C. J., Schooler, J. W., et al. 2014. Free will and punishment: A mechanistic view of human nature reduces retribution. *Psychological Science* 25, 1563–1570.

Shea, N. 2018. *Representation in Cognitive Science*. Oxford University Press.

Shea, N., and Frith, C. D. 2016. Dual-process theories and consciousness: The case for "type zero" cognition. *Neuroscience of Consciousness* 2016, niw005, https://doi.org/10.1093/nc/niw005.

Shea, N. J., Boldt, A., Bang, D., Yeung, N., Heyes, C., and Frith, C. D. 2014. Supra-personal cognitive control and metacognition. *Trends in Cognitive Sciences* 18, 186–193.

Sheng, F., and Han, S. 2012. Manipulations of cognitive strategies and intergroup relationships reduce the racial bias in empathic neural responses. *NeuroImage* 61, 786–797.

Shepherd, J. 2012. Free will and consciousness: Experimental studies. *Consciousness and Cognition* 21, 915–927.

Sherif, M., and Sherif, C. W. 1956. *An Outline of Social Psychology*. Harper & Bros.

Sherman, B. E., Graves, K. N., and Turk-Browne, N. B. 2020. The prevalence and importance of statistical learning in human cognition and behavior. *Current Opinion in Behavioral Sciences* 32, 15–20.

Shettleworth, S. J. 2010. *Cognition, Evolution, and Behavior*. Oxford University Press.

Shoda, Y., Mischel, W., and Peake, P. 1990. Predicting adolescent cognitive and social competence from preschool delay of gratification: Identifying diagnostic conditions. *Developmental Psychology* 26, 978–986.

Shteynberg, G. 2018. A collective perspective: Shared attention and the mind. *Current Opinion in Psychology* 23, 93–97.

Silani, G., Bird, G., Brindley, R., Singer, T., Frith, C., and Frith, U. 2008. Levels of emotional awareness and autism: An fMRI study. *Social Neuroscience* 3, 97–112.

Silbert, L. J., Honey, C. J., Simony, E., Poeppel, D., and Hasson, U. 2014. Coupled neural systems underlie the production and comprehension of naturalistic narrative speech. *Proceedings of the National Academy of Sciences* 111, E4687–E4696.

Sime, J. D. 1983. Affiliative behaviour during escape to building exits. *Journal of Environmental Psychology* 3, 21–41.

Singer, T., and Klimecki, O. M. 2014. Empathy and compassion. *Current Biology* 24, R875–R878.

Singer, T., Kiebel, S. J., Winston, J. S., Dolan, R. J., and Frith, C. D. 2004. Brain responses to the acquired moral status of faces. *Neuron* 41, 653–662.

Singer, T., Seymour, B., O'Doherty, J., Kaube, H., Dolan, R. J., and Frith, C. D. 2004. Empathy for pain involves the affective but not sensory components of pain. *Science* 303, 1157–1162.

Skinner, B. F. 1938. *The Behavior of Organisms: An Experimental Analysis*. Appleton-Century.

Skyrms, B. 2003. *The Stag Hunt and the Evolution of Social Structure*. Cambridge University Press.

Sliwa, J., and Freiwald, W. A. 2017. A dedicated network for social interaction processing in the primate brain. *Science* 356, 745–749.

Sloman, S., and Fernbach, P. (2018). *The Knowledge Illusion: Why We Never Think Alone*. Penguin.

Smilansky, S. 2002. Free will, fundamental dualism, and the centrality of illusion. In *The Oxford Handbook of Free Will*, ed. R. Kane, 425–441. Oxford University Press.

Smith, A. 1759. *The Theory of Moral Sentiments*. Liberty Classics, 1982.

Smith, D., Schlaepfer, P., Major, K., Dyble, M., Page, A. E., Thompson, J., et al. 2017. Cooperation and the evolution of hunter-gatherer storytelling. *Nature Communications* 8, 1853, https://doi.org/10.1038/s41467-017-02036-8.

Smith, M. L., Asada, N., and Malenka, R. C. 2021. Anterior cingulate inputs to nucleus accumbens control the social transfer of pain and analgesia. *Science* 371, 153.

Sober, E., and Wilson, D. S. 1998. *Unto Others*. Harvard University Press.

Soltani, A., and Izquierdo, A. 2019. Adaptive learning under expected and unexpected uncertainty. *Nature Reviews Neuroscience* 20, 635–644.

Sommerfeld, R. D., Krambeck, H. J., and Milinski, M. 2008. Multiple gossip statements and their effect on reputation and trustworthiness. *Proceedings of the Royal Society B: Biological Sciences* 275, 2529–2536.

Sommerfeld, R. D., Krambeck, H. J., Semmann, D., and Milinski, M. 2007. Gossip as an alternative for direct observation in games of indirect reciprocity. *Proceedings of the National Academy of Sciences* 104, 17435–17440.

Song, H.-J., Onishi, K. H., Baillargeon, R., and Fisher, C. 2008. Can an agent's false belief be corrected by an appropriate communication? Psychological reasoning in 18-month-old infants. *Cognition* 109, 295–315.

Sorrel, W. E. 1978. Cults and cult suicide. *International Journal of Group Tensions* 8, 96–105.

Southgate, V. 2020. Are infants altercentric? The other and the self in early social cognition. *Psychological Review* 127, 505–523.

Southgate, V., Senju, A., and Csibra, G. 2007. Action anticipation through attribution of false belief by 2-year-olds. *Psychological Science* 18, 587–592.

Sparenberg, P., Topolinski, S., Springer, A., and Prinz, W. 2012. Minimal mimicry: Mere effector matching induces preference. *Brain and Cognition* 80, 291–300.

Sperber, D. 1996. *Explaining Culture: A Naturalistic Approach*. Wiley-Blackwell.

Sperber, D. 2020. Why a deep understanding of cultural evolution is incompatible with shallow psychology. In *Roots of Human Sociality: Culture, Cognition and Interaction*, eds. N. J. Enfield and S. C. Levinson, 431–449. Routledge.

Sperber, D., and Wilson, D. 1986. *Relevance: Communication and Cognition.* Blackwell.

Sperber, D., Clement, F., Heintz, C., Mascaro, O., Mercier, H., Origgi, G., and Wilson, D. 2010. Epistemic vigilance. *Mind & Language* 25, 359–393.

Sroufe, L. A. 1977. Wariness of strangers and the study of infant development. *Child Development* 48, 731–746.

Stanley, J., Gowen, E., and Miall, R. C. 2010. How instructions modify perception: An fMRI study investigating brain areas involved in attributing human agency. *NeuroImage* 52, 389–400.

Stasser, G., and Titus, W. 1985. Pooling of unshared information in group decision-making: Biased information sampling during discussion. *Journal of Personality and Social Psychology* 48, 1467–1478.

Stel, M., and Harinck, F. 2010. Being mimicked makes you a prosocial voter. *Journal of Experimental Psychology* 58, 79–84.

Stel, M., van Baaren, R. B., Blascovich, J., van Dijk, E., McCall, C., Pollmann, M. M., et al. 2010. Effects of a priori liking on the elicitation of mimicry. *Journal of Experimental Psychology* 57, 412–418.

Stel, M., van Dijk, E., and Olivier, E. 2009. You want to know the truth? Then don't mimic! *Psychological Science* 20, 693–699.

Stenzel, A., Dolk, T., Colzato, L. S., Sellaro, R., Hommel, B., and Liepelt, R. 2014. The joint Simon effect depends on perceived agency, but not intentionality, of the alternative action. *Frontiers in Human Neuroscience* 8, 595, https://doi.org/10.3389/fnhum.2014.00595.

Stephens, D. W., Brown, J. S., and Ydenberg, R. C. 2008. *Foraging: Behavior and ecology.* University of Chicago Press.

Stephenson, L. J., Edwards, S. G., and Bayliss, A. P. 2021. From gaze perception to social cognition: The shared-attention system. *Perspectives on Psychological Science* 16, 553–576.

Stirrat, M., and Perrett, D. I. 2010. Valid facial cues to cooperation and trust: Male facial width and trustworthiness. *Psychological Science* 21, 349–354.

Stirrat, M., and Perrett, D. I. 2012. Face structure predicts cooperation. *Psychological Science* 23, 718–722.

Stolk, A., Noordzij, M. L., Verhagen, L., Volman, I., Schoffelen, J.-M., Oostenveld, R., et al. 2014. Cerebral coherence between communicators marks the emergence of meaning. *Proceedings of the National Academy of Sciences* 111, 18183.

Sul, S., Tobler, P. N., Hein, G., Leiberg, S., Jung, D., Fehr, E., and Kim, H. 2015. Spatial gradient in value representation along the medial prefrontal cortex reflects individual differences in prosociality. *Proceedings of the National Academy of Sciences* 112, 7851–7856.

Sulik, J., Bahrami, B., and Deroy, O. 2021. The diversity gap: When diversity matters for knowledge. *Perspectives on Psychological Science* 17, 752–767.

Sumpter, D. J. 2010. *Collective Animal Behaviour.* Princeton University Press.

Surtees, A., Apperly, I., and Samson, D. 2016. I've got your number: Spontaneous perspective-taking in an interactive task. *Cognition* 150, 43–52.

Susskind, J. M., Lee, D. H., Cusi, A., Feiman, R., Grabski, W., and Anderson, A. K. 2008. Expressing fear enhances sensory acquisition. *Nature Neuroscience* 11, 843–850.

Sutton, R. S., and Barto, A. G. 1998. *Reinforcement Learning: An Introduction.* MIT Press.

Swets, J. A., Tanner, W. P. J., and Birdsall, T. G. 1961. Decision processes in perception. *Psychological Review* 68, 301–340.

Sylwester, K., and Roberts, G. 2010. Cooperators benefit through reputation-based partner choice in economic games. *Biology Letters* 6, 659–662.

Sznycer, D., Xygalatas, D., Agey, E., Alami, S., An, X.-F., Ananyeva, K. I., et al. 2018. Cross-cultural invariances in the architecture of shame. *Proceedings of the National Academy of Sciences* 115, 9702–9707.

Tausch, A., and Heshmati, A. 2009. Asabiyya: Re-interpreting value change in globalized societies. *IZA Discussion Paper No. 4459*, https://ssrn.com/abstract=1484630.

Tauzin, T., and Gergely, G. 2018. Communicative mind-reading in preverbal infants. *Scientific Reports* 8, 9534, https://doi.org/10.1038/s41598-018-27804-4.

Tauzin, T., and Gergely, G. 2019. Variability of signal sequences in turn-taking exchanges induces agency attribution in 10.5-mo-olds. *Proceedings of the National Academy of Sciences* 116, 15441–15446.

Tavares, Rita M., Mendelsohn, A., Grossman, Y., Williams, Christian H., Shapiro, M., Trope, Y., and Schiller, D. 2015. A map for social navigation in the human brain. *Neuron* 87, 231–243.

Tazelaar, M. J. A., van Lange, P. A. M., and Ouwerkerk, J. W. 2004. How to cope with "noise," in social dilemmas: The benefits of communication. *Journal of Personality and Social Psychology* 87, 845–859.

Tenney, E. R., MacCoun, R. J., Spellman, B. A., and Hastie, R. 2007. Calibration trumps confidence as a basis for witness credibility. *Psychological Science* 18, 46–50.

Tennie, C., Frith, U., and Frith, C. D. 2010. Reputation management in the age of the world-wide web. *Trends in Cognitive Sciences* 14, 482–488.

Thaler, R. H., and Sunstein, C. R. 2008. *Nudge: Improving Decisions about Health, Wealth, and Happiness.* Yale University Press.

Thomas, A. J., Saxe, R., and Spelke, E. S. 2022. Infants infer potential social partners by observing the interactions of their parent with unknown others. *Proceedings of the National Academy of Sciences* 119, e2121390119, https://doi.org/10.1073/pnas.2121390119.

Thorndike, E. L. 1898. Animal intelligence: An experimental study of the associative processes in animals. *Psychological Review: Monograph Supplements* 2, i–109.

Thornton, A., and Clutton-Brock, T. 2011. Social learning and the development of individual and group behaviour in mammal societies. *Philosophical Transactions of the Royal Society B: Biological Sciences* 366, 978–987.

Thornton, A., and McAuliffe, K. 2006. Teaching in wild meerkats. *Science* 313, 227–229.

Ting, F., He, Z., and Baillargeon, R. 2019. Toddlers and infants expect individuals to refrain from helping an ingroup victim's aggressor. *Proceedings of the National Academy of Sciences* 116, 6025–6034.

Todorov, A., Olivola, C. Y., Dotsch, R., and Mende-Siedlecki, P. 2015. Social attributions from faces: Determinants, consequences, accuracy, and functional Significance. *Annual Review of Psychology* 66, 519–545.

Todorov, A., Pakrashi, M., and Oosterhof, N. N. 2009. Evaluating faces on trustworthiness after minimal time exposure. *Social Cognition* 27, 813–833.

Tomasello, M. 2010. *Origins of Human Communication.* MIT Press.

Tomasello, M., and Carpenter, M. 2007. Shared intentionality. *Developmental Science* 10, 121–125.

Tomasello, M., and Vaish, A. 2013. Origins of human cooperation and morality. *Annual Review of Psychology* 64, 231–255.

Tomasello, M., Carpenter, M., and Liszkowski, U. 2007. A new look at infant pointing. *Child Development* 78, 705–722.

Tompkins, V., Benigno, J. P., Kiger Lee, B., and Wright, B. M. 2018. The relation between parents' mental state talk and children's social understanding: A meta-analysis. *Social Development* 27, 223–246.

Travers, E., Fairhurst, M. T., and Deroy, O. 2020. Racial bias in face perception is sensitive to instructions but not introspection. *Consciousness and Cognition* 83, 102952, https://doi.org/10.1016/j.concog.2020.102952.

Treadway, M. T., Buckholtz, J. W., Martin, J. W., Jan, K., Asplund, C. L., Ginther, M. R., et al. 2014. Corticolimbic gating of emotion-driven punishment. *Nature Neuroscience* 17, 1270–1275.

Tremoulet, P. D., and Feldman, J. 2000. Perception of animacy from the motion of a single object. *Perception* 29, 943–951.

Trivers, R. 2000. The elements of a scientific theory of self-deception. *Annals of the New York Academy of Sciences* 907, 114–131.

Trouche, E., Sander, E., and Mercier, H. 2014. Arguments, more than confidence, explain the good performance of reasoning groups. *Journal of Experimental Psychology: General* 143, 1958–1971.

Trouche, E., Johansson, P., Hall, L., and Mercier, H. 2016. The selective laziness of reasoning. *Cognitive Science* 40, 2122–2136.

Truesdell, C. A. 1966. Thermodynamics of visco-elasticity. In *Six Lectures on Modern Natural Philosophy,* 35–52. Springer-Verlag.

Trujillo, J. P., Simanova, I., Bekkering, H., and Ozyurek, A. 2018. Communicative intent modulates production and comprehension of actions and gestures: A Kinect study. *Cognition* 180, 38–51.

Tsai, C. C., and Brass, M. 2007. Does the human motor system simulate Pinocchio's actions? Coacting with a human hand versus a wooden hand in a dyadic interaction. *Psychological Science* 18, 1058–1062.

Tunçgenç, B., El Zein, M., Sulik, J., Newson, M., Zhao, Y., Dezecache, G., and Deroy, O. 2021. Social influence matters: We follow pandemic guidelines most when our close circle does. *British Journal of Psychology* 112, 763–780.

Tuomela, R. 2006. Joint intention, We-mode and I-mode. In *Midwest Studies in Philosophy: Shared Intentions and Collective Responsibility*, eds. P. A. French and H. K. Wettstein, 35–58. Wiley-Blackwell.

Tversky, A., and Kahneman, D. 1974. Judgment under uncertainty: Heuristics and biases. *Science* 185, 1124–1131.

Ule, A., Schram, A., Riedl, A., and Cason, T. N. 2009. Indirect punishment and generosity toward strangers. *Science* 326, 1701–1704.

Ullsperger, M., Volz, K. G., and von Cramon, D. Y. 2004. A common neural system signaling the need for behavioral changes. *Trends in Cognitive Sciences* 8, 445–446.

Umiltà, M. A., Escola, L., Intskirveli, I., Grammont, F., Rochat, M., Caruana, F., et al. 2008. When pliers become fingers in the monkey motor system. *Proceedings of the National Academy of Sciences* 105, 2209–2213.

Urgen, B. A., and Saygin, A. P. 2020. Predictive processing account of action perception: Evidence from effective connectivity in the action observation network. *Cortex* 128, 132–142.

Uzzi, B., Amaral, L. A. N., and Reed-Tsochas, F. 2007. Small-world networks and management science research: A review. *European Management Review* 4, 77–91.

Vale, G. L., Flynn, E. G., Kendal, J., Rawlings, B., Hopper, L. M., Schapiro, S. J., et al. 2017. Testing differential use of payoff-biased social learning strategies in children and chimpanzees. *Proceedings of the Royal Society B: Biological Sciences* 284, 20171751, http://doi.org/10.1098/rspb.2017.1751.

Vallortigara, G. 2021. *Born Knowing: Imprinting and the Origins of Knowledge.* MIT Press.

Vallortigara, G., Regolin, L., and Marconato, F. 2005. Visually inexperienced chicks exhibit spontaneous preference for biological motion patterns. *PLoS Biology* 3, e208, https://doi.org/10.1371/journal.pbio.0030208.

van Baaren, R., Janssen, L., Chartrand, T. L., and Dijksterhuis, A. 2009. Where is the love? The social aspects of mimicry. *Philosophical Transactions of the Royal Society B: Biological Sciences* 364, 2381–2389.

van Baaren, R. B., Holland, R. W., Kawakami, K., and van Knippenberg, A. 2004. Mimicry and prosocial behavior. *Psychological Science* 15, 71–74.

van Bergen, Y., Coolen, I., and Laland, K. N. 2004. Nine-spined sticklebacks exploit the most reliable source when public and private information conflict. *Proceedings of the Royal Society B: Biological Sciences* 271, 957–962.

van den Berg, R., Zylberberg, A., Kiani, R., Shadlen, M. N., and Wolpert, D. M. 2016. Confidence is the bridge between multi-stage decisions. *Current Biology* 26, 3157–3168.

van der Hulst, H., ed. 2010. *Recursion and Human Language*. Walter de Gruyter.

van der Plas, E., Mason, D., Livingston, L. A., Craigie, J., Happé, F., and Fleming, S. M. 2021. Computations of confidence are modulated by mentalizing ability. *PsyArXiv*, July 7, https://doi.org/10.31234/osf.io/c4pzj.

van der Wel, R. R. D., and Fu, E. 2015. Entrainment and task co-representation effects for discrete and continuous action sequences. *Psychonomic Bulletin & Review* 22, 1685–1691.

van Lange, P. A. M., and Columbus, S. 2021. Vitamin S: Why is social contact, even with strangers, so important to well-being? *Current Directions in Psychological Science* 30, 267–273.

van Lange, P. A. M., Otten, W., DeBruin, E. M. N., and Joireman, J. A. 1997. Development of prosocial, individualistic, and competitive orientations: Theory and preliminary evidence. *Journal of Personality and Social Psychology* 73, 733–746.

van Lange, P. A. M., Ouwerkerk, J. W., and Tazelaar, M. J. A. 2002. How to overcome the detrimental effects of noise in social interaction: The benefits of generosity. *Journal of Personality and Social Psychology* 82, 768–780.

van Leeuwen, M. L., van Baaren, R. B., Martin, D., Dijksterhuis, A., and Bekkering, H. 2009. Executive functioning and imitation: Increasing working memory load facilitates behavioural imitation. *Neuropsychologia* 47, 3265–3270.

Van Overwalle, F., and Vandekerckhove, M. 2013. Implicit and explicit social mentalizing: Dual processes driven by a shared neural network. *Frontiers in Human Neuroscience* 7, 560, https://doi.org/10.3389/fnhum.2013.00560.

van Schie, H. T., van Waterschoot, B. M., and Bekkering, H. 2008. Understanding action beyond imitation: Reversed compatibility effects of action observation in imitation and joint action. *Journal of Experimental Psychology: Human Perception and Performance* 34, 1493–1500.

Vesper, C., and Richardson, M. J. 2014. Strategic communication and behavioral coupling in asymmetric joint action. *Experimental Brain Research* 232, 2945–2956.

Vesper, C., Abramova, E., Bütepage, J., Ciardo, F., Crossey, B., Effenberg, A., et al. 2017. Joint action: Mental representations, shared information and general mechanisms for coordinating with others. *Frontiers in Psychology* 7, 2039, https://doi.org/10.3389/fpsyg.2016.02039.

Vesper, C., Van Der Wel, R. P. R. D., Knoblich, G., and Sebanz, N. 2011. Making oneself predictable: Reduced temporal variability facilitates joint action coordination. *Experimental Brain Research* 211, 517–530.

Vettin, J., and Todt, D. 2004. Laughter in conversation: Features of occurrence and acoustic structure. *Journal of Nonverbal Behavior* 28, 93–115.

Vicary, S., Sperling, M., von Zimmermann, J., Richardson, D. C., and Orgs, G. 2017. Joint action aesthetics. *PLoS One* 12, e0180101, https://doi.org/10.1371/journal.pone.0180101.

Viding, E., and McCrory, E. 2019. Towards understanding atypical social affiliation in psychopathy. *The Lancet Psychiatry* 6, 437–444.

Viviani, P., and Flash, T. 1995. Minimum-jerk, two-thirds power law, and isochrony: Converging approaches to movement planning. *Journal of Experimental Psychology: Human Perception and Performance* 21, 32–53.

Vogeley, K., May, M., Ritzl, A., Falkai, P., Zilles, K., and Fink, G. R. 2004. Neural correlates of first-person perspective as one constituent of human self-consciousness. *Journal of Cognitive Neuroscience* 16, 817–827.

Vohs, K. D., and Schooler, J. W. 2008. The value of believing in free will: Encouraging a belief in determinism increases cheating. *Psychological Science* 19, 49–54.

von Helmholtz, H. 1866/1962. Concerning the perceptions in general. In his *Treatise on Physiological Optics*. 24–44. Dover.

von Helmholtz, H. 1878/1971. The facts in perception. In *Selected Writings of Hermann Helmholtz*, ed. R. Kahl. Wesleyan University Press.

Vuillaume, L., Martin, J.-R., Sackur, J., and Cleeremans, A. 2020. Comparing self- and hetero-metacognition in the absence of verbal communication. *PLoS One* 15, e0231530, https://doi.org/10.1371/journal.pone.0231530.

Vukovic, N., and Shtyrov, Y. 2017. Cortical networks for reference-frame processing are shared by language and spatial navigation systems. *NeuroImage* 161, 120–133.

Vullioud, C., Clément, F., Scott-Phillips, T., and Mercier, H. 2017. Confidence as an expression of commitment: Why misplaced expressions of confidence backfire. *Evolution and Human Behavior* 38, 9–17.

Vygotsky, L. S. 1978. *Mind in Society*. Harvard University Press.

Wadge, H., Brewer, R., Bird, G., Toni, I., and Stolk, A. 2019. Communicative misalignment in autism spectrum disorder. *Cortex* 115, 15–26.

Walton, M. E., and Mars, R. B. 2007. Probing human and monkey anterior cingulate cortex in variable environments. *Cognitive Affective & Behavioral Neuroscience* 7, 413–422.

Wang, L., Hemmer, P., and Leslie, A. M. 2019. A Bayesian framework for the development of belief-desire reasoning: Estimating inhibitory power. *Psychonomic Bulletin & Review* 26, 205–221.

Wang, Y., and Hamilton, A. 2012. Social top-down response modulation (STORM): A model of the control of mimicry in social interaction. *Frontiers in Human Neuroscience* 6, 153, https://doi.org/10.3389/fnhum.2012.00153.

Ward, A. F. 2021. People mistake the Internet's knowledge for their own. *Proceedings of the National Academy of Sciences* 118, e2105061118, https://doi.org/10.3389/fnhum.2012.00153.

Ward, J. 2013. Synesthesia. *Annual Review of Psychology* 64, 49–75.

Warneken, F., and Tomasello, M. 2006. Altruistic helping in human infants and young chimpanzees. *Science* 311, 1301–1303.

Warneken, F., Hare, B., Melis, A. P., Hanus, D., and Tomasello, M. 2007. Spontaneous altruism by chimpanzees and young children. *PLoS Biology* 5, 1414–1420.

Wason, P. C. 1968. Reasoning about a rule. *Quarterly Journal of Experimental Psychology* 20, 273–281.

Watson-Jones, R. E., Whitehouse, H., and Legare, C. H. 2016. In-group ostracism increases high-fidelity imitation in early childhood. *Psychological Science* 27, 34–42.

Watts, D. J., and Strogatz, S. H. 1998. Collective dynamics of "small-world" networks. *Nature* 393, 440–442.

Watts, I., Nagy, M., Burt de Perera, T., and Biro, D. 2016. Misinformed leaders lose influence over pigeon flocks. *Biology Letters* 12, 20160544, https://doi.org/10.1098/rsbl.2016.0544.

Webster, M. A., Kaping, D., Mizokami, Y., and Duhamel, P. 2004. Adaptation to natural facial categories. *Nature* 428, 557–561.

Wedekind, C., and Milinski, M. 2000. Cooperation through image scoring in humans. *Science* 288, 850–852.

Wegner, D. M. 1987. Transactive memory: A contemporary analysis of the group mind. In *Theories of Group Behavior,* eds. B. Mullen and G. R. Goethals, 185–208. Springer New York.

Wegner, D. M. 2003. *The Illusion of Conscious Will.* MIT Press.

Wellman, H. M., Cross, D., and Watson, J. 2001. Meta-analysis of theory-of-mind development: The truth about false belief. *Child Development* 72, 655–684.

Werchan, D., and Amso, D. 2021. All contexts are not created equal: Social stimuli win the competition for organizing reinforcement learning in 9-month-old infants. *Developmental Science* 24, e13088, https://doi.org/10.1111/desc.13088.

Werker, J. F., John, H. V. G., Humphrey, K., and Tees, R. C. 1981. Developmental aspects of cross-language speech perception. *Child Development* 52, 349–355.

Whalen, P. J., Rauch, S. L., Etcoff, N. L., McInerney, S. C., Lee, M. B., and Jenike, M. A. 1998. Masked presentations of emotional facial expressions modulate amygdala activity without explicit knowledge. *Journal of Neuroscience* 18, 411–418.

Wheatley, T., Boncz, A., Toni, I., and Stolk, A. 2019. Beyond the isolated brain: The promise and challenge of interacting minds. *Neuron* 103, 186–188.

Wheatley, T., Milleville, S. C., and Martin, A. 2007. Understanding animate agents: distinct roles for the social network and mirror system. *Psychological Science* 18, 469–474.

White, S. J. 2013. The Triple I hypothesis: Taking another('s) perspective on executive dysfunction in Autism. *Journal of Autism and Developmental Disorders* 43, 114–121.

Whiten, A., Caldwell, C. A., and Mesoudi, A. 2016. Cultural diffusion in humans and other animals. *Current Opinion in Psychology* 8, 15–21.

Whiten, A., McGuigan, N., Marshall-Pescini, S., and Hopper, L. M. 2009. Emulation, imitation, over-imitation and the scope of culture for child and chimpanzee. *Philosophical Transactions of the Royal Society B: Biological Sciences* 364, 2417–2428.

Whitfield, J. 2002. Nosy neighbours. *Nature* 419, 242–243.

Wicker, B., Keysers, C., Plailly, J., Royet, J. P., Gallese, V., and Rizzolatti, G. 2003. Both of us disgusted in My insula: The common neural basis of seeing and feeling disgust. *Neuron* 40, 655–664.

Wiederman, S. D., Fabian, J. M., Dunbier, J. R., and O'Carroll, D. C. 2017. A predictive focus of gain modulation encodes target trajectories in insect vision. *eLife* 6, e26478, https://doi.org/10.7554/eLife.26478.

Wikan, U. 2004. Deadly distrust: Honor killings and Swedish multiculturalism. In *Distrust*, ed. R. Hardin, 192–204. Russell Sage Foundation.

Wild, B., Rodden, F. A., Grodd, W., and Ruch, W. 2003. Neural correlates of laughter and humour. *Brain* 126, 2121–2138.

Wilkinson, A., Kuenstner, K., Mueller, J., and Huber, L. 2010. Social learning in a non-social reptile (Geochelone carbonaria). *Biology Letters* 6, 614–616.

Wilks, M., Kirby, J., and Nielsen, M. 2018. Children imitate antisocial in-group members. *Developmental Science* 21, e12675, https://doi.org/10.1111/desc.12675.

Willems, R. M., Benn, Y., Hagoort, P., Toni, I., and Varley, R. 2011. Communicating without a functioning language system: Implications for the role of language in mentalizing. *Neuropsychologia* 49, 3130–3135.

Wilson, D. S., Philip, M. M., MacDonald, I. F., Atkins, P. W. B., and Kniffin, K. M. 2020. Core design principles for nurturing organization-level selection. *Scientific Reports* 10, 13989, https://doi.org/10.1038/s41598-020-70632-8.

Wilson, D. S., Wilczynski, C., Wells, A., and Weiser, L. 2000. Gossip and other aspects of language as group-level adaptations. *The Evolution of Cognition*, 347–365.

Wimmer, H., and Perner, J. 1983. Beliefs about beliefs: Representation and constraining function of wrong beliefs in young children's understanding of deception. *Cognition* 13, 103–128.

Wing, A. M., Endo, S., Bradbury, A., and Vorberg, D. 2014. Optimal feedback correction in string quartet synchronization. *Journal of the Royal Society Interface* 11, 20131125, http://doi.org/10.1098/rsif.2013.1125.

Wing, L., Gould, J., Yeates, S. R., and Brierly, L. M. 1977. Symbolic play in severely mentally retarded and in autistic children. *Journal of Child Psychology and Psychiatry* 18, 167–178.

Winkielman, P., Berridge, K. C., and Wilbarger, J. L. 2005. Unconscious affective reactions to masked happy versus angry faces influence consumption behavior and judgments of value. *Personality and Social Psychology Bulletin* 31, 121–135.

Wittenbaum, G. M., Hubbell, A. P., and Zuckerman, C. 1999. Mutual enhancement: Toward an understanding of the collective preference for shared information. *Journal of Personality and Social Psychology* 77, 967–978.

Wittgenstein, L. 1953. *Philosophical Investigations.* Blackwell.

Wittmann, M. K., Lockwood, P. L., and Rushworth, M. F. S. 2018. Neural mechanisms of social cognition in primates. *Annual Review of Neuroscience* 41, 99–118.

Wohltjen, S., and Wheatley, T. 2021. Eye contact marks the rise and fall of shared attention in conversation. *Proceedings of the National Academy of Sciences* 118, e2106645118, https://doi.org/10.1073/pnas.2106645118.

Wolpert, D. M., Ghahramani, Z., and Jordan, M. I. 1995. An internal model for sensorimotor integration. *Science* 269, 1880–1882.

Wood, L. A., Kendal, R. L., and Flynn, E. G. 2012. Context-dependent model-based biases in cultural transmission: Children's imitation is affected by model age over model knowledge state. *Evolution and Human Behavior* 33, 387–394.

Woodruff, G., and Premack, D. 1979. Intentional communication in the chimpanzee: The development of deception. *Cognition* 7, 333–362.

Woodward, A. L. 1998. Infants selectively encode the goal object of an actor's reach. *Cognition* 69, 1–34.

Wu, J., Balliet, D., and van Lange, P. A. M. 2016. Gossip versus punishment: The efficiency of reputation to promote and maintain cooperation. *Scientific Reports* 6, 23919, https://doi.org/10.1038/srep23919.

Wyman, E., Rakoczy, H., and Tomasello, M. 2012. Non-verbal communication enables children's coordination in a "Stag Hunt" game. *European Journal of Developmental Psychology* 10, 597–610.

Xiang, T., Lohrenz, T., and Montague, P. R. (2013). Computational substrates of norms and their violations during social exchange. *Journal of Neuroscience* 33, 1099–1108.

Xiao, N. G., Wu, R., Quinn, P. C., Liu, S., Tummeltshammer, K. S., Kirkham, N. Z., et al. 2018. Infants rely more on gaze cues from own-race than other-race adults for learning under uncertainty. *Child Development* 89, e229–e244.

Xu, X., Zuo, X., Wang, X., and Han, S. 2009. Do you feel my pain? Racial group membership modulates empathic neural responses. *Journal of Neuroscience* 29, 8525–8529.

Xygalatas, D., Mitkidis, P., Fischer, R., Reddish, P., Skewes, J., Geertz, A. W., et al. 2013. Extreme rituals promote prosociality. *Psychological Science* 24, 1602–1605.

Yager, D. D., May, M. L., and Fenton, M. B. 1990. Ultrasound-triggered, flight-gated evasive maneuvers in the praying mantis Parasphendale agrionina. I. Free flight. *Journal of Experimental Biology* 152, 17–39.

Yang, F., Choi, Y.-J., Misch, A., Yang, X., and Dunham, Y. 2018. In defense of the commons: young children negatively evaluate and sanction free riders. *Psychological Science* 29, 1598–1611.

Yang, Y., Tian, T. Y., Woodruff, T. K., Jones, B. F., and Uzzi, B. 2022. Gender-diverse teams produce more novel and higher-impact scientific ideas. *Proceedings of the National Academy of Sciences* 119, e2200841119.

Yon, D., and Frith, C. D. 2021. Precision and the Bayesian brain. *Current Biology* 31, R1026–R1032.

Yoon, J. M. D., Johnson, M. H., and Csibra, G. 2008. Communication-induced memory biases in preverbal infants. *Proceedings of the National Academy of Sciences* 105, 13690–13695.

Yoshida, K., Saito, N., Iriki, A., and Isoda, M. 2011. Representation of others' action by neurons in monkey medial frontal cortex. *Current Biology* 21, 249–253.

Yuniarto, L. S., Gerson, S. A., and Seed, A. M. 2020. Better all by myself: Gaining personal experience, not watching others, improves 3-year-olds' performance in a causal trap task. *Journal of Experimental Child Psychology* 194, 104792, https://doi.org/10.1016/j.jecp.2019.104792.

Zahavi, D. 2014. Empathy and other-directed intentionality. *Topoi* 33, 129–142.

Zaki, J., and Mitchell, J. P. 2013. Intuitive prosociality. *Current Directions in Psychological Science* 22, 466–470.

Zaki, J., and Ochsner, K. N. 2012. The neuroscience of empathy: Progress, pitfalls and promise. *Nature Neuroscience* 15, 675–680.

Zawidzki, T. W. 2008. The function of folk psychology: Mind reading or mind shaping? *Philosophical Explorations* 11, 193–210.

Zebrowitz, L. A., White, B., and Wieneke, K. 2008. Mere exposure and racial prejudice: exposure to other-race faces increases liking for strangers of that race. *Social Cognition* 26, 259–275.

Zuberbühler, K. 2008. Gaze following. *Current Biology* 18, R453–R455.

Zuckerman, M., DePaulo, B. M., and Rosenthal, R. 1981. Verbal and nonverbal communication of deception. In *Advances in Experimental Social Psychology*, ed. L. Berkowitz, 1–59. Academic Press.

Zylberberg, A., Wolpert, D. M., and Shadlen, M. N. 2018. Counterfactual reasoning underlies the learning of priors in decision making. *Neuron* 99, 1083–1097.

Index

The Jean Nicod Lectures

Francois Recanati, editor

The Elm and the Expert: Mentalese and Its Semantics, Jerry A. Fodor (1994)

Naturalizing the Mind, Fred Dretske (1995)

Strong Feelings: Emotion, Addiction, and Human Behavior, Jon Elster (1999)

Knowledge, Possibility, and Conscsiousness, John Perry (2001)

Rationality in Action, John R. Searle (2001)

Varieties of Meaning, Ruth Garrett Millikan (2004)

Sweet Dreams: Philosophical Obstacles to a Science of Consciousness, Daniel C. Dennett (2005)

Things and Places: How the Mind Connects with the World, Zenon W. Pylyshyn (2007)

Reliable Reasoning: Induction and Statistical Learning Theory, Gilbert Harman and Sanjeev Kulkarni (2007)

Origins of Human Communication, Michael Tomasello (2008)

The Evolved Apprentice: The Nicod Lectures 2008, Kim Sterelny (2011)